Modularity

Modularity

Understanding the Development and Evolution of Natural Complex Systems

edited by Werner Callebaut and Diego Rasskin-Gutman
Foreword by Herbert A. Simon

The MIT Press
Cambridge, Massachusetts
London, England

The MIT Press is pleased to keep this title available in print by manufacturing single copies, on demand, via digital printing technology.

First MIT Press paperback edition, 2009

© 2005 Massachusetts Institute of Technology

This book was set in Times by SNP Best-set Typesetter Ltd., Hong Kong.

Library of Congress Cataloging-in-Publication Data

Modularity : understanding the development and evolution of natural complex systems / edited by Werner Callebaut and Diego Rasskin-Gutman.
     p. cm.—(The Vienna series in theoretical biology)
  Includes bibliographical references and index.
  ISBN 978-0-262-03326-8 (hc.: alk. paper)—978-0-262-51326-5 (pb.)
    1. Evolution (Biology). 2. Natural selection. 3. Chaotic behavior in systems. 4. Modularity (Psychology). I. Callebaut, Werner. II. Rasskin-Gutman, Diego. III. Series.

QH366.2.M63 2005
576.8—dc22

                                              2004055907

Contents

Biology promises to be the leading science in the twenty-first century. As in all other sciences, progress in biology depends on interactions between empirical research, theory building, and modeling. But whereas the techniques and methods of descriptive and experimental biology have evolved dramatically in recent years, generating a flood of highly detailed empirical data, the integration of these results into useful theoretical frameworks has lagged behind. Driven largely by pragmatic and technical considerations, research in biology continues to be guided less by theory than it is in other fundamental sciences. By promoting the discussion and formulation of new theoretical concepts in the biosciences, this series intends to help fill conceptual gaps in our understanding of some of the major open questions of biology, such as the origin and organization of organismal form, the relationship between development and evolution, and the biological bases of cognition and mind.

Theoretical biology is firmly rooted in the experimental biology movement of early twentieth-century Vienna. Paul Weiss and Ludwig von Bertalanffy were among the first to use the term *theoretical biology* in a modern scientific context. In their understanding the subject was not limited to mathematical formalization, as is often the case today, but extended to the general theoretical foundations of biology. Their synthetic endeavors aimed at connecting the laws underlying the organization, metabolism, development, and evolution of organisms. It is this commitment to a comprehensive, cross-disciplinary integration of theoretical concepts that the present series intends to emphasize. A successful integrative theoretical biology must encompass not only genetic, developmental, and evolutionary components—the major connective concepts in modern biology—but also relevant aspects of computational biology, semiotics, and cognition, and should have continuities with a modern philosophy of the sciences of natural systems.

The series, whose name reflects the location of its initiating meetings and commemorates the seminal work of the aforementioned scientists, grew out of the yearly Altenberg Workshops in Theoretical Biology, held near Vienna at the Konrad Lorenz Institute for Evolution and Cognition Research (KLI), a private, nonprofit institution closely associated with the University of Vienna. KLI fosters research projects, seminars, workshops, and symposia on all aspects of theoretical biology, with an emphasis on the developmental, evolutionary, and cognitive sciences. The workshops, each organized by leading experts in their fields, concentrate on new conceptual advances originating in these disciplines, and are meant to facilitate the formulation of integrative, cross-disciplinary models. Volumes on emerging topics of crucial theoretical importance not directly related to any of the workshops will also be included in the series. The series editors welcome suggestions for book projects on theoretical advances in the biosciences.

Gerd B. Müller, University of Vienna, KLI
Günter P. Wagner, Yale University, KLI
Werner Callebaut, Limburg University Center, KLI

Foreword
The Structure of Complexity in an Evolving World: The Role of Near Decomposability

In today's world, we pay a great deal of attention to complex systems, and with good reason. The phenomena we wish to understand, in all of the sciences, are increasingly complex, and the problems we need to solve in order to guarantee the future of our planet have complexity at their core.

Complex systems are usually big, but complexity means something more fundamental than bigness. A balloon, as big as you like, filled with hydrogen or helium gas would not usually be regarded as complex. We could build a theory of the balloon, quite satisfactory for many purposes, on the basis of the simple, fundamental law that relates pressure and volume to temperature of gases and the other laws of thermodynamics.

Of course, if we set the balloon loose in the atmosphere, all sorts of complexity may be observed in its behavior. But the source of this complexity is the balloon's interaction with a complex atmosphere; it is not intrinsic to the balloon.

Complexity arises, then, when we have a large system *and* when the system divides into a number of components that interact with each other in ways that amount to something more than uniform, frequent elastic collisions. Interactions among components can lead to all kinds of nonlinearities in behavior, and very often carry the system into the especially complex regions of phase space that we call chaos. For us humans, a principal symptom of complexity is that the complex system becomes harder and harder to understand and to predict. Our ordinary mathematical methods no longer lead to solutions in closed form, and the complexity can even carry us beyond the simulation capacities of our largest computers, present or prospective.

Even in the absence of chaos, the phase spaces of our complex systems may, and usually do, exhibit irregular surfaces of local minima and maxima of the variables of interest, preventing us from using our favorite mathematical tools to find the global maxima or minima that we often set as the goals of our search.

These difficulties of understanding and predicting the behavior of complex systems confront not only us human beings; they also confront, in a manner of speaking, the forces of inorganic and organic evolution. If we take the phrase "survival of the fittest" literally, then our theories of change and evolution, organic and inorganic, have force only in a world where maxima are attainable and the paths toward them are discoverable, conditions that are seldom satisfied in the real world. If Nature is unaware of these difficulties that confront it as it moves forward in time, we human beings who are observing the course of events are only too well aware of them.

The universe before the big bang apparently was not too unlike our hydrogen balloon—big, but perhaps not very complex—the universe immediately afterward, and from then on, became progressively more complex, with the rapid growth of components that were themselves increasingly complex. It becomes a fundamental problem of science, then, to explain why this growth in complexity came about and to observe and explain any commonalities of form that the resulting complex subsystems of the universe exhibit.

It has been noted for a long time that most of the complex systems we observe in the world, beginning with atoms and going on to stellar galaxies or galaxy clusters, have distinctly hierarchical structures: that is, they consist of (complex) subsystems, that consist of subsystems, and so forth, through many levels. The subsystems interact with each other (else they would not form systems), but the frequencies (and energy densities) associated with the interactions drop steadily as we go upward in the hierarchy—typically by an order of magnitude or two for each level that we ascend. Thus, quarks and other atomic particles jostle each other at speeds that are very hard to observe (the steady rise in cost of leading-edge accelerators being testimony to that), whereas the stars and even the planets revolve about each other at a majestic, solemn pace.

Another way to describe this structure is to state that the frequencies of interaction among elements in any particular subsystem of a system are an order of magnitude or two greater than the frequencies of interaction between the subsystems. We call systems with this property nearly completely decomposable systems, or for short, nearly decomposable (ND) systems (Simon and Ando, 1961).

The question, then, is, Why does complexity in our universe, at virtually all levels, generally take this hierarchical, nearly decomposable form? The first step toward answering the question is to observe that the complex systems had to evolve from simpler systems by processes of evolution. This leads to the second question: What is the connection between speed of evolution and near decomposability?

The answer to this second question is still far from complete, and the incomplete answer we have comes in several parts. First, it is easy to show that if large systems form by the agglomeration of smaller systems, and if the agglomeration process involves chance meetings between components that can form new, more or less stable, systems, then the probability is very high that large systems will consist of many layers, produced by successive meetings of components—and earlier still, of components of those components—and not by a single "reverse big bang" that instantly assembles a massive system from numerous tiny components.

Many of you are familiar with this argument of "The Two Watchmakers" (Simon, 1996, pp. 188–190), one of whom never succeeds in assembling a watch before the many independent components fall apart again, and the other of whom builds his watches in stages, putting together smaller, stable components into larger ones until the whole is complete.

But not all of the large systems we encounter in the world, especially biological organisms, appear to have been assembled by evolution from diverse smaller components. In particular, the formation of multicelled organisms has been quite different, based mostly on specialization of identical or similar units that are generated by cell division, but retain a lifetime's mutual attachment. Yet, the same ND architecture appears in these organisms, with their division into organs and tissues, and finally into (specialized) cells. What kind of evolutionary process could lead to the ubiquity of this particular hierarchical scheme?

Here we must return to the complexity of the fitness landscape, especially the proliferation of local maxima that make it extremely difficult to reach global maxima by any local hill-climbing, survival-of-the-fittest strategy. The solution to the dilemma appears when we recognize that those which survive need not be fittest, in the sense of occupying global maxima in the fitness landscape. The winners need not be "optimal" organisms. For an organism to survive and flourish requires only that it be *fitter* (in some niche) than the other extant organisms with which it competes. "Potential" organisms that have not yet been introduced by mutation or crossover or immigration offer no competition until they actually appear, as hares finally did in Australia. So local maxima, or even uplands that are not maxima, are all that the evolving organism need aspire to.

With this recognition, the process of evolution takes a new shape. Suppose that a large, complex system, with many interacting parts, is changing through the usual evolutionary mechanisms. Natural selection only evaluates the fitness of the organism as a whole, not the fitness of individual organs except as they contribute to the whole. But if the effectiveness of design of each organ depends on the design of the organs with which it interacts, then there is no guarantee that improvements of one organ will not worsen the performance of others. Natural selection will have to depend on the lucky chance that simultaneous, mutually beneficial changes in a number of different organs will occur to improve overall fitness. The probability of this coordination of events decreases rapidly with the amount of interdependence of design of the organs.

Suppose, instead, that the effectiveness of each organ depends very little on the design of the others, provided that the inputs each requires are supplied by the others. Then, up to a scale factor, the design of each organ can be improved independently of what is happening to the others; and it is easy to show that fitness will rise much more rapidly than when there is mutual dependence of design. The most convincing demonstration of this is the simulation by Frenken, Marengo, and Valente of the University of Trento (1999), using a genetic algorithm and simulated organisms, some of which were ND and some of which were not. The ND organisms soon reached higher fitnesses than the others, and displaced them.

I have omitted important details from this brief sketch. In particular, organs inhabiting a single organism would require some mutual balance of their relative capacities. However, one can describe a variety of allometric mechanisms that can balance capacities through an entire organism by size adjustments of component organs without damaging their basic independence of design (Simon, 1996, pp. 204–205).

The evolutionary scenario I have just described is not limited to organisms. Exciting explorations, some of them by the same group at Trento who demonstrated the superior growth of fitness of ND systems, have deepened our understanding of the evolutionary development of business firms with changing technology. The same ND property that facilitates rapid evolution provides the key to rapid adaptation to new technological opportunities. This is of particular interest in a society that uses a combination of markets and organizations to perform its economic functions.

Markets are basically simple systems, in the sense that our hydrogen balloon is simple. In fact, most of the arguments for their economic efficiency rest on their simplicity. However, historians of economics have noticed that when Adam Smith was writing, markets dominated the economy of Britain, but in the two succeeding centuries markets have been steadily and rapidly displaced by large organizations. Today, most people engaged in economically productive activity in the developed world work within organizations and not in the marketplace. Whether e-commerce will substantially shift that balance remains to be seen, but I, for one, see no compelling reason why it should. By lowering market costs, e-commerce may actually reduce the fraction of total productive effort required to operate the economy's markets.

Economic theory about the relative roles of markets and business firms in a modern economy is in a state of great flux. The neoclassical theory that had placed markets in the central role is being seriously challenged by the New Institutional Economics, by the theory of bounded rationality, and by evolutionary theories—the latter three viewpoints being complementary in complicated ways. I would not like to try to forecast how this will all work out, but I offer at least the conjecture that a theory of ND systems will be important to the outcome.

Near decomposability also plays a major role in the understandability of complex systems when we humans seek to investigate their properties and behavior. Because of the (incomplete) separation of frequencies between levels of an ND system, one can usually obtain good approximations to the short-run and middle-run behavior at any given level without considering the details of the higher-frequency movements at the levels below, and while taking the situation at the levels above as constant over the interval of interest. We have been using this divide-and-conquer strategy in science for many generations, and it is a principal basis for specialization of the sciences into levels, from sociology and economics through biology and chemistry, and ultimately to particle physics.

Improvements in our methods for approximating the behavior of ND systems (for example, renormalization in quantum electrodynamics, which depends on near decomposability) will continue to be an important task for applied mathematics, certainly requiring advances in pure mathematics as well.

I have touched on only a few aspects of complexity and complex systems that happen to have attracted my interest over the years. There are many other stories to be told about complexity, and I am sure that many of them will be told at this workshop. For example, we are still far from digesting all of the lessons of chaos, advancing our computational theories to explore it, or designing feedback devices that can constrain and manage it. I will not presume to speculate about all of the forms that complexity will take in the world and in our thoughts, or to enlarge further the brief account I have given of one of its very visible aspects, the near decomposability of most of the world's complex systems.

—Herbert A. Simon†

References

Frenken K, Marengo L, Valente M (1999) Interdependencies, nearly-decomposability and adaptation. In: Computational Techniques for Modelling Learning in Economics (Brenner T, ed), 145–165. Dordrecht: Kluwer.

Simon HA (1996) The Sciences of the Artificial, 3rd ed. Cambridge, MA: MIT Press.

Simon HA, Ando A (1961) Aggregation of variables in dynamic systems. Econometrica 29: 111–138.

The quest to apprehend the world in terms of the modular organization of its parts has an impressive track record in the sciences. Its roots reach back at least to the Scientific Revolution (the heliocentric conception of the cosmos). It includes Leibnizian and Kantian faculty psychology, Gall's phrenology, and nineteenth- and early twentieth-century debates in physiology and genetics, among many other relevant developments. This is just what one would expect, since modularity abounds in nature. Moreover, human art and engineering, as well as mathematics, crucially rely on modular *design* principles in their constructions. The instances of modules that the *New Webster's Dictionary and Thesaurus of the English Language* (1992) lists are from architecture, electronics, rocketry, and mathematics.

At present, modularity is a prominent theme mainly in the life sciences (including the neurosciences), cognitive science, and computer science (modular architectures). In biology, structural and functional modules are now recognized at many levels of organization. Their diversity urges for precise characterizations in their respective domains as well as the identification of commonalities and differences. In cognitive science and linguistics, a lively debate on "mind modules" was spurred by Jerry Fodor, and the interdisciplinary field of evolutionary psychology is grounded on the controversial "Swiss army knife" model of perception and cognition. Modularity also remains a central concept in the neurosciences, where new anatomical, imaging, and experimental methods make possible the identification of brain modules at various levels of granularity.

With some rare exceptions, discussions of modularity to date remain largely confined to the "home disciplines" (developmental and evolutionary biology, evolutionary psychology, etc.) in which they originated. The editors felt the time was ripe to bring together experts in the fields of—in alphabetical order—artificial life, cognitive science, developmental and evolutionary biology, economics, evolutionary computation, linguistics, mathematics, morphology, paleontology, physics, psychology, and theoretical chemistry, as well as philosophers of biology and mind, to try to

1. Survey the variety of disciplinary contexts in which "modular thinking" in general (e.g., hierarchical organization, near decomposability, quasi-independence, recursion) or more specialized concepts (e.g., character complex, gene family, encapsulation, mosaic evolution) play a role

2. Clarify, against this background, what modules are, why and how they originate and change (develop, evolve), and what this implies for the respective research agendas in the disciplines involved

3. Bring about a useful knowledge transfer between diverse fields regarding the broad topic of modularity wherever this appears useful and feasible.

To achieve these aims, the fifth Altenberg Workshop in Theoretical Biology, "Modularity: Understanding the Development and Evolution of Complex Natural Systems," was convened in the Lorenz mansion in Altenberg, Austria, in October 2000. Twenty-one participants discussed these and related issues intensively during four days. At the end of the meeting, the general feeling was that it had been very successful in that the first and third of our ambitious objectives were largely met, and that substantial progress was made regarding the second, and arguably the most difficult, objective as well. The (often multiauthored) papers that had been prepared for the workshop were rewritten, often substantially, in the light of our sometimes heated debates. This volume presents the results to the reader. A companion Web site containing a plethora of graphical materials that could not be included in the book is available at www.kli.ac.at/mit/modularity.

The editors thank the board of directors of the Konrad Lorenz Institute for Evolution and Cognition Research (KLI) in Altenberg, Austria, for its financial support of the workshop. We owe special thanks to the chairman, Prof. Gerd B. Müller, and the general manager, Dr. Astrid Juette, without whose logistic and moral support our task would have been much less agreeable. We warmly thank our 26 contributors and the staff at MIT Press, in particular Robert Prior, Valerie Geary, and Katherine Almeida for all their work and patience with the long gestation of the final manuscript.

Finally, with much pain in our hearts we learned in February 2001, three months after the workshop was held, that Prof. Herbert A. Simon had died in Pittsburgh at the age of 84. Because of his careful thoughts on the near decomposability of complex systems as well as much of his other work, Herb undoubtedly deserves to be called the master of modularity thinking. He had declined our invitation to participate in the workshop, not because of the physical stress involved in traveling but because he felt he had so much important work to finish. Instead, he gently offered to send us a manuscript to read, assuring us that his thoughts would be with us. After his death, Herb's family kindly granted us permission for this manuscript (dated September 29, 2000) to be included as the foreword in the present volume. We dedicate this book, imperfect as it is, to the memory of this great man, in the hope that as a "satisficer," not as an optimizer, he would have enjoyed it.

I INTRODUCTION

1 The Ubiquity of Modularity

Werner Callebaut

In our world, modular systems, both natural and artificial (in Herbert Simon's sense[1]), abound. The majority of the contributions to this volume deal with modularity as uncovered and specified in a number of biological disciplines. Part II deals with the challenge modularity poses for evolutionary biology, developmental biology, and the emerging interfield between evolution and development usually referred to as evolutionary developmental biology or *evo-devo*.[2] Part III considers the implications of modularity for macroevolution, morphology, and paleobiology. While the focus in part II is on process ("The Making of a Modular World"), the emphasis in part III is more structural ("Working Toward a Grammar of Forms"). Part III also includes chapters on modularity in art and at the boundary between art and science. The fourth and last part of the book deals with mind and culture. The vexed question of the modularity of the human mind is framed in the context of advancements in artificial intelligence, neuroscience, and the cognitive sciences in general, with a particular emphasis on connectionism. It is also argued here that although "modularity" is not normally part of the terminology of social scientists (but see Baldwin and Clark, 1997; Hurley, 1999), the realm of economic interactions provides almost ideal field and laboratory settings to study modularity.

Because the editors of this volume view minds and cultures as well as their exosomatic products (technologies) as naturally evolved and naturally developing systems, we insist on subsuming them all under "natural complex systems," as the subtitle of the book indicates. (The view that modularity is a prerequisite for adaptive technological evolution as well as for biological evolution has been popularized by the influential work of the economic historian Paul David and others on the robustness of the QWERTY system.[3]) The final chapter in the book is concerned with the "natural logic" of communicative possibilities, extending the concept of morphospace that was central to part III to the (human and animal) psychological and cultural realms.

This introductory chapter is concerned primarily with providing some conceptual foundations for the enterprise that follows. I will first suggest that there is an intimate connection between the ubiquity of modular organization in the world and the circumstance that Western science has *historically* been so successful at "the knowledge game" (Hull, 1988)—which is not to deny that science may also be successful at understanding, say, nonmodular deterministic systems or chaotic systems (see Agre, 2003 on the evolutionary significance and generalizability of Simon's parable of the two watchmakers). Moreover, with Simon, I would argue that we already do possess more—and better—than "transcendental" arguments for the ubiquity of evolved modules, and thus can at least begin to genuinely *explain* the ubiquity of modularity.

Next I will look at a number of contexts in which scientists deem it necessary or useful to invoke modularity, try to disentangle various meanings and uses of the word (e.g., as *explanans* or *explanandum*), and attempt to provide as general a definition of modularity as possible, with a view to the wide range of applications considered in this book. I then will look in some detail at the biological uses of modularity, in particular in the contexts of development and evolution. I will round off this chapter by discussing some aspects of the issue of the modularity of the mind/brain.

"Fortune Smiled upon Kepler and Newton"

The success of modern science depends on plausible simplification. The number of actual or conceivable interactions among the parts of a system greatly exceeds the number of interactions that must actually be taken into account to yield system descriptions that are good enough for most theoretical or practical purposes. This happy condition was critical for the articulation of classical mechanics—the paradigmatic science of *simplicity*. Imagine a system that grows by multiplying its parts (or by the agglomeration of smaller systems; see Foreword in this volume). The number of potential interactions between the system's parts will increase much more quickly than its cardinality (number of parts). To illustrate this point, Simon (1977b) described an episode that from the perspective of the history of science has been of dramatic importance in that it consolidated the Scientific Revolution substantially. Because this example provides the starting point for my argument concerning the ubiquity of modularity, I think it is worthwhile to discuss it in some detail.

Consider our solar system as it was known in Kepler's and Newton's day. It consisted of a sun, six planets (including Earth), and about ten visible satellites belonging to three of the planets. (Comets are left out of the picture.) If one considers only pairs of heavenly bodies, there are already $17 \times 16 = 272$ potential interactions between the elements of the system (half as many if one assumes that $I(a,b)$ and $I(b,a)$ represent identical interactions), resulting in a rather complex description of the system. Note that there is no a priori reason to assume that the system's behavior is determined only by pairs of bodies, although Newtonian physics did make this assumption. It turns out that in practice, one does not have to take into account most of these potential interactions:

Kepler and Newton did no such thing. Kepler detected three observational regularities: the orbits of the planets about the sun are ellipses with the sun at the focus; equal areas are swept out in a given planetary orbit in equal times; and the periodic times of the planets vary with the 3/2 power of their distances from the sun. Newton showed that these phenomena, together with the analogous ones for the satellites, could be deduced from his Laws of motion, taken in combination with the gravitational law, gravitational force varying inversely with the square of the distance between a pair of bodies. But the observed regularities, and the derivation as well, depended upon the assumption that each particular orbit under consideration was determined by the interaction between the sun and a a single central body (a planet or satellite). In each case, interactions with all other components of the system were ignored, and yet an excellent fit was obtained between the theoretical derivations and the observations. (Simon, 1977b, pp. 508–509)

Why were such elementary calculations sufficient?[4] The explanation, Simon argued, is to be sought in the circumstance that our solar system, in comparison with the kinds of solar systems one would expect on the basis of purely statistical considerations, turns out to be a very special case indeed. "If the deck of cards that Nature dealt to Kepler and Newton was not stacked, it was at least a very lucky deal." Why was this so? Simon continues:

First there was a single body, the sun, that was larger by three orders of magnitude than any other body in the system. Second, there were six bodies, the planets, that were several orders of magnitude larger than their satellites. Third, the distances of the planets from each other were of the same order of magnitude as their distances from the sun, while the distances of the satellites from their planets were orders of magnitude smaller than their distances from the sun. *None of these distributional facts follows from the laws of mechanics but had they not been true of the solar system, Kepler's regularities and Newton's derivations would not have described that system.* Although Newton's Laws are generally valid (up to the classical, non-relativistic approximation) for systems of masses, the relatively simple calculations used to test those laws would not have sufficed had the system been more "general." Because of these relative sizes and distances, each planet orbited around the sun almost exactly as if it had been attracted by the sun alone, and each satellite around its planet in a similar way. (Simon, 1977b, p. 509; italics added)

These simplifications depended not only on the distribution of masses and distances; they were amplified by the *forms of the laws* themselves (Simon, 1997b, p. 509). For these several reasons, our solar system is simpler—*and hence can be treated as such*[5]—than one would expect on the basis of 272 pairwise interactions. Most interesting is the fact that this "fortune" also extends, albeit often less spectacularly, to the natural complex systems scientists are now beginning to understand (see, e.g., Székely, 2001 for an application of "Simon's theorems" to the complexity of the brain). The possibility of scientific understanding crucially depends, then, on the *near decomposability* (Simon) of modular systems, which allows the subdivision of the explanatory task into manageable chunks. Generally speaking, a system may be characterized as modular to the extent that each of its components operates primarily according to its own, intrinsically determined principles. Modules within a system or process are tightly integrated but relatively independent or "dissociable" (Needham, 1933) from other modules (e.g., Simon, 1969, 1973, 1977b, 1995; and Foreword in this volume; Raff, 1996; Wagner, 1996). Because the strength or weakness of interactions is a matter of degree, modularity should itself be seen as a gradual property (see Wagner and Altenberg, 1996 and chapters 2 and 9 in this volume).

Explaining the Ubiquity of Modularity

Let us suppose that most of the natural complex systems that science encounters do display the nearly decomposable organization—characterized by "meager" (i.e., thinly populated) interaction patterns that Simon and other systems theorists have described.[6] As Simon has emphasized, philosophers and other skeptics could easily object that this "fact" tells us little or nothing about the structure of reality but could be due, entirely or in part, to the perceptual and/or analytical biases of limited human epistemic subjects (see my discussion of Brandon below, as well as chapter 13 in this volume). On this quite influential view, the universe could well be ultra-complex, but will remain barred to us, presumably forever.

My reply to this is threefold. First, there are general reasons to resist "global skepticism," which as far as science is concerned has been shown to be a doomed heuristic (see Shapere, 1984). Second, pace evolutionary epistemologists and evolutionary psychologists who claim that our evolutionary heritage has inescapably "bleak implications" for human rationality (but see Samuels et al., 1999), the progress of science hitherto suggests no principal reasons whatsoever, and certainly no a priori reasons, to doubt that our science, through the improvement of its observational, experimental, and computational techniques, will be able to transcend any such limitations (see, most forcefully, Levinson, 1982; see also Callebaut and Stotz, 1998). Although, say, evidence from the anthropology of science suggests that work in taxonomy to date remains tributary to folk-biological categories (Atran, 1998) and that "psychological essentialism" (Barrett, 2001) may be the result of a history of natural selection on human representation and inference systems,[7] the very same work also indicates that scientists *are* able to reflexively overcome their remnant anthropocentrism.

Third, in the vein of Simon's a posteriori arguments concerning classical mechanics, it is very possible to look for, say, the *biological* reasons why "the bodies of higher organisms are so obviously built in a modular way such that apparently natural units are often easy to recognize" (Wagner, 1996, p. 36). As Wagner puts it in his discussion of homologues, "Homologues, if they are natural kinds, do not exist in order to serve the needs of comparative anatomists" (1996, p. 36; see also Wagner and Laubichler, 2000 on the role of the organism in character identification).[8]

A perhaps more serious objection concerns the logic of influential arguments for the ubiquity of (evolutionary-developmental) modularity by Lewontin and Bonner. Lewontin (1974) stressed the "quasi independence" of characters, by which he meant that there are at least some developmental trajectories that allow one character to be changed without affecting others. Bonner (1988) considered "gene nets," groupings of networks of gene actions and their products into discrete units during the course of development. As Brandon (chapter 3 in this volume) makes clear, Bonner's and Lewontin's arguments are "transcendental" in that they claim that modularity is *necessary* for the very existence of the phenomena of adaptation: "Adaptive evolution, which produces the phenomena of adaptation, requires quasi independence/gene nets. The phenomenon of adaptation is real. Therefore, quasi independence/gene nets exist." The problem with transcendental arguments, Brandon states, is that although they are perfectly valid from a logical point of view, they are not explanatory (see Brandon, 1999, p. 178, n. 11). Also, and perhaps more important, they may unwittingly reflect limitations on our understanding of the world rather than a limitation on how the world works.

Brandon's diagnosis seems to me to be convincing as far as the cases at hand are concerned. But is this grounds to worry about the feasibility of the general enterprise of trying to explain the ubiquity of modularity—and, conversely, its limitations—as, say, a variational principle? Brandon himself points the way when he writes that belief in the existence of evolutionary modules may be inferred "indirectly" from phylogenetic data or based on the "direct" observation of modules.

This is in fact what current practice tries to do. For instance, Jablonka (2001), in her comparison of the genetic inheritance system (GIS) with epigenetic (EIS), behavioral (BIS), and symbolic (SIS) inheritance systems, identifies both modular and nonmodular ("holistic") types of information and modes of transmission of information. Only the GIS, methylation (one of the several EISs she describes), and imitation and teaching (two of the several BISs) concern purely modular information that is also transmitted in a purely modular way; symbolic systems, which are transmitted by social learning, are transmitted both modularly and holistically.

Or, to take another example, the question whether genetic modularity is necessary for evolvability (i.e., the ability to respond to a selective challenge by producing the right kind of variation; see, e.g., Gerhart and Kirschner, 1998; Raff and Raff,

2000; cf. Wagner and Laubichler, 2004, pp. 100–101) can be tackled by putting it in the context of other conceivable principles of evolvability. Hansen (2003) lists quite a few alternative candidates: co-optation, cryptic variation, dissociability, duplication and divergence, "the edge of chaos," "evolutionary cranes" (Dennett, 1995), "extradimensional bypass" (Conrad, 1990), recombination, redundancy, robustness, symmetry, and—if this can be viewed as different from modularity itself—the emergence of new hierarchical levels of organization.

Reminding us of François Jacob's metaphor of evolution as tinkering, Hansen cautions that while genetic modularity "may indeed be a simple, logical and efficient way of achieving evolvability," it does not follow that it *is* the biological basis of evolvability. (But see chapter 9 in this volume, where Rasskin-Gutman postulates that the space of modular design is the only available pool for the evolutionary arrow to proceed.) Nor is it to be excluded that genetic evolvability is achieved "in ways that appear complex and illogical to our minds" (Hansen, 2003, p. 85). In the scenario Hansen considers, the most evolvable genetic architectures are typically those with an intermediate level of integration among characters, and in particular those where pleiotropic effects are variable and able to compensate for each other's constraints. Several of the chapters in this volume probe other such scenarios. This and related work seem to indicate that the question "Why does complexity in our universe, at virtually all levels, generally take this hierarchical, nearly decomposable form?" (Simon, foreword to this volume) is clearly amenable to theoretical and empirical investigation that can lead to genuinely explanatory answers (see chapter 11 in this volume). At least *some* transcendental arguments can be "naturalized"! Since chapter 2 deals entirely with the origin of modules, this important but vexed issue will not be further pursued here.

Dimensions and Kinds of Modularity

At this juncture I will introduce a number of conceptual distinctions in order to prepare the reader for the somewhat bewildering panorama of modularities awaiting her. Modules are invoked in many different contexts with different purposes, some of which have little in common with our preoccupations in the present volume. Yet it seems fair to say that there is a sense that runs through *any* ascription of modularity, from the art motifs discussed in chapters 12 and 13 of this volume to chapter 17's autonomous and anonymous economic agents who take decisions independently from one another and interact only through the price system. It is that "of a unit that is a component part of a larger system and yet possessed of its own structural and/or functional identity" (Moss, 2001, p. 91). In addition to the criteria of tight internal integration and relative independence from other modules introduced above, this characterization suggests two further criteria: that modules must *persist* as identifiable units for long enough time spans (or, in the case of evolutionary modules, generations), and that they must be more or less identical, *repetitive*, and reusable "building blocks" of larger wholes and/or different systems (e.g., Müller and Wagner, 1996; Müller and Newman, 2003).

Both of the latter criteria indicate that modules may be subject to a certain amount of change within them (Raff, 1996, p. 322; Gilbert and Bolker, 2001, p. 10). For instance, the economic agents of classical and neoclassical economic theory discussed in chapter 17 of this volume must live long enough to participate in market transactions such as working for a salary, buying or selling goods, and the like; and there must be large enough populations of these agents for markets to be able to function properly. Or, to take another example, developmental modules can be deployed repeatedly in the same organism, as in the case of the left and right forelimb buds. The two forelimbs are two *different developmental modules* of the organism, but they are also parts of the *same evolutionary module* (chapter 2 in this volume).

Still at this most general level, "ontologically" speaking, modularity comes in two varieties: "It may be a primary property of the way organisms are built, for instance due to organizational principles of self-maintaining systems" (Fontana and Buss, 1994), or it may be an "evolved property" (Wagner, 1996, p. 38). If modular organization is the product of evolution by natural selection—the only evolutionary force capable of explaining adaptation on the standard, neo-Darwinian view (e.g., Ridley, 1993, part 3)—it can result either from *parcellation* (i.e., the differential elimination of pleiotropic effects among characters belonging to different character complexes) or from the differential *integration* of independent characters serving a common functional role. The relative frequency of either is an empirical question (Wagner, 1996, pp. 38–39). I will return to the issue of evolutionary modules below.

Structure, Process, and Function

Still at this level of greatest generality, it seems useful and even imperative to distinguish modularity of *structure* from modularity of *process*. Whereas at least the identification of structural or architectural modules is often a straightforward matter (Bolker, 2000), many biologists have been reluctant to talk about process modules because they would seem to be much more ephemeral. In biology at least, the issue is further complicated by the circumstance that modularity and *homology* have a common (recent) history (Wagner, 1995, 1996; Moss, 2001). As late as 1971, De Beer "drew a clear line between structure, which he viewed as the only appropriate thing to be homologized, and function: 'An organ is homologous with another because of what it is, not because of what it does' " (Gilbert and Bolker, 2001, p. 1).

Recent progress in developmental genetics has led to remarkable insights into the molecular mechanisms of morphogenesis, but has at the same time blurred the clear distinction between structure and function that De Beer was relying on. Gilbert and Bolker (2001, p. 10) are confident that "[i]dentifying the ways in which homologous processes are regulated, replicated and changed over time will enable us to better understand how changes in development generate changes in morphology and, ultimately, the evolution of new groups of animals." More generally, process modularity is required to make sense of modular *functions*, which are behaviors, not structures.[9] (Recall Moss's definition of a module as "a unit that is a component part of a larger system and yet possessed of its own structural and/or functional identity," introduced above.)

Following Bechtel and Richardson, whose work on decomposition and localization as research strategies in the biological and cognitive sciences is in many ways an elaboration and refinement of Simon's view on near decomposability, I want to frame this issue in the context of sound—which for me means *mechanistic*—explanation (see Callebaut, 1989, 1995):

Simple localization differentiates tasks performed by a system, localizing each in a structural or functional component. Complex localization requires a decomposition of systemic tasks into subtasks, localizing each of these in a distinct component. Showing how systemic functions are, or at least could be, a consequence of these subtasks is an important element in a fully mechanistic explanation. Confirming that the components realize those functions is also critical. Both are necessary for a sound mechanistic explanation. (Bechtel and Richardson, 1993, p. 125)

It is important to note that structures often do not map neatly one to one onto functions (and vice versa), making functions indispensable. As Bechtel and Richardson show, the route to complex localization frequently begins with direct localization, which then develops into a more complex localization in which functional decomposition of tasks becomes more central (see Star, 1989). This is common in psychology, where research often begins by dividing psychological activities into broad performance categories such as perception, memory, language, reasoning, and emotion. Bechtel and Richardson note that "Noam Chomsky has provided one of the clearest expressions of this approach in his own 'organology', strikingly reminiscent in tone of phrenology" (p. 126). Fodor has generalized Chomsky's organology, or modularity, beyond the domain of language to modular cognitive systems, which he claims are "domain specific, innately specified, hardwired, autonomous, and not assembled" (Fodor, 1983, p. 37). I will return to the issue of cognitive modules later on. Again with a view to mechanistic explanatory concerns, Von Dassow and Munro (1999, pp. 307–308) write:

The experimental study of development assumes that one may meaningfully isolate (physically or conceptually) and study individual processes independent from one another. Functional decomposability is thus a necessary presumption to considering developmental mechanisms either as units of explanation within development or as units of evolutionary change.

Top-Down and Bottom-Up Research Strategies

Quite often, modules have been or are posited "top-down," beginning the investigation with the phenomenal properties of a system, and then attempting to explain its working on the basis of one or several modules. Cognitive science abounds with examples (Bechtel and Richardson, 1993). Such modules, like the "Darwinian" modules postulated in evolutionary psychology and related "massive modularity" accounts (see below), have not been empirically observed but, in straightforward Popperian fashion, speculatively brought forth as *explanans* (Moss, 2001; see also chapter 9 in this volume). Modules of this sort are quite often associated with a preformationist stance, as in Chomsky's and Fodor's view or, more recently, in evolutionary psychology, where

The cognitive capacity/phenotype (whether still of adaptive value or not) is . . . construed to be the expression of a developmentally invariant, preformationistically transmitted module that has been passed along from generation to generation ever since [the Pleistocene]. (Moss, 2001, p. 92)

Yet, as Moss points out, "the concept of module itself does not specify its place along a preformationism–epigenesis axis" (2001, p. 92). Clear-cut examples of the converse, "bottom-up" research strategy may be found in, say, computer programming and neurocomputing (e.g., Barbuti et al., 1993; Husken et al., 2002) or in the ab initio calculations of artificial life (e.g., Adami 2002; see also Fontana and Buss, 1994). As Moss (2001, p. 92) notes, "the 'genetic revolution' of the twentieth century did not result in a search for any form of subcellular modules, nor any expectation of finding such. Rather, the recognition of modularity came as a surprise." At least in subcellular biology, modularity has arisen as an *explanandum* in the first place, but by now, new and promising module-centered explanatory approaches have begun to emerge which usefully complement many developmental accounts (see below).[10] The bottom-up versus top-down distinction should not be pushed too far, however. In the end, modularity becomes "all-around" as the modules are recognized, characterized, and used empirically. Once a module has been established, its constituent parts become irrelevant, so to speak (see Simon's "pragmatic holism" as discussed in chapter 15 of this volume). What matters most from now on is the interaction among modules.

Biological Modules

For the purposes of my discussion of biological modularity, it will be convenient to distinguish three aspects: development, morphology, and evolution.

Developmental Modularity

Although many of the structures and processes with which developmental biologists have been traditionally concerned are readily referred to as "developmental modules," it is not always clear what this is supposed to mean or imply. Von Dassow and Munro (1999, p. 308) warn that at present we have only the "rudiments of a developmental modularity concept," which comprise many intuitive notions about modularity (Raff, 1996, esp. chap. 10), including morphogenetic fields (Gilbert et al., 1996), gene networks (Bonner, 1988), and the several notions of homologues (Hall, 1992). One way to define developmental modules operationally is to state that any subsystem manifesting some quasi-autonomous behavior qualifies (Von Dassow and Munro, 1999, p. 313).

According to the current "interactionist consensus" that emerged from nature/nurture debates (Kitcher, 2001; Oyama et al., 2001), developmental modules are viewed as "phenotypic expressions of genes in an environment" (Sperber, 2002). But at least since the hardening of the Modern Synthesis, in practice the environment typically has been left out of the picture (see Robert et al., 2001).[11] "In genetic experiments, variability that was associated with flexibility and condition-sensitive development came to be regarded as noise, a factor to be controlled and not studied for its own sake" (West-Eberhard, 1998, p. 8417). Worse, common practice also suppresses the important roles that *epigenetic* factors play in development as well as in evolution, although "epigenesis is a primary factor directing morphological evolution, even in evolved developmental systems" (Newman and Müller, 2000, p. 312; see also Griesemer, 2002; Müller and Olson, 2003). Thus the tendency to black-box development that had been inaugurated by Darwin (Amundson, 1994) is continued.[12] "Molecular developmental systems" is one among several recent attempts to counteract this: "A necessary molecular concomitant of organismal complexity appears to be that of great developmental versatility in the resources available for constructing cell-to-cell and cell-to-matrix linkages" (Moss, 2001, p. 93).

As Sterelny (2001, p. 341) emphasizes, an important feature of developmental modules is their *reusability*. Müller and Wagner describe the "machinery of development" as follows:

> The more we learn about molecular mechanisms of development in widely different organisms, the higher the number of conserved mechanisms that become known. Some of them do indicate homology of morphologically divergent characters. . . . Still others illustrate that highly conserved molecular mechanisms may be used in radically different developmental contexts, indicating that the machinery of development consists of modular units that become recombined during evolution. (Müller and Wagner, 1996, p. 11)

At the very least, such insights suggest that biological reality is much too complex to be captured by a linear mapping of genes onto developmental schedules, and of developmental schedules onto phenotypes (Minelli, 1998). The extent and functional basis of developmental modularity will need to be investigated in much greater detail, however (see Griffiths and Gray, 2001, p. 215).

Morphological Modularity

At the *morphological* or architectural level, the structure and function of specific parts or elements of organisms like the mammalian forelimb or the modular structures of animal skeletons are characterized (see, e.g., Riedl, 1978). The contributions of modularity in art in part III and some of the chapters on neurocognitive modularity in part IV also concentrate on architectural aspects. At this level, a part is to be viewed as a module in what might be called the "operation" of an organism, for example, in its physiology or behavior, rather than its development (chapter 8 in this volume). Alternatively, morphological modules may be seen as preserving the functional integrity of the part but not its purposive function (chapter 9 in this volume).

As Thomas's discussion (chapter 11 in this volume) of animal skeletons as nearly decomposable systems shows, individual elements may have "a relatively high degree of local, short-run integrity of structure and function, while being interdependent at the level of operation of the organism as a whole." The structural elements defined by the parameters of Thomas's *skeleton space* may be skeletons in themselves, parts of skeletons, or parts of tightly integrated skeletal complexes with specific functions. At this level, biologists are increasingly interested in the way in which organisms and their parts can be viewed as an articulation of more or less autonomous mechanisms. See, for example, chapter 8 in this volume on the "remodularization" of organisms in the evolutionary transition from single-cell to multicellular organisms, or chapter 9 in this volume, where modularity is characterized as integration on four different morphological levels: proportions, orientations, connections, and articulations.

Eble (chapter 10 in this volume) notes that "the parts and characters routinely identified by the morphologist reflect hypotheses of modularity based on observational or quantitative criteria, without reference to the generative mechanisms or the theoretical contexts to which modules relate." However, a notion of developmental modularity in terms of mechanisms of genetic and epigenetic specification of units of phenotypic evolution is now being advanced (see below). Since morphological patterns of organization emerge in ontogeny, "morphological modularity might thus be seen as an aspect of developmental modularity" (chapter 10 in this volume). See also chapter 9 in this volume on morphological modularity as a consequence of binary division in multicellular organisms.

Evolutionary Modularity

As elaborated in various ways in part II of this volume, the concept of modularity provides a powerful nexus between developmental and evolutionary questions (see esp. chapter 4 in this volume). Perhaps most important, there turns out to be an intimate connection between continued *evolutionary plasticity* in a lineage and developmental modularity (Wagner, 1995; Müller and Wagner, 1996; Raff, 1996; Wagner and Altenberg, 1996; Brandon, 1999 and chapter 3 in this volume; Bolker, 2000; see also Dawkins, 1996, on "kaleidoscope embryology"). I have already mentioned Lewontin's argument for the necessity of "quasi independence" of characters and Bonner's view of "gene nets"—adaptive change would be impossible if development were holistic. Wagner and Altenberg (1996) have translated this point in the language of genotype–phenotype mappings. On this view, an evolutionary module is "a set of phenotypic features that are highly integrated by phenotypic effects of the underlying genes and are relatively isolated from other such sets by a paucity of pleiotropic effects" (see chapter 2 in this volume, esp. figure 2.1).

Thus the *genetic representation* (see chapter 17 in this volume) is modular. In the same vein, a *module of selection* may be defined as "a set of genes, their products and interactions (their developmental pathways), the resulting character complex and that complex's functional effect" (Brandon, 1999, p. 177). This, Brandon suggests, is what evolution by natural selection "picks out, selects among, and transforms," implying at long last a solution to the units and levels of selection riddle: "These modules are the units of evolution by natural selection" (1999, p. 177; see also chapter 7 in this volume).

Moss (2001, pp. 87–88) usefully distinguishes between "Genes-P" and "Genes-D." A Gene-P is defined in terms of its functional relationship to a phenotype, "black-boxing," as it were, requirements in regard to specific molecular sequence, and with respect to the biology involved in producing the phenotype (e.g., a "gene for blue eyes"). A gene-D, in contrast, is "mechanistically" defined by its molecular sequence. It is a developmental resource in the sense of Developmental Systems Theory (Oyama et al., 2001) and, as such, "indeterminate" with respect to phenotype. See also Wheeler and Clark (1999) on the analogy of genes in the production of biological form and the role of neural states in the production of behavior, and in particular their discussion of "causal spread." Using this distinction, Moss describes how modularity and homology have come together as complementary themes arising out of research in subcellular biochemistry and molecular biology:

Modularity, at the level of individual genes (Gene-D), which is the rule not the exception for the eukaryotic cell and all metacellular organisms, provides for developmentally contingent flexibility in the expression and realization of "gene-products" from out of the resource base which any Gene-D represents. N-CAM is just such a modularized Gene-D resource, but it is also just one member of a "superfamily" of modularized genetic resources whose kinship is defined by the possession of homologous modules. . . . Much of the evolutionary novelty associated with increasing organismic complexity, it turns out, has been achieved through the reshuffling and mixing and matching of modular exon units to form families of homologous genetic (Gene-D) resources. This has been particularly pronounced with respect to those molecules associated with developmentally and functionally contingent associations between cells and other cells, and cells and extracellular matrices. (Moss, 2001, p. 93; more details in Moss, 2003)

In the modeling scenarios of Schank and Wimsatt, unless development is modular, phenotypes will become *generatively entrenched*, for a change in a developmental sequence is likely to ramify, having many effects on the developed phenotype, some of which would be deleterious (Schank and Wimsatt, 1988, 2001; Wimsatt and Schank, 1988).

Neural and Cognitive Modules

With one exception, all of the chapters in part IV directly or indirectly address the issue of neurocognitive modularity, or at least the question of how the brain and the mind are to be meaningfully interrelated. (The exception is chapter 17, which elaborates Simon's view with respect to the economic realm, showing neatly how classical economic theory can be seen to display modularity in its purest form, and offering a new take on problems we have encountered, such as the problem of genetic representation, along the way.) Chapter 16, on the attractiveness and pitfalls of cognitive modularity in general, and the more specialized treatments of evolutionary connectionism and mind/brain modularity in chapter 14, of modularity in "classical" information-processing models in chapter 15, and of the modular elements that are an integral part of Oller's "natural logic of communicative capabilities" in chapter 18 are all remarkably self-contained, and as such would not require much by way of conceptual preparation here. However, since they are also critical of the computational approach that continues to dominate the cognitive sciences to this day, it should be worthwhile to critically survey some of the developments and views they are responding to—in particular, evolutionary psychology.

Neural Modularity

The modular conception of the human brain goes back at least to the efforts to explain the uniqueness of our species by such pioneers as Pierre-Paul Broca (1824–1880), Carl Wernicke (1848–1904), and—yes—Sir Russell Brain (1895–1966), who sought to discover the neurological "magic module" (Merlin Donald) that might explain human language and symbolic thought. The same motivation was still very much present in most twentieth-century research on patients with impaired brains (see, e.g., Geschwind, 1974) or, say, in Chomsky's battle for the view that the unique properties of human language require a built-in brain device for its generation (contrast Deacon, 1997, who offers an alternative view in which language and the brain coevolve, and Oller, chapter 18 in this volume, for both a critique of the Chomskyan nativist view and a forceful defense of a "self-organizational" alternative). However, the results have been inconclusive at best, if not largely negative:

Every conceivable anatomical comparison has been made between chimpanzees and humans, in the hope of finding the critical structure that explains the gulf between us and our closest relatives. But this has yielded very little. Essentially every structure we can describe in the human brain has an equivalent, or homologue, in the chimpanzee. It is thus virtually certain that our common ancestor five million years ago must also have had the same brain architecture. This in turn implies that no radical modular redesign of the human nervous system has occurred during our evolution. If we are looking for a modular "table of elements" to explain our uniqueness, we had better look somewhere else. It is not there. (Donald, 2001, p. 111)[13]

This caveat having been issued, it seems uncontroversial today that progress in neuroscience, enabled by ever improving anatomical, imaging, and experimental data, has allowed us to identify a number of brain modules at various levels of granularity (e.g., Churchland and Sejnowski, 1992; see also chapters 15 and 16 in this volume). Neuropsychology typically links behavioral data with regions of the brain, using mainly brain-damaged patients and brain-imaging techniques. In this context, Kosslyn and Koenig's (1992) notion of "weak modularity" is relevant: "Even though networks compute input–output mappings, the same network may belong to several processing systems; and, while there is a good measure of localization in the brain, it is also often the case that neurons participating in the same computation belong to different regions" (chapter 15 in this volume).

Cognitive Modularity

Although they are much "softer" in comparison to neural accounts of modularity, much more heat has been generated by various *modularity of mind* hypotheses. These originated in the 1980s on a wave of skepticism about the possibility of a "grand design" for different cognitive phenomena (Turner et al., 1997; see also chapter 16 in this volume) on which evolutionary psychologists continue to surf.

Fodor (1983) argued that there are only two major classes of cognitive entities in the brain: (1) domain-specific mental modules, which include the (unconscious) computations underlying vision and our other input systems as well as the output systems that account for behavior (see also Rozin, 1976), and (2) a domain-general (and conscious) central processor that is barred access to the details of whatever modules do (there must be some domain-general "central systems" that interface with the modules; Fodor, 1983, pp. 101–103).

Echoing Chomsky, Fodor thought that language is one of the modules of mind rather than part of the central processor. (One argument for this stance concerns the *involuntary* dimension of the acquisition, generation, and perception of language.) The essence of Fodorian modularity, then, is "information encapsulation": some of the information outside the module is not accessible from within (Fodor, 1983, p. 71). The restrictions on information flow engender several other symptoms. Modules are "mandatory" (one cannot control whether or not a module applies to a given input); they are typically fast in comparison to nonmodular processes (see Gigerenzer, 1997, and Gigerenzer and Todd, 1999 on "fast and frugal heuristics"); they are "computationally shallow" in that they provide only a preliminary characterization of output; modular mechanisms are associated with fixed neural architecture and, as a consequence, possess characteristic breakdown patterns.

Whereas Fodor's view, like Chomsky's, was and remains clearly anti-Darwinian (chapter 14 in this volume), the evolutionary psychologists who radicalized the modularistic stance (e.g., Barkow et al., 1992; see also Carroll, 1988; Garfield, 1991; Hirschfeld and Gelman, 1994; Sperber, 1994, 2002; Charland, 1995, Segal, 1996) and their philosophical associates, such as Steven Pinker (1997), typically embrace the adaptationist reading of the Modern Synthesis due to Williams (1966) and Dawkins (1976). According to their "massive modularity hypothesis" (MMH), the human mind is composed largely, if not entirely, of innate, special-purpose computational mechanisms or "modules." (Paradoxical as it may seem for scholars who claim to take evolution seriously, evolutionary psychologists tend to be remarkably silent on the issue of apes-to-humans continuity: Heyes, 2000.) The four central tenets of evolutionary psychology (EP) are (1) *computationalism*: the human mind is an information-processing device that can be likened to "a computer made out of organic components rather than silicon chips" (Cosmides et al., 1992, p. 7); (2) *nativism*: much of the human mind is taken to be innate; (3) *adaptationism*, as suggested above: our minds are the mosaic, evolutionary product (see the Swiss army knife metaphor) of a great number of adaptations to challenges posed by the "environment of evolutionary adaptation" in our Pleistocene past; and (4) *massive modularity*, according to which the human mind contains a (very) large number—hundreds, if not thousands—of "Darwinian modules," comprising both peripheral systems and central capacities such as reasoning (see Samuels 1998, 2000; Sperber, 2002). Contrary to Fodor, evolutionary psychologists have had little to say about the neural constraints on their cognitive modules (Scholl, 1997; Panksepp and Panksepp, 2000; chapter 16 in this volume).

EP is sometimes presented as simply "psychology that is informed by the additional knowledge that evolutionary biology has to offer" (Cosmides et al., 1992, p. 3). Its advocates suggest that the very existence of modularity and of the specific modules it postulates begs for an evolutionary explanation. They wonder why this is uncontroversial in the case of nonpsychological modular components of the organism (e.g., the liver or the eyes), "which are generally best understood as adaptations" (Sperber, 2002), but raises eyebrows as soon as it comes to psychology. To the extent that EP aims to complete our causal account of mental capacities by including the phylogenetic dimension (the *explanatory* project in Grantham and Nichols's 1999 terms), it should be rather uncontroversial. As Sperber (2002, p. 49) views it, the evolutionary perspective is especially relevant to psychology, and in particular to the study of cognitive architecture, because *we know so little about the mind*: "Apart from input and output systems, which, being linked to sensory and motor organs, are relatively discernible, there is nothing obvious about the organization of the mind into parts and sub-parts. Therefore all sources of insight and evidence are welcome."

But EP's range of usefulness is supposed to extend beyond explaining why we have certain mechanisms once other branches of psychology have discovered them. Its more ambitious goal is to use the tools of evolutionary biology to predict which mechanisms ought to constitute the brain (Grantham and Nichols call this the *predictive* project). Drawing on the theory of natural selection is appealing here because one of its primary objectives is to explain the functional organization of organisms (see note 1). However, there is no consensus on the kinds of *constraints* evolutionary concepts place on psychological inquiry. The stand one takes on such constraints is formed in large part by how one understands the operation of natural selection. On one account, exemplified by EP sensu stricto, selection must produce highly specialized products tailored to fit the specific environmental conditions considered to cause adaptive problems. Consequently, the brain should be composed of a number of dedicated modules, each outfitted to deal with an adaptive problem.

In contrast, Millikan (1993) and Rozin (1976), among others, argue that natural selection *could* produce general-purpose cognitive devices. Dennett's pragmatic view, although in principle closer to EP, ultimately boils down to the same: "Learning is *not* a general-purpose process, but human beings have so many special-purpose gadgets, and learn to harness them with such versatility, that learning *often* can be treated as if it were an entirely medium-neutral and content-neutral gift of non-stupidity" (Dennett, 1995, p. 491). The philosopher Brandon (1990), for one, has suggested that flexible phenotypes prove advantageous in rapidly changing environments whose fluctuations are difficult to predict. Extending this idea to human cognitive evolution, overly specialized devices might not be able to cope with the rapid changes—and, more to the point, natural selection would not have the time to build an array of specialized devices to contend with the conditions.

Inspired in part by connectionism, Karmiloff-Smith (1992) combines a minimal nativism, which she redefines within a "truly epigenetic perspective of genetic expression rather than genetic unfolding," with Piagetian constructivism (Piaget's view was basically antimodularist). She argues that domain-specific predispositions give development "a small but significant kickstart" by focusing the young infant's attention on proprietary inputs. The early period is then followed by intricate interaction with the environment, which crucially affects brain development in return as subsequent learning takes place.

In chapter 14 of this volume, on mind/brain modularity in an evolutionary-connectionist framework, Calabretta and Parisi likewise argue for a form of connectionism that is neither antimodularist nor antinativist. In their discussion of "theory of mind" (ToM)—the (meta)theory of how people or animals attribute mental states to each other and use them to predict others' behavior—Scholl and Leslie (1999) also address the seeming tension between developmental and "static," nondevelopmental, cognitive-modular accounts of ToM. They explore how ToM may be grounded in a cognitive module, yet still afford development, and conclude that a modular capacity such as ToM may be acquired in at least four distinct ways: (1) the innate capacity is fixed but needs to be appropriately triggered in order to develop fully; (2) the essential character of the capacity is determined by environmental parameter setting; (3) it has an innate basis which is later fixed by module-*internal* development, making use only of information "allowed" past the module's informational boundaries; and finally (4) some of the properties and contents of the capacity or skill may not have an innate basis at all—the capacity may be "cognitively penetrable" and learnable by induction (see Karmiloff-Smith, 1992). This classification is not necessarily exhaustive. Obviously, much more systematic connectionist modeling and fine-grained neurodevelopmental and neurogenetic evidence will be required to settle this rapprochement.

Closing the Gap Between Mind and Brain

Gobet (chapter 15 in this volume) distinguishes three meanings of modularity in psychology: the biological (see above), the functional (à la Fodor or EP), and the knowledge meaning. The latter refers to the modular organization of knowledge ("representation") and has some kinship with the notion of modularity used in computer science and artificial intelligence.

Evolutionary psychologists are not always very clear as to where their "Darwinian" modules belong in terms of this threefold distinction. Samuels (2000) usefully distinguishes between "computational" and "Chomskyan" modules. In his terminology, a Chomskyan module is a domain-specific body of mentally represented knowledge or information that accounts for a cognitive capacity, whereas computational modules are specific computational devices. As systems of *representations*, "inert" Chomskyan modules play a role that differs importantly from that of computational modules, which often "manipulate" the former. The "Darwinian modules" of EP are typically domain-specific computational *mechanisms*, and hence not Chomskyan modules in Samuels' sense.

However, evolutionary psychologists do typically assume that (many) Darwinian modules utilize domain-specific systems of knowledge (i.e., Chomskyan modules). Samuels further distinguishes between strong massive modularity, which assumes that *all* cognitive mechanisms are Darwinian modules in the aforementioned sense, and weak massive modularity, which maintains only that the human mind, including its parts that are responsible for central processing, is *largely* modularly structured. At least some evolutionary psychologists reject the strong MMH in Samuels' sense.

These distinctions are relevant if one wants to assess the empirical evidence for the role of Darwinian as opposed to just Chomskyan modules in central cognition. ToM is quite generally regarded as "the most well-developed experimental case for a computational or Darwinian module that is not peripheral in character (Samuels, 2000, p. 38). Evidence for a computational ToM module comes mainly from dissociative studies (selective impairment). Thus, Williams syndrome subjects with wide ranges of cognitive impairment typically pass false belief tasks (i.e., tasks evaluating whether or not subjects understand when one might hold a false belief), whereas autistic adolescents and adults with normal IQs typically fail them. However, the available evidence does not allow one to decide in favor of an impaired *computational* ToM module as opposed to a specialized body of ToM *knowledge* (Samuels, 1998).[14] Data from experiments on normal subjects, such as the Wason selection task, are similarly inconclusive. These and similar problems concerning the discrimination between functional (say, Darwinian) modules and knowledge or Chomskyan modules add to the general problem of interrelating mind modules and their neural correlates (Scholl, 1997).

My aim in this introductory chapter has not been to spell out the views of the evolutionary psychologists in any more detail than necessary for a proper understanding of part IV—they are extremely well publicized, especially in the more popular media. Nor is this the place to survey the various lines of criticism that have been addressed to them in addition to those included in this volume (see, among many other sources, Carroll, 1988; Sterelny, 1995; Looren de Jong and van der Steen, 1998; Shapiro and Epstein, 1998; Lloyd, 1999; Buller and Hardcastle, 2000; Fodor, 2000; Panksepp and Panksepp 2000, 2001; Rose and Rose, 2000). Just one final remark: Wagner et al. (chapter 2 in this volume), in their discussion of the origin of modularity, point to "mechanistic plurality" as a real possibility. Maybe taking this message to heart can alleviate the frustration of those among us who feel that too much arbitrariness is involved in the current evolutionary-psychological debates.

Acknowledgments

I am grateful to my coeditor Diego Rasskin-Gutman for carefully commenting on the draft version of this chapter. Although I did my best to take his advice to heart wherever I could endorse it, the tension between his Platonist weltanschauung and my more down-to-earth naturalistic philosophy of science remains considerable, so that I cannot humanly expect him to embrace the final product. Our conversation will thus have to be continued. Meanwhile, I have to take full responsibility for any remaining imbalances or worse errors. I also thank James Phelps, whose thoughtful interpretation of the program of EP has inspired some of the views expounded in the section on cognitive and neural modularity, and Sahotra Sarkar for allowing me to read parts of an unpublished book manuscript on evolution from a developmental point of view.

Notes

1. According to Simon (1969, pp. 5–6), artificial things "are synthesized (though not always or usually with full forethought) by man"; they "may imitate appearances in natural things while lacking, in one or many respects, the reality of the latter"; they "can be characterized in terms of functions, goals, adaptation"; and they "are often discussed, particularly when they are being designed, in terms of imperatives as well as descriptives." An important fact about functional explanation, Simon specified, is that it demands an understanding mainly of the outer environment (see Godfrey-Smith, 1996). "Analogous to the role played by natural selection in evolutionary biology is the role played by rationality in the sciences of human behavior. If we know of a business organization only that it is a profit-maximizing system, we can often predict how its behavior will change if we change its environment. . . . We can make this prediction . . . without any detailed assumptions about the adaptive mechanisms, the decision-making apparatus that constitutes the inner environment of the business firm" (Simon, 1969, p. 8). At the most general level, Simon (1973, p. 3) maintained, there are properties related to hierarchy that are "common to a very broad class of complex systems, independently of whether those systems are physical, chemical, biological, social, or artificial. The existence of these commonalities is a matter of empirical observation; their explanation is, in a broad sense, Darwinian—they concern properties that facilitate the evolution and survival of complexity."

2. See, e.g., Hall (1992); Gilbert et al. (1996); the editorial by Raff et al. (1999) in the first issue of the journal *Evolution and Development*; Müller and Newman (2003); and Robert (2004).

3. See, e.g., Dennett (1995); David (2000); and Langlois and Savage (2001). See also Wagner and Altenberg (1996); Wagner et al. (chapter 2 in this volume); and Marengo et al. (chapter 17 in this volume) on decomposability in genetic algorithms.

4. In their work on the heuristics used in both original scientific discovery and novice learning, Simon and his coworkers have shown that calculations such as these can quite easily be automated; see in particular Simon (1977a); Simon et al. (1981); and Langley et al. (1987) on the BACON programs.

5. Supposing that the more exact a system's description is, the more complex it is, one could say that the inherent complexity of an object in principle constitutes the floor for the complexity of an exact description.

6. Although on some of its interpretations, modularity comes conceptually close to Simon's concept of near decomposability of hierarchical systems (e.g., chapter 15 in this volume), hierarchy does not conceptually imply modularity. Thus Simon's (1969) original example of rooms connected by corridors shows a modular design but no hierarchy. In biology this is even more evident: a molecule with different domains shows modularity, but not hierarchy (the hierarchy here is manifested differently, as linear chain → secondary and tertiary structure). The limb is modular, but there is no hierarchy of bones; in this case the hierarchy is one of molecules-cells-tissues, etc. (Diego Rasskin-Gutman, personal communication; see also Agre, 2003). Yet many of the complex systems we encounter, whether assembled from diverse smaller components or through specialization of identical or similar units, display a ND structure (Simon, foreword to this volume) *and* do happen to be modular as well (e.g., Blume and Appel, 1999, and chapters 2 and 11 in this volume). Simon's suggestion that *any* complex, naturally evolved system is constituted by a decomposable hierarchy is challenged by Zawidzki (1998), who takes Kauffman's (1993) models of genetic regulatory networks to provide counterexamples.

7. Not without irony, the experimental work reported in Atran (1998) "supports a modular view of folk biology as a core domain of human knowledge."

8. In the conclusion of his *Inductive Inference and Its Natural Ground*, Hilary Kornblith has this to say on the human ability to cope with natural kinds: "[W]e are quite adept at detecting the very features of natural kinds which are essential to them, and our conceptual structure places these essential features in the position of driving inductive inference. . . . [W]e typically project the properties of natural kinds which are universally shared by their members. It is thus that our inductive inferences are tailored to the causal structure of the world, and thus that inductive understanding of the world is possible" (Kornblith, 1995, p. 107).

9. An important question in this context, which can only be mentioned here, concerns the nature of the "glue" that holds the components of a system together. In Simon's example of the solar system, the "distributional facts" remain unexplained; they call for further explanation beyond the Newtonian framework, possibly in the realm of cosmogony. In living systems—especially if one is interested in their origination (Müller and Newman, 2003)—physical (as well as chemical) *forces* that vary with distance continue to be important *explanantia* of form. But as soon as some sort of *scarce energy* enters the scene, the issue of differential allocation arises (Marengo et al., chapter 17 in this volume). Here Simon's (1969) evolutionary argument of the two watchmakers would seem to gain its full force.

10. Additional conceptual distinctions are provided in Bolker (2000) and Winther (2001).

11. As Brandon (1990) argues, by avoiding the ecological process of selection, genic selectionism—the idea that all of evolution can be understood in terms of selection acting at the level of genes (Williams, 1966; Dawkins, 1976)—cannot possibly explain what makes an adaptation adaptive. The reasons why modern *developmental* biology has come to ignore the environment ultimately can be traced back to Weismann's influential proposal that development was merely the segregation of entities residing within the nucleus (Gilbert and Bolker, 2003, p. 4).

12. West-Eberhard (1998, p. 8417) relates this suppression to the opposition between Darwin's gradualism and the saltationist views that developmental biologists have continued to hold: "The large variants sometimes produced by development . . . invite explanation of adaptive form in terms of accident or divine creation. Darwin was uncompromising on this point and cleverly explained developmentally mediated heterochrony as involving complex traits first established by gradual change in ancestral juveniles or adults. . . ."

13. Donald acknowledges Terry Deacon's claim that in humans, certain parts of the frontal cortex expanded considerably and extended their range of interconnections over evolutionary time (see Deacon, 1997). But he considers this a quibble over small facts, insisting that "The Big Fact is one that should be inscribed on every cognitive theorist's door: NO NEW MODULES" (2001, p. 112).

14. Connectionist neuropsychology poses additional problems for modularists invoking dissociation; see, e.g., Plaut (1995).

References

Adami C (2002) Ab initio modeling of ecosystems with Artificial Life. Nat Res Model 15: 133–145.

Agre PE (2003) Hierarchy and history in Simon's "Architecture of complexity." J Learn Sci 12: 413–426.

Amundson R (1994) Two concepts of constraint: Adaptationism and the challenge from developmental biology. Phil Sci 61: 556–578.

Atran S (1998) Folk biology and the anthropology of science: Cognitive universals and cultural particulars. Behav Brain Sci 21: 547–609.

Baldwin CY, Clark KB (1997) Managing in an age of modularity. Harvard Bus Rev 75(5): 84–93.

Barbuti R, Giacobazzi R, Levi G (1993) A general framework for semantics based on bottom-up abstract interpretation of logic programs. ACM Trans Program Lang Sys 15(1): 133–181.

Barkow JH, Cosmides L, Tooby J (eds) (1992) The Adapted Mind: Evolutionary Psychology and the Generation of Culture. New York: Oxford University Press.

Barrett HC (2001) On the functional origins of essentialism. Mind Soc 2: 1–30.

Bechtel W, Richardson RC (1993) Discovering Complexity: Decomposition and Localization as Strategies in Scientific Research. Princeton, NJ: Princeton University Press.

Blume M, Appel AW (1999) Hierarchical modularity. ACM Trans Program Lang Systems 21: 513–547.

Bolker JA (2000) Modularity in development and why it matters to Evo-Devo. Amer Zool 40: 770–776.

Bonner JT (1988) The Evolution of Complexity by Means of Natural Selection. Princeton, NJ: Princeton University Press.

Brandon RN (1990) Adaptation and Environment. Princeton, NJ: Princeton University Press.

Brandon RN (1999) The units of selection revisited: The modules of selection. Biol Philos 14: 167–180.

Brooks RA (1997) Intelligence without representation. In: Mind Design II (Haugeland J, ed), 395–420. Cambridge, MA: MIT Press.

Buller DJ, Hardcastle VG (2000) Evolutionary psychology meets developmental neurobiology: Against promiscuous modularity. Mind Brain 1: 307–325.

Callebaut W (1989) Post-positivist views of scientific explanation. In: Explanation in the Social Sciences: The Search for Causes in Demography (Duchêne J, Wunsch G, Vilquin E, eds), 141–196. Brussels: Ciaco.

Callebaut W (1995) Réduction et explication mécaniste en biologie. Rev Philos Louvain 91: 33–66.

Callebaut W, Stotz K (1998) Lean evolutionary epistemology. Evol Cognit 4: 11–36.

Carroll, JM (1988) Modularity and naturalness in cognitive science. Metaphor Symbol Activ 3(2): 61–86.

Charland L (1995) Feeling and representing: Computational theory and the modularity of affect. Synthese 105: 273–301.

Churchland PS, Sejnowski TJ (1992) The Computational Brain. Cambridge, MA: MIT Press.

Conrad M (1990) The geometry of evolution. BioSystems 24: 61–81.

Cosmides L, Tooby J, Barkow JH (1992) Introduction: Evolutionary psychology and conceptual integration. In: The Adapted Mind (Barkow JH, Cosmides L, Tooby J, eds), 3–15. New York: Oxford University Press.

David, PA (2000) Path dependence and varieties of learning in the evolution of technological practice. In: Technological Innovation as an Evolutionary Process (Ziman J, ed), 118–133. Cambridge: Cambridge University Press.

Dawkins R (1976) The Selfish Gene. Oxford: Oxford University Press.

Dawkins R (1996) Climbing Mount Improbable. New York: Norton.

Deacon TW (1997) The Symbolic Species: The Co-evolution of Language and the Brain. New York: Norton.

Dennett DC (1995) Darwin's Dangerous Idea: Evolution and the Meanings of Life. New York: Simon & Schuster.

Donald M (2001) A Mind So Rare: The Evolution of Human Consciousness. New York: Norton.

Fodor JA (1983) The Modularity of Mind: An Essay on Faculty Psychology. Cambridge, MA: MIT Press.

Fodor JA (2000) The Mind Doesn't Work That Way: The Scope and Limits of Computational Psychology. Cambridge, MA: MIT Press.

Fontana W, Buss LW (1994) "The arrival of the fittest": Toward a theory of biological organization. Bull Math Biol 56: 1–64.

Garfield, JL (1991) Modularity in Knowledge Representation and Natural-Language Understanding. Cambridge, MA: MIT Press.

Gerhart J, Kirschner M (1998) Evolvability. Proc Natl Acad Sci USA 95: 8420–8427.

Geschwind N (1974) Selected Papers on Language and the Brain. Dordrecht: Reidel.

Gigerenzer G (1997) The modularity of social intelligence. In: Machiavellian Intelligence II: Extensions and Evaluations (Byrne RW, Whiten A, eds), 264–288. Cambridge: Cambridge University Press.

Gigerenzer G, Todd PM (eds) (1999) Simple Heuristics That Make Us Smart. New York: Oxford University Press.

Gilbert SF, Bolker JA (2001) Homologies of process and modular elements of embryonic construction. J Exp Zool (Mol Dev Evol) 291: 1–12.

Gilbert SF, Bolker JA (2003) Ecological developmental biology: Preface to the symposium. Evol Dev 5(1): 3–8.

Gilbert SF, Opitz JM, Raff RA (1996) Resynthesizing evolutionary and developmental biology. Dev Biol 173: 357–372.

Godfrey-Smith P (1996) Complexity and the Function of Mind in Nature. Cambridge: Cambridge University Press.

Grantham TA, Nichols S (1999) Evolutionary psychology: Ultimate explanations and Panglossian predictions. In: Where Biology Meets Psychology: Philosophical Essays (Hardcastle VG, ed), 47–66. Cambridge, MA: MIT Press.

Griesemer JR (2002) What is "epi" about epigenetics? In: From Epigenesis to Epigenetics: The Genome in Context (Van Speybroeck L, Van de Vijver G, De Waele D, eds), Ann NY Acad Sci 981: 97–110.

Griffiths PE, Gray RD (2001) Darwinism and developmental systems. In: Cycles of Contingency: Developmental Systems and Evolution (Oyama S, Griffiths PE, Gray RD, eds), 195–218. Cambridge, MA: MIT Press.

Hall BK (1992) Evolutionary Developmental Biology. London: Chapman and Hall.

Hansen TF (2003) Is modularity necessary for evolvability? Remarks on the relationship between pleiotropy and evolvability. BioSystems 69: 83–94.

Heyes CM (2000) Evolutionary psychology in the round. In: Evolution of Cognition (Heyes CM, Huber L, eds), 1–21. Cambridge, MA: MIT Press.

Hirschfeld LA, Gelman SA (eds) (1994) Mapping the Mind: Domain Specificity in Cognition and Culture. New York: Cambridge University Press.

Hull DL (1988) Science as a Process: An Evolutionary Account of the Social and Conceptual Development of Science. Chicago: University of Chicago Press.

Hurley SL (1999) Rationality, democracy, and leaky boundaries: Vertical vs. horizontal modularity. In: Democracy's Edges (Shapiro I, Hacker-Cordón C, eds), 273–294. Cambridge: Cambridge University Press.

Husken M, Igel C, Toussaint M (2002) Task-dependent evolution of modularity in neural networks. Connect Sci 14: 219–229.

Jablonka E (2001) The systems of inheritance. In: Cycles of Contingency: Developmental Systems and Evolution (Oyama S, Griffiths PE, Gray RD, eds), 99–116. Cambridge, MA: MIT Press.

Karmiloff-Smith A (1992) Beyond Modularity: A Developmental Perspective on Cognitive Science. Cambridge, MA: MIT Press.

Kauffman S (1993) The Origins of Order. Oxford: Oxford University Press.

Kitcher P (2001) Battling the undead: How (and how not) to resist genetic determinism. In: Thinking About Evolution: Historical, Philosophical, and Political Perspectives (Singh RS, Krimbas CB, Paul DB, Beatty J, eds), 396–414. New York: Cambridge University Press.

Kornblith H (1995) Inductive Inference and Its Natural Ground: An Essay in Naturalistic Epistemology. Cambridge, MA: MIT Press.

Kosslyn SM, Koenig O (1992) Wet Mind. New York: Free Press.

Langley P, Simon HA, Bradshaw GL, Zytkow JM (1987) Scientific Discovery: Computational Explorations of the Creative Processes. Cambridge, MA: MIT Press.

Langlois RN, Savage DA (2001) Standards, modularity, and innovation: The case of medical practice. In: Path Dependence and Creation (Garud R, Karnøe P, eds), 149–168. Mahwah, NJ: Lawrence Erlbaum.

Levinson P (1982) Evolutionary epistemology without limits. Knowl: Creat, Diff, Util 4: 465–502.

Lewontin RC (1974) The Genetic Basis of Evolutionary Change. New York: Columbia University Press.

Lloyd EA (1999) Evolutionary psychology: The burdens of proof. Biol Philos 14: 211–233.

Looren de Jong H, Van der Steen WJ (1998) Biological thinking in evolutionary psychology: Rockbottom or quicksand? Phil Psych 11: 183–205.

Millikan RG (1993) White Queen Psychology and Other Essays for Alice. Cambridge, MA: MIT Press.

Minelli A (1998) Molecules, developmental modules, and phenotypes: A combinatorial approach to homology. Molec Phylogenet Evol 9: 340–347.

Moss L (2001) Deconstructing the gene and reconstructing molecular developmental systems. In: Cycles of Contingency: Developmental Systems and Evolution (Oyama S, Griffiths PE, Gray RD, eds), 85–97. Cambridge, MA: MIT Press.

Moss L (2003) What Genes *Can't* Do. Cambridge, MA: MIT Press.

Müller GB, Newman SA, eds (2003) Origination of Organismal Form: Beyond the Gene in Developmental and Evolutionary Biology. Cambridge, MA: MIT Press.

Müller GB, Olsson L (2003) Epigenesis and epigenetics. In: Keywords in Evolutionary Developmental Biology (Hall BK, Olson W, eds), 114–123. Cambridge, MA: Harvard University Press.

Müller GB, Wagner GP (1996) Homology, Hox genes, and developmental integration. Amer Zool 36: 4–13.

Needham J (1933) On the dissociability of the fundamental processes in ontogenesis. Biol Rev 8: 180–223.

Newman SA, Müller GB (2000) Epigenetic mechanisms of character organization. J Exp Zool (Mol Dev Evol) 288: 304–317.

Oyama S, Griffiths PE, Gray RD (2001) Introduction: What is Developmental Systems Theory? In: Cycles of Contingency: Developmental Systems and Evolution (Oyama S, Griffiths PE, Gray RD, eds), 1–11. Cambridge, MA: MIT Press.

Panksepp J, Panksepp JB (2000) The seven sins of evolutionary psychology. Evol Cognit 6: 108–131.

Panksepp J, Panksepp JB (2001) A continuing critique of evolutionary psychology: Seven sins for seven sinners, plus or minus two. Evol Cognit 7: 56–80.

Pinker S (1997) How the Mind Works. London: Penguin.

Plaut DC (1995) Double dissociation without modularity: Evidence from connectionist neuropsychology. J Clin Exp Neuropsychol 17: 291–321.

Raff EC, Raff RA (2000) Dissociability, modularity, evolvability. Evol Dev 2(5): 235–237.

Raff RA (1996) The Shape of Life. Chicago: University of Chicago Press.

Raff RA, Arthur W, Carroll SB, Coates MI, Wray G (1999) Editorial: Chronicling the birth of a discipline. Evol Dev 1: 1–2.

Ridley M (1993) Evolution. Boston: Blackwell Scientific.

Riedl RJ (1978) Order in Living Organisms: A Systems Analysis of Evolution (Jefferies RPS, trans). New York: Wiley. German orig. 1975.

Robert JS (2004) Embryology, Epigenesis, and Evolution: Taking Development Seriously. Cambridge: Cambridge University Press.

Robert JS, Hall BK, Olson WM (2001) Bridging the gap between developmental systems theory and evolutionary developmental biology. BioEssays 23: 1–9.

Rose H, Rose S, eds (2000) Alas, Poor Darwin: Arguments Against Evolutionary Psychology. New York: Harmony Books.

Rozin P (1976) The evolution of intelligence and access to the cognitive unconscious. In: Progress in Psychology vol. 6, (Sprague JM, Epstein AN, eds), 245–280. New York: Academic Press.

Samuels R (1998) Evolutionary psychology and the massive modularity hypothesis. Brit J Phil Sci 49: 575–602.

Samuels R (2000) Massively modular minds: Evolutionary psychology and cognitive architecture. In: Evolution and the Human Mind: Modularity, Language and Meta-Cognition (Carruthers P, Chamberlain A, eds), 13–46. Cambridge: Cambridge University Press.

Samuels R, Stich S, Tremoulet P (1999) Rethinking rationality: From bleak implications to Darwinian modules. In: What Is Cognitive Science? (Lepore E, Pylyshyn Z, eds), 74–120. Oxford: Blackwell.

Schank JC, Wimsatt WC (1988) Generative entrenchment and evolution. In: PSA 1986, vol. 2 (Fine A, Machamer PK, eds), 33–60. East Lansing, MI: Philosophy of Science Association.

Schank JC, Wimsatt WC (2001) Evolvability, adaptation and modularity. In: Thinking About Evolution, vol. 2: Historical, Philosophical, and Political Perspectives (Singh RS, Krimbas CB, Paul DB, Beatty J, eds), 322–335. New York: Cambridge University Press.

Scholl BJ (1997) Neural constraints on cognitive modularity? Behavi Brain Sci 20: 575–576.

Scholl BJ, Leslie AM (1999) Modularity, development and "theory of mind". Mind Lang 14: 131–153.

Segal G (1996) The modularity theory of minds. In: Theories of Theories of Mind (Carruthers P, Smith PK, eds), 141–157. Cambridge: Cambridge University Press.

Shapere D (1984) Reason and the Search for Knowledge: Investigations in the Philosophy of Science. Dordrecht: Reidel.

Shapiro L, Epstein W (1998) Evolutionary theory meets cognitive psychology: A more selective perspective. Mind Lang 13: 171–194.

Simon HA (1969) The Sciences of the Artificial. Cambridge, MA: MIT Press. 2nd ed.

Simon HA (1973) The organization of complex systems. In: Hierarchy Theory (Pattee HH, ed), 3–27. New York: Braziller.

Simon HA (1977a) Models of Discovery. Dordrecht: Reidel.

Simon HA (1977b) How complex are complex systems? In: PSA 1976 (Suppe F, Asquith PD, eds), 507–522. East Lansing, MI: Philosophy of Science Association.

Simon HA (1995) Near decomposability and complexity: How a mind resides in a brain. In: Mind, the Brain, and Complex Adaptive Systems (Morowitz HJ, Singer JL, eds), 25–44. Santa Fe Institute Studies in the Sciences of Complexity, Proceedings vol. 22. Reading, MA: Addison-Wesley.

Simon HA, Langley P, Bradshaw G (1981) Scientific discovery as problem solving. Synthese 47: 1–27.

Sperber D (1994) The modularity of thought and the epidemiology of representations. In: Mapping the Mind: Domain Specificity in Cognition and Culture (Hirschfeld LA, Gelman SA, eds), 39–67. New York: Cambridge University Press.

Sperber D, ed (2000) Metarepresentations: A Multidisciplinary Perspective. New York: Oxford University Press.

Sperber D (2002) In defense of massive modularity. In: Language, Brain and Cognitive Development. Essays in Honor of Jacques Mehler (Dupoux E, ed), 47–57. Cambridge, MA: MIT Press.

Star LS (1989) Regions of the Mind: Brain Research and the Quest for Scientific Certainty. Stanford, CR: Stanford University Press.

Sterelny K (1995) Review of (Barkow et al. 1992). Biol Psychol 10: 365–380.

Sterelny K (2001) Niche construction, developmental systems, and the extended replicator. In: Cycles of Contingency: Developmental Systems and Evolution (Oyama S, Griffiths PE, Gray RD, eds), 333–349. Cambridge, MA: MIT Press.

Székely G (2001) An approach to the complexity of the brain. Brain Res Bull 55(1): 11–28.

Turner JH, Borgerhoff Mulder M, Cosmides L, Giesen B, Hodgson G, Maryanski AM, Shennan SJ, Tooby J, Velichkovsky BM (1997) Looking back: Historical and theoretical context of present practice. In: Human by Nature: Between Biology and the Social Sciences (Weingart P, Mitchell SD, Richerson PJ, Maasen S, eds), 17–64. Malwah, NJ: Erlbaum.

Von Dassow G, Munro E (1999) Modularity in animal development and evolution: Elements of a conceptual framework for EvoDevo. J Exp Zool (Mol Dev Evol) 285: 307–325.

Wagner GP (1995) The biological role of homologues: A building block hypothesis. N Jb Geol Paläont Abh 195: 279–288.

Wagner GP (1996) Homologues, natural kinds, and the evolution of modularity. Amer Zool 36: 36–43.

Wagner GP, Altenberg L (1996) Perspective: Complex adaptations and the evolution of evolvability. Evolution 50: 967–976.

Wagner GP, Laubichler MD (2000) Character identification in evolutionary biology: The role of the organism. Theory Biosci 119: 20–40.

Wagner GP, Laubichler MD (2004) Rupert Riedl and the re-synthesis of evolutionary and developmental biology: Body plans and evolvability. J Exp Zool (Mol Dev Evol) 302B: 92–102.

West-Eberhard MJ (1998) Evolution in the light of developmental and cell biology, and vice versa. Proc Natl Acad Sci USA 95: 8417–8419.

Wheeler M, Clark A (1999) Genic representation: Reconciling content and causal complexity. Brit J Phil Sci 50: 103–135.

Williams GC (1966) Adaptation and Natural Selection. Princeton, NJ: Princeton University Press.

Wimsatt WC, Schank JC (1988) Two constraints on the evolution of complex adaptations and the means of their avoidance. In: Evolutionary Progress (Nitecki MH, ed), 231–275. Chicago: University of Chicago Press.

Winther RG (2001) Varieties of modules: Kinds, levels, origins and behaviors. J Exp Zool (Mol Dev Evol) 291: 116–129.

Zawidzki TW (1998) Competing models of stability in complex, evolving systems: Kauffman vs. Simon Biol Phil 13: 541–554.

II EVO-DEVO: THE MAKING OF A MODULAR WORLD

The mutual relationship that holds between development and evolution was recognized early on by Ernst Haeckel, and is the main theme of the now fashionable field of evolutionary developmental biology (evo-devo), which did not participate in the mainstream until only about 2000. Brian Hall puts it this way:

> the challenge taken up by evolutionary developmental biologists is to integrate development with genomic, organismal, and population approaches to evolution. In Gabriel Dover's words: "to Dobzhansky's 'nothing in biology makes sense except in the light of evolution,'" I would add that nothing in evolution makes sense except in the light of the processes, starting within genomes, that affect developmental operations and ultimately spread through a population. . . . Wed this to natural selection . . . and we have the hierarchical, yet integrative approach required. (Hall, 1999, p. 14)

This having been said, making the connection between development and evolution is no straightforward matter. Attempts to provide development with an evolutionary link have emphasized population dynamics, fitness, functional optimization, and natural selection as pivotal connections between both disciplines. Gene sequences shared by such distant organisms as worms, flies, and vertebrates readily suggest a relationship between these groups and prompt the search for an explanation of both the evolutionary origin of their divergent sequences and their shared developmental mechanisms. Remarkably, both the same molecules and the same mechanisms are involved in building quite different organs (even in the same organism), suggesting that the evolutionary process has co-opted them in order to generate structural and metabolic innovations in different groups. This is the fuel of evolution.

Wagner, Chiu, and Laublicher (2000) have argued that evo-devo has a twofold importance in broadening our understanding of biological problems: (1) by providing new ways to assess homological relations and the genotype–phenotype map in well-established fields, such as systematics and evolutionary theory, and (2) by providing a unique research program that is able to tackle questions such as developmental constraints and mechanistic explanations of evolutionary innovations.

We open this book with a section devoted to evo-devo, to find that the concept of modularity can be a powerful nexus between developmental and evolutionary questions. This is perhaps the most biological section of the book, and a logical starting point.

Günter Wagner, Jason Mezey, and Raffaele Calabretta (chapter 2) start this section with an overview of ideas about and models of the evolutionary origin of modules. They identify up to eight different and mechanistically independent possible mechanisms, and further discuss the problems that are involved when trying to provide a causal explanation for the origin of modularity. Their view on modularity rests on the structure of the genotype–phenotype map, implying that specific sets of genes are strongly integrated, affecting some parts of the body but not others. They suggest that the origin of modules is a special case of the evolution of genetic architecture and, as such, any evolutionary model for the origin of modules has to explain how natural selection could produce this distribution of genetic effects. They end on the cautionary note that it would be illusory to assume that one mechanism will explain the origin of modularity in all circumstances; hence, "mechanistic plurality" is a real possibility.

Continuing the discussion of the origin of evolutionary modules (or "modules of selection," as he calls them), Robert Brandon (chapter 3) first reviews earlier arguments by Lewontin (quasi independence, allowing selection to act on a characteristic without altering others) and Bonner (discrete "gene nets"). He argues that these arguments were sound but "transcendental," in philosophical parlance (claiming that modularity is necessary for the very existence of adaptation), and as such are nonexplanatory. Alternatively, belief in the existence of evolutionary modules may be inferred "indirectly" from phylogenetic data or based on the "direct" observation of modules. Brandon then argues that an evolutionary module must be both functionally and developmentally modular. Finally, he contrasts his own "conceptual analysis" with the "empirical hypothesis" of the Wagner–Altenberg model by means of a comparison with an earlier controversy over the proper characterization of teleonomic systems.

To tackle the question "Do modules mechanistically interact, or selectively compete, or both?," Rasmus Winther (chapter 4) surveys research on cells and social insect (particularly hymenopteran) organisms as parts of a whole to explore two perspectives on modular processes. The *integration* view (parts of organisms are intermediate-level modules involved mainly in mechanistic processes) is embraced by evo-devo; the *competition* view (parts are interactor modules engaged primarily in selective processes) is prominent in the "levels of selection" debate. There turns out to be significant overlap between the two perspectives. Genes are important mechanistic modules in the integration view and are, generally, replicator models in the competition view; and partitioning strategies in both perspectives tend to focus on the context-independent properties and powers of genes. In the terminology of Lenny Moss, the goal of the integration perspective is to unravel the mechanisms involving "genes-D," whereas the competition perspective is concerned with the change in frequencies of "genes-P."

In chapter 5, Lee Altenberg offers an account of mutational kinetics and provides a provocative view of the nature of modularity in the genotype–phenotype mapping. Modularity here refers to a genotype–phenotype map that can be (nearly) decomposed into the product of independent genotype–phenotype maps of smaller dimension. For Altenberg, the evolutionary advantages that are attributed to modular design do not derive from modularity per se; what matters is the "alignment" between the spaces of variation of a phenotypic trait and the selective forces that are available to the organism. This determines what Alfenberg calls "constructional selection," by which evolution favors the appearance of new loci that are in alignment with selective gradients, and rejects those that can be detrimental for the overall fitness of the population. Modular genotype–phenotype mapping facilitates such an alignment but is not sufficient: the appropriate phenotype fitness map is also necessary for evolvability.

Lauren Ancel Meyers and Walter Fontana (chapter 6) elegantly tackle the issue of modularity, using computational models of RNA folding, a particularly straightforward case of the genotype–phenotype mapping. In their view, RNAs can be decomposed into modules or subunits that are independent with respect to their thermodynamic environment, genetic context, and folding kinetics. Modules in RNA are stretches of contiguous ribonucleotides held together by a covalent backbone. A test for modularity in RNA is suggested: by looking at their melting profiles, modules can be identified as those subunits that dissolve discretely without perturbing the remaining structure as temperature increases. Ancel Meyers and Fontana finally show that modularity arises as a necessary by-product when natural selection acts to reduce the plasticity of molecules by stabilizing their shapes. Modules resist change; yet by enabling variation "at a new syntactical level," modularity may also provide an escape from the evolutionary dead end that produced it in the first place.

To end this section, Gerhard Schlosser (chapter 7) offers a discussion of modules with respect to the concept of unit of selection and, by showing several examples in the evolution of amphibians, sets out the conditions that are necessary to positively identify a module. These conditions are based on putting things in a proper phylogenetic framework so that the correlation in their variation can be readily established and on identifying the structure as a "true" module (i.e., by showing that it is independent from its surroundings, both structurally and functionally). Modules often coincide with units of evolution. Yet, Schlosser argues, not all modules need to act as units of evolution, and not all units of evolution need to be modules. Most important, Schlosser suggests both empirical and experimental procedures to analyze this two-step test for modularity.

References

Hall BK (1999) Evolutionary Developmental Biology. London: Chapmann and Hall.

Wagner GP, Chiu C-H, Laubichler MD (2000) Developmental evolution as a mechanistic science: The inference from developmental mechanisms to evolutionary processes. Amer Zool 40: 819–831.

2 Natural Selection and the Origin of Modules

Günter P. Wagner, Jason Mezey, and Raffaele Calabretta

There is an emerging consensus about the existence of developmental and evolutionary modules and their importance for understanding the evolution of morphological phenotypes (Bolker, 2000; Raff, 1996; Wagner and Altenberg, 1996). Modules are considered important for the evolvability of complex organisms (Bonner, 1988; Wagner and Altenberg, 1996) and for the identification of independent characters (Houle, 2001; Kim and Kim, 2001; Wagner, 1995) and necessary for heterochrony (Gould, 1977). Methods to recognize and test for modularity have been developed (Cheverud et al., 1997; Mezey et al., 2000) and comparative developmental data have been reinterpreted in the context of the modularity concept (Nagy and Williams, 2001; Schlosser, chapter 7 in this volume; Stock, 2001). In contrast to the progress made in these areas, there has been very little research on the origin of modules, and the few results published about models for the origin of modules point in widely different directions (Altenberg, 1994; Ancel and Fontana, 2000; Calabretta et al., 2000; Rice, 2000). As of 2004 no unitary explanation has emerged for the evolution of modularity. This is surprising, since modularity seems to be so common among higher organisms that one might expect a robust and unitary mechanism behind its origin.

In this chapter we want to review the current models and ideas for the evolutionary origin of modules. The majority of the models discussed below were published in 2000 or 2001, and we thus feel that an overview might be useful. Another goal of this chapter is to identify the range of open problems we face in explaining the ultimate causes of modularity.

Kinds of Modules

While the intuitive idea of modularity is pretty simple, the distinction between different types of modularity and their operational definition stimulates ongoing conceptual development (Brandon, 1999; von Dassow and Munro, 1999; Nagy and Williams, 2001; Sterelny, 2000; and chapters 4, 7, and 8 in this volume). In this chapter, however, we do not want to enter the discussion about the more subtle aspects of the modularity concept but, rather, use a few fairly simple and perhaps robust distinctions and definitions sufficient to communicate about models for the origin of modularity.

The biological modularity concept has several largely independent roots. In developmental biology the modularity concept is based on the discovery of semiautonomous units of embryonic development (Raff, 1996). The empirical basis for *developmental modules* is the observation that certain parts of the embryo can develop largely independent of the context in which they occur. Examples are limb buds and tooth germs (Raff, 1996), developmental fields (Gilbert et al., 1996), and clusters of interacting molecular reactions (Abouheif, 1999; Gilbert and Bolker, 2001; Wray 1999). On the other hand, *evolutionary modules* are defined by their variational independence from each other and the integration among their parts, either in interspecific variation or in mutational variation (Wagner and Altenberg, 1996).

The preliminary definition of an evolutionary module used in this chapter is a set of phenotypic features that are highly integrated by pleiotropic effects of the underlying genes and are relatively isolated from other such sets by a paucity of pleiotropic effects (figure 2.1). This preliminary definition is also the basis for attempts to measure and test for modularity in genetic data (Cheverud et al., 1997; Mezey et al., 2000). *Functional modules*, on the other hand, are parts of organisms that are independent units of physiological regulation (Mittenthal et al., 1992), such as biomechanical units (Schwenk, 2001), or an isolated part of the metabolic network (Rohwer et al., 1996). The precise definition of all these concepts is somewhat difficult and still controversial. The real challenge, however, is to determine how these different kinds of modules relate to each other. For instance, are evolutionary and developmental modules the same? If not, why and in what respects are they different?

Intuitively, developmental and evolutionary modules should be very closely related. The developmental process determines how a gene influences the

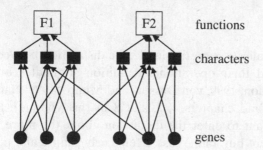

Figure 2.1
Variational modularity of a set of phenotypic characteristics is defined as integration due to the presence of many pleiotropic effects of genes and relative independence from other phenotypic characters due to a relative lack of pleiotropic effects. It is also often the case that a phenotypic module also is dedicated primarily to a specific function. In this case the variational module is also an adaptive character, or an evolutionary module.

phenotype, and hence the existence of developmental modules should influence the structure of the genotype–phenotype map. This is a largely correct argument, but it fails to show that developmental modules map one-to-one to evolutionary modules. One of the reasons is that developmental modules can be deployed repeatedly, as in the case of the left and right forelimb buds. Each of the two forelimb buds is an independent developmental module because each is a self-contained developmental unit with its own capacity for self-differentiation. From a variational point of view, however, the left and right forelimbs are not independent, because they express the same genetic information. Mutations are thus expected to affect both forelimbs simultaneously, and the genetic variation is correlated. Hence the two forelimbs indeed are two *different developmental modules* of the organism, and also are parts of the *same evolutionary module*.

The distinction between developmental and evolutionary modules may be critical for the question of how evolutionary modules originate. One of the most common modes for the origin of evolutionary modules (i.e., phenotypic units of variation) is the differentiation of repeated developmental modules (Raff, 1996; Riedl, 1978; Weiss, 1990). One example is the evolutionary differentiation of teeth. Each individual tooth germ is a developmental module, but each differentiated tooth class is an evolutionary module (Stock, 2001). Another example is arthropod segments, which are potential developmental modules, and tagmata like thorax and abdomen as the evolutionary modules derived from the differentiation of a set of segments (Nagy and Williams, 2001). This fact may be relevant for the origin of evolutionary modules. The main problem is to explain the suppression of pleiotropic effects among genetically coupled parts of the body (i.e., the evolution of individuality of primitively integrated units). The implications of this fact have not been explored systematically, but may hold the key to one of the problems in the origin of modules discussed in the next section.

Mechanisms for the Origin of Modules

In this section we review models for the evolutionary origin of modules. The objective is to understand how natural selection may have acted on the phenotype so as to produce evolutionary modules. As defined above, evolutionary modularity is a statement about the statistical structure of the genotype–phenotype map (Mezey et al., 2000). It implies that certain sets of phenotypic features are affected by the same set of genes, and thus are highly integrated, but these genes have few pleiotropic effects affecting other parts of the body. An evolutionary model for the origin of modules has to explain how natural selection could produce this distribution of genetic effects. Hence the origin of modules is a special case of the evolution of genetic architecture. So far we recognize two classes of models. In one class of models there is a more or less direct selective advantage associated with evolutionary modularity. Within this class, different models differ with respect to the kind of connection assumed between modularity and fitness. In the second class there is no direct selection for modularity, which arises more indirectly through the dynamics of evolution (Calabretta et al., 2000; Force et al., 2004).

Direct Selection for Modularity

For natural selection to cause modularity, there has to be a connection between a selective advantage and modularity. One of the most frequently noted effects of modularity is its potential impact on evolvability (Altenberg, 1995; Galis, 1999, 2001; Gerhart and Kirschner, 1997; Holland, 1992; Liem, 1973; Riedl, 1978; Vermeij, 1970; Wagner and Altenberg, 1996). Hence it is tempting to suggest that modularity evolves as a result of selection for evolvability (Gerhart and Kirschner, 1997; Riedl, 1978). We will explore this possibility first. The other possibility is that modularity is a result of mutations that break developmental constraints due to nonadaptive linkages between characters (Leroi, 2000).

Selection for Evolvability   The question of whether modularity can be explained as an adaptation for evolvability has to be discussed in the broader context of whether selection for evolvability can be a factor in the evolution of genetic architecture. This question is unresolved. In principle, selection for evolvability is possible, particularly in asexual species. The mechanism is a simple Darwinian selection process based on a difference in mean fitness caused by differences in the rate of adaptation among clones (Wagner, 1981). Experimentally it has been shown that alleles that increase the mutation rate get selected in bacterial populations if the population faces a new environment, a situation which is consistent with models for the selection for evolvability (Cox and Gibson, 1974).

However, the mechanism works well only if there is either no recombination or there is a strong linkage disequilibrium between, say, the mutator locus and the genes which mutate to advantageous alleles. With recombination, the mutator gene can no longer ride to fixation on the coattails of the other genes, a process that has been called "hitchhiking" (Maynard-Smith and Haigh, 1974). The reason is that recombination will separate the mutator from the advantageous mutations. The same argument holds for any other mechanism that may influence the rate of adaptation, such as differential epistasis that may suppress deleterious pleiotropic effects (see below). Consequently, with recombination, selection for evolvability becomes a very weak force.

At this point, we want to report the results of a study that aimed at modeling the evolution of pleiotropic effects (Mezey, 2000). Let us consider two characters, one under directional selection and one under stabilizing selection. This model represents a fairly generic scenario for a complex organism. Whenever natural selection acts to change a character, many other characters of the same organism will remain under stabilizing selection. It has been shown that pleiotropic effects among these two characters decrease the rate of evolution of the character under directional selection (Baatz and Wagner, 1997). Pleiotropic effects decrease evolvability. The question then is whether natural selection could fix a modifier allele which suppresses the pleiotropic, and thus increases evolvability (figure 2.2). We used an individual-based model to investigate this question and estimated the selection coefficients of the modifier allele by measuring the time to fixation. The result was that there was quite a strong selection for the modifier (a sample of the results is given in table 2.1).

However, the selection coefficient alone does not tell us whether we are dealing with selection for evolvability. The mean fitness of genotypes with different modifier alleles is influenced by at least two factors: (1) the amount of variation in the character under stabilizing selection, and (2) the relative location of the genotypes along the direction of directional selection (figure 2.3). Only the second factor can be called selection for evolvability, since it derives from differential rates of adaptation. We determined the relative contributions of these two factors to the selection coefficient of the modifier and found that in all cases where we checked, the

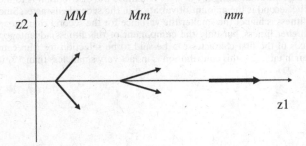

Figure 2.2
A modifier model in which the genotype at a modifier locus determines the relative size of pleiotropic effects between two characters. With MM the effects on the two characters of a mutation are of the same magnitude; with Mm the effects on $z2$ are smaller than the effects on $z1$; and with mm the mutations have no effect on $z2$. The modifier allele m suppresses the pleiotropic effects on the character that is under stabilizing selection. Selection of this allele increases evolvability (Baatz and Wagner, 1997).

Table 2.1
Selection coefficient of a modifier that suppresses pleiotropic effects and the percentage of the selection coefficient explained by selection for evolvability

	s	% explained by evolvability
Vs = 2	0.08	2.0%
Vs = 10	0.08	4.5%

Note that the selection coefficient is quite high with 8%, but only 2%–5% of that can be attributed to selection for evolvability. Vs is the strength of stabilizing selection; the directional selection was 0.1; and the population size was 100.

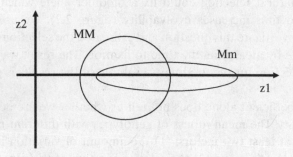

Figure 2.3
Comparison of the distribution of genotypic values of two classes of genotypes. The MM genotypes have equal mutational effects on the two characters, and the distribution is thus circular in this model. The other class of genotypes, Mm, has smaller effects on the second character, and the distribution of genotypic values is thus more extended along the axis of the first character. If the first character is under directional selection for larger character values and the second is under directional selection, the relative mean fitness of these two classes of genotypes is influenced by two factors. The first is the relative location of the genotype distribution along the z1 axis. The more the distribution is to the right, the higher is the mean fitness. The second is the amount of variation in the second character. Since z2 is under stabilizing selection, the fitness is higher the smaller the variance for the second character. In this case the Mm distribution has higher fitness, but only the component of this fitness advantage that is due to the location along the axis of the first character can be said to be selection for Mm caused by selection for evolvability. As seen in table 2.1, this contribution is in fact very small, less than 5% in most cases tested.

fraction of the selection advantage due to selection for evolvability was much less than 10%. In other words, more than 90% of the selective advantage of suppressing pleiotropic effects was due to direct selective advantages rather than advantages related to evolvability per se. Hence, we concluded that even if natural selection can be effective in removing pleiotropic effects, the resulting increase in evolvability is not explained by direct selection for the rate of evolution.

Another study on the evolution of evolvability had a similar result (Turney, 2000). The model considered mutations which increased the dimensionality of the phenotype and thus the number of degrees of freedom for adaptive variation. It was shown that evolvability increases during the simulation runs. The evolutionary mechanism was a direct selective advantage to the mutations that increased evolutionary versatility. Mutations that increased versatility led directly to higher fitness phenotypes that were previously inaccessible.

Hence evolvability can evolve and even improve, but evolvability per se is perhaps not the target of selection. From that we conclude that evolution of modularity is unlikely to result from direct selection in favor of evolvability. One caveat in this argument, however, is that we are not aware of any work on selection for evolvability in populations with spatial structure. Spatial structure may make selection for evolvability more likely than selection in panmictic populations.

These results suggest two possible mechanisms for the origin of modules. One is that the genotype–phenotype map has a direct impact on mean fitness, in particular if the population is far from equilibrium (see also Rice, 1990). Hence it is conceivable that modularity results from the fact that pleiotropic effects can decrease the mean fitness of a population if the population experiences directional selection. The other possibility is that mutations that produce modularity break genetic constraints on adaptation and thus would be selected because they make advantageous phenotypes accessible.

Direct Selection on Pleiotropic Effects   Based on the results reported above, we attempted to evolve modularity in a quantitative genetic model by alternating directional selection and differential epistasis (figure 2.4). The rationale was that directional selection on a single character selects against pleiotropic effects on other characters. If two characters never experience directional selection simultaneously, a modular genetic architecture for the two characters may arise (i.e., one set of genes with most of their effects focused on one character and another set of genes with most of their effects focused on the other character). The results, however, showed that alternating selection does not lead to a separation of genes into two

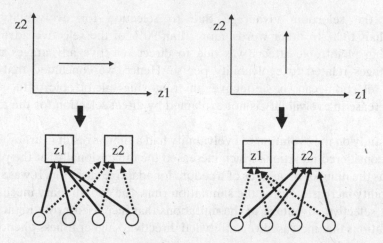

Figure 2.4
The effect of directional selection on one character on the genetic architecture of a two-character phenotype. Whenever there is directional selection on any one character, all the genes increase their contribution to this character. Even directional selection strictly alternating between the two characters does not lead to a segregation of genes into a modular pattern.

character-specific sets, one with effects on one character and the other with effects on the other character. The distribution of gene effects did not settle into a modular pattern; rather, any episode of directional selection tended to recruit genes into the selected character (Mezey, 2000). From this we concluded that alternating selection alone cannot account for the origin of evolutionary modularity.

Modularity as an Escape from Adaptive Constraints   The second alternative mentioned above is that modularity may result from mutations which overcome constraints among adaptive traits. This idea is related to the fact that structural and functional decoupling can facilitate adaptation (Galis, 2001; Liem, 1973) and was proposed as a mechanisms for the origin of modularity by Leroi (2000), but to our knowledge it has not been explicitly modeled, and thus is hard to evaluate at this time. Perhaps the most relevant, but still limited, model is that of Turney (2000) on the evolution of evolutionary versatility discussed above.

Constructional Selection   The oldest model for the origin of modularity that in fact works is constructional selection, proposed by Altenberg (1994; see also chapter 5 in this volume). It is based on the assumption that genes with fewer pleiotropic effects have a higher probability of establishing duplicated copies of themselves in the genome. This model is thus based on intragenomic competition among genes with different degrees of pleiotropy. It predicts the evolution of lower and lower average degrees of pleiotropy. The problematic aspect of this model, however, is the assumption that the degree of pleiotropy is heritable among copies of genes, in particular if the genes acquire new functions. In fact there is evidence for lower pleiotropy among duplicated gene copies, but this fact is better explained by subspecialization of the gene copies due to degeneration of and complementation among modular enhancer elements (Force et al., 1999).

Phenotypic Stability   In an important computational study on the evolution of RNA secondary structure, Ancel and Fontana (2000) found that selection for phenotypic stability also leads to modularity (see also chapter 6 in this volume). Ancel and Fontana found that in RNA there is a three-way correlation among phenotypic stability in the sense of robustness against thermal noise, mutational robustness, and modularity of the molecule. Of these three properties, phenotypic stability is most effectively selected, that is, is best "seen" by natural selection (Wagner et al., 1997). The evolution of mutational robustness and modularity is a correlated response to selection on phenotypic robustness. Since the correlations are not coincidental, but are intrinsic to the biophysics of RNA, Ancel and Fontana call this phenomenon "plasto-genetic congruence."

Similar principles have been found to hold for protein structure (Bornberg-Bauer and Chan, 1999). These results suggest the intriguing possibility that modularity and other properties of the genetic architecture may evolve as a side effect of the evolution of phenotypic robustness against environmental perturbations. It is thus of greatest importance to investigate whether similar congruence principles may hold for organismal characters as well.

In the older literature about genetic and environmental canalization, the question of whether there might be a correlation between these two forms of robustness was addressed (reviewed in Scharloo, 1991). In general, however, the conclusion was negative. There is no simple relationship between genetic and environmental canalization of a character. The methods available at the time, however, were quite limited, and the question requires new studies with better experimental techniques. One set of papers which supports the notion of a correlation between genetic and environmental robustness for organismal characters consists of the studies on the canalization of life history characters of *Drosophila melanogaster* (Stearns et al., 1995; Stearns and Kawecki, 1994). Stearns and his collaborators found a three-way correlation among fitness sensitivity and mutational and environmental robustness. The results, however, do not address the question of whether genetic and environmental robustness are two independent characters or are variationally correlated.

Modularity Facilitates Physiological Adaptation   Another intriguing model is one with which Calabretta and collaborators simulated the evolution of an artificial neural network dedicated to two functional tasks, the "where and what" task (Di Ferdinando et al., 2001; see also chapter 14 in this volume). The network was expected to produce two kinds of outputs. One indicated the location of an object and the other, its identity. The model led to the evolution of a modular neural architecture and had two components. The neural architecture (i.e., the question of which neurons are connected with each other) was genetically determined, and evolved by mutation and selection. On the other hand, the strength of the neural connection was determined by a learning algorithm based on back propagation (i.e., was acquired by each individual during its ontogeny).

This model, but none of the others investigated by Calabretta and his colleagues, led to the evolution of modularity. The reason is that the effectiveness of the learning algorithm depended on the neuronal architecture. Only a modular architecture provides the basis for successful learning. Hence modularity, which was genetically determined, had a direct fitness advantage mediated through its influence on the effectiveness of individual learning. In addition, the modular neural architectures are also genetically modular with respect to certain mutations. However, the genetic modularity quite evidently did not evolve in this model because of its variational (genetic) consequences. All attempts to evolve modularity without learning (i.e., only with genetic mutations) failed.

This scenario is similar to the one described by Ancel and Fontana (2000) in that there is an interaction between genetic modularity and plasticity or learning, but the selective mechanism is quite different. In the study of Di Ferdinando and collaborators (2001), the highest fitness phenotype could not develop without modular architecture. In the RNA example the highest fitness phenotype was attainable, but at a lower frequency than with modularity. In addition, in the RNA example it was not clear whether there was any causality from modularity to phenotypic robustness at all, while in the Di Ferdinando model there was a clear causal connection from neuronal modularity to high fitness phenotypes due to plasticity.

Modularity from "Frustration"   In a study on the general mathematical theory of gene interactions, Sean Rice (2000) discovered an unexpected mechanism for the origin of modules. Rice found that positive correlations are expected to evolve if the effects of two characters on fitness are synergistic (i.e., if the increase of one character value increases directional selection on the other character). On the other hand, the evolution of a negative correlation is predicted if the characters are antagonistic with respect to fitness. If we consider more than two characters with pairwise antagonistic interactions on fitness, however, something unexpected happens. It is impossible to have negative correlations among three or more characters simultaneously. The evolution of negative correlations is said to be "frustrated."

The only stable solution is that the characters evolve variational independence. It is surprisingly simple to find a scenario for this phenomenon. For instance, assume that three characters contribute to a composite of characters $C = x + y + z$, and in addition assume that the composite character C is under stabilizing selection. Then there is antagonism among all three characters, and Rice's theory predicts selection for independence among the characters. Hence, modularity (i.e., character independence) can result from antagonistic fitness interaction among three or more characters.

Modularity as a Dynamical Side Effect

In all the models discussed above, modularity is assumed to be connected to some sort of selective advantage. In a study on the evolution of functional modularity using an artificial life model, Calabretta and collaborators (2000) discovered a mechanism which cannot be classified as direct or indirect selection for modularity per se. Functional modularity arises from subspecialization of duplicated structural modules without any intrinsic benefit in terms of level of performance or rate of advantage. Modularity arises entirely as a side effect of the evolutionary dynamics.

Calabretta et al. (2000) investigated an artificial life model in which a genetic algorithm had the task of developing both the architecture and the connection weights for a population of neural networks controlling the behavior of a mobile robot. Each robot lived in a walled arena and had the task of exploring the arena and picking up objects and dropping them outside the arena. The robot had infrared sensors that informed it of the presence of objects and walls. It had two wheels for moving forward and backward and for turning in the environment, and a gripper for picking up one object at a time and transporting it outside the environment. The task of the robot was to move in the environment by differentially rotating the two wheels, to find an object, to pick up the object with the gripper and transport it near one of the walls, and finally to release the gripper in such a way that the object was placed outside the environment. To do this, the robot had four motor systems: the two wheels, the motor that controlled the opening and rising of the gripper, and the motor controlling the lowering and opening of the gripper.

As one can easily see, this was a difficult task to learn. The neural network must be able to control the correct sequence of subbehaviors: to explore the environment, to find an object by discriminating it from the wall, to pick up the object by lowering and then raising the gripper, to find the wall while avoiding the other

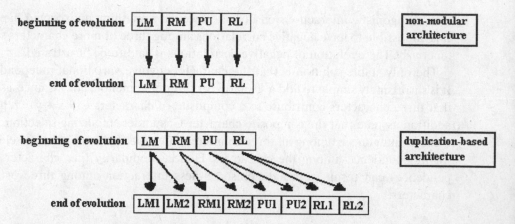

Figure 2.5
Schematic representation of the genomes of the nonmodular and duplication-based modular architectures. LM, genetic encoding for the connection weights of the left motor; RM, right motor; PU, pickup motor; RL, release motor. (Modified from Calabretta et al., 2000).

objects, and to release the object correctly outside the wall. Hence there were a number of behavioral tasks that required different neuronal control over the motor output. Basically there were two types of behavior: searching for a new object and removing the object from the arena. The absence of an object in the gripper should lead to searching and pickup behavior, and the presence of an object in the gripper should lead to a behavioral sequence resulting in the removal of the object from the arena. The question is whether these two behavioral sequences were represented by different neuronal substrates (i.e., functional modularity).

In a study by Nolfi and collaborators (Nolfi 1997) it was shown that functional modularity is not necessary for solving this adaptation problem. Nolfi provided the robot with duplicated neuronal elements to control the output to the motor units of the robot. He found that the genetic algorithm could solve the problem but that the behaviors were not represented by different neuronal elements. No functional modularity evolved. This result shows that functional modularity is not necessary for solving a complex adaptive challenge consisting of a number of different tasks.

Calabretta and colleagues conducted slightly modified simulations. The robots started out with only one neuronal control element per output unit (i.e., motor). During the evolution of the neuronal network, however, a new form of mutation was allowed, the duplication of these control units (figure 2.5).

By analyzing the behavior exhibited by the robots, the authors showed that duplication-based modular networks possess a high degree of specialization

(Calabretta et al., 2000). Some neural modules are specialized for some subtasks (e.g., controlling the robot's movements when the robot is exploring the environment in search of objects), and other neural modules are specialized for other subtasks (e.g., picking up an object). It is important to note that the populations which evolved functional modularity reached the same level of performance as the populations which did not. Furthermore, modular neural networks did not reach the solution faster than others. Hence there was no intrinsic adaptive benefit to functional modularity.

But what was the mechanism that produced functional modularity in these simulations? Various observations point to an evolutionary scenario like the following. First, the duplication of a neuronal control unit. This step was neutral in this model, since the two duplicates were identical. Second, the acquisition of a neutral change in the regulation of the duplicated modules which made one unit more likely to be deployed in one situation—for instance, while searching for another object rather than in object removal. Finally, the accumulation of mutations that adapt the neuronal control unit to the functional context in which it is employed more frequently. This step led to a coadaptation between the regulatory and the functional parts of the control units that locked the system into the functionally specialized state.

From a population genetic point of view, the evolution of functional specialization in this model was caused by epistatic interactions among genes that influence in what situation a control unit is active and genes which control the motor output that is produced. There was a ratchet between a bias in deployment of a control unit and the specialization of the output to the behavioral context in which it was used. One can think of this process as being like a dynamical bifurcation which leads to increasing specialization between control units.

Conclusions

The above overview of possible mechanisms for the origin of modularity identified eight different evolutionary mechanisms, each mechanistically independent from all the others. The majority of them have been proposed since about 2000, and none is understood well enough to be excluded as a candidate. A massive amount of research is necessary to sort out these various possibilities and perhaps even discover additional ones. In our understanding of the situation, there are a number of pressing research questions that need to be addressed to make progress in this area.

Evolution of Evolvability
As summarized above, the results on the possibility of direct selection for evolvability are mixed, but we still lack important results for it to be dismissed entirely.

The most glaring gap in our knowledge is a lack of studies including subdivided populations. There is the possibility that in structured populations, selection of genetic traits influencing the rate of adaptation is more likely than in unstructured populations (Joshua Mitteldorf, pers. communication, 2000).

Congruence Principles

Many of the models that have been shown to create modularity imply some sort of congruence between modularity and some directly selectable property. The best example is the study on modularity in RNA secondary structure by Ancel and Fontana (2000), in which a correlation exists between the degree of modularity and phenotypic stability against environmental noise. But other models can be understood along similar lines. For instance, the model of Di Ferdinando and collaborators on the "where and what" task points to a congruence between physiological and genetic modularities that leads to a selective advantage. Constructional selection assumes a congruence between variational pleiotropy and probability of fixation of a duplicated gene. And the simulations regarding the evolution of pleiotropic effects point to a congruence between evolvability and mean fitness in nonequilibrium populations.

We think that there are sufficient grounds to consider congruence principles as an important component of many scenarios for the evolution of modularity, and that they should therefore be the focus of future investigations. Congruence principles have been discovered in models of molecular dynamics and neuronal networks rather than being modeled themselves. We think that it is time to develop abstract models of congruence principles in order to incorporate them into population genetic theory. In addition, it will be important to find new examples of congruence principles in models of physiology or development and in empirical research. Empirical research is not mentioned first here because it is expensive and should be done only for good reasons. For instance, measuring the mutational variability of a trait is a serious effort, and comparing it to phenotypic stability is even more so.

Mechanistic Plurality Is a Real Possibility

It would be a mistake to assume that we will discover one and only one mechanism that explains the origin of modularity in all circumstances. It is clear that for the origin of species there are many population biological and genetic mechanisms that can lead to the origin of a new species (Otte and Endler, 1989). There is no unitary "speciation mechanism." Similarly, there might be a multitude of mechanisms acting in nature to produce modular genetic architectures. Hence, it might not be productive to try to identify one mechanism among proposed models as "the Solution." Each model needs to be judged on its own merits, and it may be that we end up with an array of mechanisms, each of which may play a role in a variety of different circumstances.

References

Abouheif E (1999) Establishing homology criteria for regulatory gene networks: Prospects and challenges. In: Homology (Bock GR, Cardew G, eds), 207–221. New York: Wiley.

Altenberg L (1994) The evolution of evolvability in genetic programming. In: Advances in Genetic Programming (Kinnear JKE, ed), 47–74. Cambridge, MA: MIT Press.

Altenberg L (1995) The schema theorem and Price's theorem. In: Foundations of Genetic Algorithms 3 (Whitley D, Vose MD, eds), 23–49. Cambridge, MA: MIT Press.

Ancel LW, Fontana W (2000) Plasticity, evolvability and modularity in RNA. J Exp Zool (Mol Dev Evol) 288: 242–283.

Baatz M, Wagner GP (1997) Adaptive inertia caused by hidden pleiotropic effects. Theor Pop Biol 51: 49–66.

Bolker JA (2000) Modularity in development and why it matters in Evo-Devo. Amer Zool 40: 770–776.

Bonner JT (1988) The Evolution of Complexity by Means of Natural Selection. Princeton, NJ: Princeton University Press.

Bornberg-Bauer E, Chan HS (1999) Modelling evolutionary landscapes: Mutational stability, topology, and superfunnels in sequence space. Proc Natl Acad Sci USA 96: 10689–10694.

Brandon RN (1999) The units of selection revisited: The modules of selection. Biol Philos 14: 167–180.

Calabretta RS, Nolfi S, Parisi D, Wagner GP (2000) Duplication of modules facilitates the evolution of functional specialization. Art Life 6: 69–84.

Cheverud J, Routman EJ, Irschick DK (1997) Pleiotropic effects of individual gene loci on mandibular morphology. Evolution 51: 2004–2014.

Cox EC, Gibson TC (1974) Selection for high mutation rates in chemostats. Genetics 77: 169–184.

Di Ferdinando AR, Calabretta RS, Parisi D (2001) Evolving modular architectures for neuronal networks. In: Connectionist Models of Learning, Development, and Evolution (French RM, Sougné JP, eds), 253–262. London: Springer.

Force A, Lynch M, Pickett FB, Amores A, Yan YL, Postlethwait, J (1999) Preservation of duplicate genes by complementary, degenerate mutations. Genetics 151: 1531–1545.

Force AG, Cresko WA, Pickett FB (2004) Informational accretion, gene duplication, and the mechanisms of genetic module parcellation. In: Modularity in Development and Evolution (Schlosser G, Wagner GP, eds), 315–337. Chicago: Chicago University Press.

Galis F (1999) Why do almost all mammals have seven cervical vertebrae? Developmental constraints, Hox genes, and cancer. J Exp Zool (Mol Dev Evol) 285: 19–26.

Galis F (2001) Key innovations and radiations. In: The Character Concept in Evolutionary Biology (Wagner GP, ed), 581–605. San Diego: Academic Press.

Gerhart J, Kirschner M (1997) Cells, Embryos, and Evolution. Malden, MA: Blackwell Science.

Gilbert SF, Bolker JA (2001) Homologies of process and modular elements of embryonic construction. In: The Character Concept in Evolutionary Biology (Wagner GP, ed), 435–454. San Diego: Academic Press.

Gilbert SF, Opitz JM, Raff RA (1996) Resynthesizing evolutionary and developmental biology. Dev Biol 173: 357–372.

Gould SJ (1977) Ontogeny and Phylogeny. Cambridge, MA: Belknap Press of Harvard University Press.

Holland JH (1992) Adaptation in Natural and Artificial Systems. Cambridge, MA: MIT Press.

Houle D (2001) Characters as the units of evolutionary change. In: The Character Concept in Evolutionary Biology (Wagner GP, ed), 109–140. San Diego: Academic Press.

Kim J, Kim M (2001) The mathematical structure of characters and modularity. In: The Character Concept in Evolutionary Biology (Wagner GP, ed), 215–236. San Diego: Academic Press.

Leroi AM (2000) The scale independence of evolution. Evol Dev 2: 67–77.

Liem KF (1973) Evolutionary strategies and morphological innovations: Cichlid pharyngeal jaws. Syst Zool 22: 425–441.

Maynard-Smith J, Haigh J (1974) The hichhiking effect of a favourable gene. Genet Res 23: 23–35.

Mezey JG (2000) Pattern and Evolution of Pleiotropic Effects: Analysis of QTL Data and an Epistatic Model. New Haven, CT: Yale University, Ph D Dissertation, Department of Ecology and Evolutionary Biology.

Mezey JG, Cheverud JM, Wagner GP (2000) Is the genotype–phenotype map modular? A statistical approach using mouse quantitative trait loci data. Genetics 156: 305–311.

Mittenthal JE, Baskin AB, Reinke RE (1992) Patterns of structure and their evolution in the organization of organisms: Modules, matching, and compaction. In: Principles of Organization in Organisms (Mittenthal JE, Baskin AB, eds), 321–332. Reading, MA: Addison-Wesley.

Nagy LM, Williams TA (2001) Comparative limb development as a tool for understanding the evolutionary diversification of limbs in arthropods: Challenging the modularity paradigm. In: The Character Concept in Evolutionary Biology (Wagner GP, ed), 455–488. San Diego: Academic Press.

Nolfi S (1997) Using emergent modularity to develop control system for mobile robots. Adapt Behav 5: 343–364.

Otte D, Endler JA (1989) Speciation and Its Consequences. Sunderland, MA: Sinauer.

Raff RA (1996) The Shape of Life. Chicago: University of Chicago Press.

Rice SH (1990) A geometric model for the evolution of development. J Theor Biol 143: 319–342.

Rice SH (2000) The evolution of developmental interactions. In: Epistasis and the Evolutionary Process (Wolf JB, Brodie EDB, III, Wade MJ), 82–98. Oxford: Oxford University Press.

Riedl R (1978) Order in Living Organisms: A Systems Analysis of Evolution (Jefferies RPS, trans). New York: Wiley.

Rohwer JM, Schuster S, Westerhoff HV (1996) How to recognize monofunctional units in a metabolic system. J Theor Biol 179: 213–228.

Scharloo W (1991) Canalization: Genetic and developmental aspects. Ann Rev Ecol Syst 22: 65–93.

Schwenk K (2001) Functional units and their evolution. In: The Character Concept in Evolutionary Biology (Wagner GP, ed), 165–198. San Diego: Academic Press.

Stearns SC, Kaiser M, Kawecki TJ (1995) The differential canalization of fitness components against environmental perturbations in *Drosophila melanogaster*. J Evol Biol 8: 539–557.

Stearns SC, Kawecki TJ (1994) Fitness sensitivity and the canalization of life history traits. Evolution 48: 1438–1450.

Sterelny K (2000) Development, evolution and adaptation. Phil Sci 67: S369–S387.

Stock DW (2001) The genetic basis of modularity in the development and evolution of the vertebrate dentition. Proc Roy Soc London B356: 1633–1653.

Turney PD (2000) A simple model of unbounded evolutionary versatility as a large-scale trend in organismal evolution. Art Life 6: 109–128.

Vermeij GJ (1970) Adaptive versatility and skeleton construction. Amer Nat 104: 253–260.

Von Dassow G, Munro E (1999) Modularity in animal development and evolution: Elements of a conceptual framework for EvoDevo. J Exp Zool (Mol Dev Evol) 285: 307–325.

Wagner GP (1981) Feedback selection and the evolution of modifiers. Acta Biotheor 30: 79–102.

Wagner GP (1995) The biological role of homologues: A building block hypothesis. N Jb Geol Paläont Abh 195: 279–288.

Wagner GP, Altenberg L (1996) Perspective: Complex adaptations and the evolution of evolvability. Evolution 50(3): 967–976.

Wagner GP, Booth G, Bagheri-Chaichian H (1997) A population genetic theory of canalization. Evolution 51: 329–347.

Weiss KM (1990) Duplication with variation: Metameric logic in evolution from genes to morphology. Yrbk Phys Anthropol 33: 1–23.

Wray GA (1999) Evolutionary dissociations between homologous genes and homologous structures. In: Homology (Bock GR, Cardew G, eds), 189–203. New York: Wiley.

3 Evolutionary Modules: Conceptual Analyses and Empirical Hypotheses

Robert N. Brandon

It has been argued that adaptive evolution requires the existence of evolutionary modules—in other words, that organisms must be decomposable into traits that can evolve independently of one another. Whether this is true depends, of course, on what is meant by an "evolutionary module." This chapter aims to show how we should go about deciding such a question. More specifically, it aims to differentiate conceptual analyses of evolutionary modularity from empirical hypotheses regarding modularity. This distinction, I will argue, is crucial but nonobvious.

In the first section of this chapter I will briefly review the arguments for the necessity of evolutionary modules. I will contrast those arguments with other reasons or other sorts of evidence we might have for believing in the existence of evolutionary modules. Having put the cart before the horse, in the second section I go back to the horse and present a conceptual analysis of evolutionary modules. In the third section I present the characterization of evolutionary modules that comes from Wagner and Altenberg. The fourth section clarifies the relationship between my conceptual analysis and the Wagner–Altenberg model by comparing this work with an earlier episode in philosophy of biology—the controversy between Mayr and Nagel over the proper characterization of teleonomic systems.

Early Arguments for the Existence of Evolutionary Modules

Richard Lewontin argued that the phenomena of adaptation

> can only be workable if both the selection between character states and reproductive fitness have two characteristics: continuity and quasi-independence. Continuity means that small changes in a characteristic must result in only small changes in ecological relations. . . . Quasi-independence means that there is a great variety of alternative paths by which a given characteristic may change, so that some of them will allow selection to act on the characteristic without altering other characteristics of the organism in a countervailing fashion. . . . Continuity and quasi-independence are the most fundamental characteristics of the evolutionary process. Without them organisms as we know them could not exist because adaptive evolution would have been impossible. (1978, p. 169)

What Lewontin calls continuity is irrelevant to our present concerns, so let us focus on quasi independence. The logic of Lewontin's argument is perfectly clear. Adaptive evolution, which produces the phenomena of adaptation, requires quasi independence. The phenomenon of adaptation is real. Therefore, quasi independence exists.

John Bonner put forward a similar argument. He argued that the evolution of adaptive complexity requires modular "gene nets." For Bonner a gene net is "a grouping of a network of gene actions and their products into discrete units during the course of development." He continues:

This [grouping of gene effects] not only was helpful and probably necessary for the success of the process of development, but it also means that genetic change can occur in one of the gene nets without influencing the others, thereby much increasing its chance of being viable. The grouping leads to a limiting of pleiotropy and provides a way in which complex developing organisms can change in evolution. (1988, p. 175)

Here Bonner is a bit more specific than Lewontin in offering some description of a gene net, and by explicitly tying his idea to developmental pathways, but the logic is basically the same. Again, adaptive evolution requires gene nets. Adaptive evolution does occur. Therefore gene nets exist.

Like Wagner and Altenberg (1996), I see Lewontin's concept of quasi independence, and Bonner's idea of gene nets, as precursors to our current concept of evolutionary modules (or "modules of selection"; Brandon, 1999). In the next section I will discuss more fully just how we should characterize these evolutionary modules. But for now the point is that both Lewontin and Bonner think they have shown that such modules do exist, because they are necessary for adaptive evolution and adaptive evolution does occur.

Let us classify this as a "transcendental" argument for the existence of evolutionary modules. (The term "transcendental" is appropriate here because these arguments claim that modularity is necessary for the very existence of the phenomena of adaptation.)

Although I largely agree with Lewontin and Bonner, let me make two cautionary comments about this sort of argument. First, such arguments are not explanatory. That is, they do not explain the existence of evolutionary modules any more than the following explains the existence of the sun: Sunlight is a necessary condition for the existence of green plants on Earth. Green plants exist on Earth. Therefore the sun exists. This is, of course, a perfectly valid argument; it just doesn't explain the existence of the sun.

Second, and perhaps more important, we need to be cautious about this sort of transcendental argument. The argument is supposed to reflect necessary relations in the world, but may unwittingly reflect limitations on our understanding of the world. We cannot imagine adaptive evolution without quasi independence and/or gene nets. But perhaps that is a limitation of our understanding, not a limitation on how the world works. The two should not be confused. As a physics professor I had in my undergraduate years was fond of saying, "We didn't get here first."

Given these cautionary remarks about this sort of argument for the existence of evolutionary modules, we should ask whether there are other arguments for, or evidence of, the existence of modules. There are at least two that are important to distinguish.

The first is a post hoc inference to the existence of certain modules from phylogenetic and/or fossil data. For example, the earliest mammal was a tetrapod with—it is reasonable to suppose—forelimbs and hind limbs that were both developmentally and functionally not much differentiated. But within the mammalian lineage the forelimb has become a flipper (whales), a wing (bats), and an arm with a highly dexterous hand (humans). What does this tell us? It tells us that the forelimb can evolve without totally changing the mammalian body plan—whales, bats, and dogs have similar circulatory systems. More specifically, we see that the forelimb can evolve relatively independently of the hind limb. Bats don't have two sets of wings, nor do humans have two sets of arms. I think that this inference is perfectly sound, but it, too, is not altogether satisfactory. It doesn't tell us how a module attains its modularity. What is it about the functional effects and genetic/developmental architecture of mammalian forelimbs that enable them to behave as evolutionary modules? We would like to know this, and the post hoc recognition that they are modules doesn't answer that question.

(I have described this sort of evidence as supplying us with a post hoc argument for the existence of modules because, in practice, I suspect this is the way it will most often be used. Which is just one more example of the fact that the evolutionary past is easier to know than the evolutionary future. But, in principle at least, nothing would prevent us from identifying an evolutionary module from its behavior in the present and using that information to make predictions.)

The second alternative way of knowing of the existence of evolutionary modules is, as it were, via direct observation. Just as in astronomy, where we can predict the existence of a body (e.g., Pluto) from the behavior of other bodies, it is more satisfying when we can actually observe the body itself. This is so, in part, because it confirms our predictions, which, however sound, are always fallible. (Though in astronomy we do have to reconcile ourselves to the fact that some objects, such as black holes, will never be directly observable.) But the direct observation of a module presupposes we know what we are looking for (i.e., presupposes what we will discuss in the next two sections).

To summarize this section, we can distinguish three reasons to believe in the existence of evolutionary modules: (1) the "transcendental" argument of Lewontin and Bonner; (2) the "indirect" inference to the existence of certain modules from phylogenetic/fossil data; and (3) the "direct" observation of modules.

Evolutionary Modules: A Conceptual Analysis

As I've argued elsewhere (Brandon, 1999), evolutionary modules, or "modules of selection," are such in virtue of being both functionally and developmentally modular. One way of thinking about this is to think about evolutionary modules as the "units" or "natural kinds" picked out by the process of evolution through natural selection. The process of evolution by natural selection is itself composed of two subprocesses: phenotypic selection and the genetic response to selection (see, e.g., Brandon, 1990, chap. 1). Corresponding to these two subprocesses are the two components of evolutionary modules, ecological function and genetic/developmental modularity.

The idea is fairly straightforward. For something to be an evolutionary module, it must have a (relatively) unitary function. Some things will fail to be modular in this way because they are dysfunctional or afunctional. The exact pattern of freckles on a human's face is, I'm supposing, afunctional. Other things will fail because they are at the wrong level of resolution relative to the pertinent function. For example, with respect to the viceroy butterfly's mimicry of the monarch's wing pattern, the color of a particular cell of the viceroy's wing will not be a functional module because it is the mimicry of the whole pattern that matters. The wing cell is at too small a scale of resolution. But with respect to the evolution of "eyespots" in some lepidoptera that ward off predators, the whole wing pattern is at too large a scale of resolution—it is the spot itself that is the functional module.

Two points should be noted here. First, by "unitary function" I mean that the part in question has, as a whole or as a *unit*, a function. That is the point of the butterfly wing examples above. However, by "unitary" I do not mean to imply that the part has exactly one function. It may have more than one, as do butterfly wings, which generally serve at least two functions—flight and thermoregulation. They can also serve other functions, as in the viceroy example above.

Second, it should be noted that functions do have a hierarchical structure. For example, the human arm is a functional module (its function presumably playing some role in the evolution of bipedalism). But the human hand, a part of the arm module, might also be a module relative to the function of the fine motor control of objects, as might be the thumb, and so on. To say something has a unitary function must be relativized to a particular function.

But functional modularity is not enough. For some trait to evolve effectively, it must be more or less disassociable from other traits. This is precisely the point of Lewontin's and Bonner's arguments above. One could not, for instance, effectively evolve an eyespot if the color of the midpart of the lower wing were inextricably connected to the color of the rest of the lower wing or, worse, to the color of all the wing surfaces. Or, if any change in mammalian forelimbs were always connected with the same change in hind limbs, then the differentiation of mammalian forelimbs mentioned above would have been impossible.

Thus an evolutionary module is some feature of an organism that has a unitary ecological function and a genetic/developmental architecture that allows it to evolve in a "quasi-independent" way from other features. An evolutionary module must be both functionally and developmentally modular. This, I think, is at least a good beginning toward a conceptual analysis of evolutionary modularity.

The Wagner–Altenberg Model of Modularity

In a series of papers Günter Wagner and Lee Altenberg have presented a plausible picture of evolutionary modules that is much more detailed than the analysis of the last section (see Altenberg, 1995; Wagner, 1995, 1996; Wagner and Altenberg, 1996). Figure 3.1 represents their model. Starting at the bottom, we see that there are two sets of genes: $\{G_1, G_2, G_3\}$ and $\{G_4, G_5, G_6\}$. What makes each a set or, in Bonner's terminology, a "gene net" is that the pleiotropic connections are largely within each set, and only rarely between them. Above the gene nets are sets of phenotypic characters, character complexes: $\{A, B, D, C\}$ and $\{E, G, F\}$. What unifies these characters is their functional effects. Character complex 1 largely effects function 1 while having only a minor effect on function 2. Character complex 2 mainly effects function 2, with only a small effect on function 1.

One might think that this is just a more specific version of the conceptual analysis of the preceding section. The idea that an evolutionary module must have a unitary ecological function is represented by C_1's primary effect being on F_1, and similarly for C_2 and F_2. The idea that an evolutionary module must have a genetic/developmental architecture that allows it to evolve relatively independently of other modules is represented by the fact that most of the arrows from genes to characters are within a module (that is, most of the pleiotropic connections are within a module). Thus the Wagner–Altenberg model just puts a bit more flesh on the bones of the conceptual analysis of the preceding section.

As tempting as this thought is, I believe it is wrong. I think that when it is properly understood, the Wagner–Altenberg model is best seen as an empirical hypothesis concerning the underlying nature of evolutionary modules, not as a bit of conceptual analysis. The proof of this is to see just how their model might

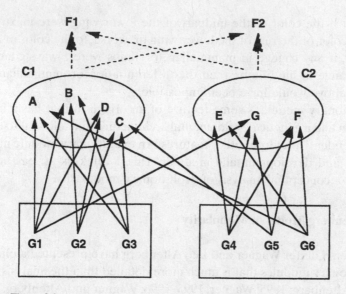

Figure 3.1
The Wagner–Altenberg model of modularity. (Reprinted with permission from Wagner and Altenberg, 1996.)

be criticized and to compare that with how my conceptual analysis might be criticized.

Conceptual Analyses and Empirical Hypotheses

Before addressing the question of the relation between my conceptual analysis of evolutionary modularity and the Wagner–Altenberg model, let me illustrate the relation by means of another example—one from the early days of philosophy of biology.

Ernest Nagel was one of the foremost philosophers of science of the second half of the twentieth century. He was a logical empiricist (which is the more general movement that started with the logical positivism of the Vienna Circle). Although certainly not a full-time philosopher of biology, Nagel was interested in functional explanation in biology precisely because he wanted to show that such explanations fit the model of scientific explanation put forward by his contemporary, Carl Hempel. He and Hempel disagreed on this matter: Hempel (1965) thought functional explanations in biology were not legitimate scientific explanations, while Nagel thought that they were legitimate. A related issue also interested Nagel: How can we characterize goal-directed systems in nature? Nagel was convinced that such systems existed in biology (as well as in the realm of human artifacts), and that they were legitimate objects of scientific investigation. Thus he sought to give a non-mysterious characterization of goal-directed systems that would facilitate their scientific investigation. The culmination of this was his paper of 1977, "Teleology Revisited."

At the same time Ernst Mayr, along with a number of other biologists, sought to remove the negative connotations of the concept of goal-directedness, or teleology, in biology. Mayr (1974) adopted Pittendridge's neologism "teleonomy" (which was meant to relate to teleology in the same way that astronomy relates to astrology). Of course, deep philosophical problems cannot be solved by the introduction of a new word. Mayr quite clearly saw that, and sought to give an analysis of teleonomy or goal-directedness. The details of Mayr's and Nagel's accounts are not relevant to the present purposes, so I will give just a cursory overview. Nagel thought that what characterized goal-directed systems was the combination of two things: *plasticity* and *persistence*. Goal-directed systems do not always achieve their goals, but they do show persistence when obstacles are put in their way, and they also exhibit plasticity in that they tend to have multiple ways of achieving their end state. Nagel gives a more formal account of this, but the informal account suffices for us.

Mayr, on the other hand, argued that goal-directed systems were ones whose behavior was directed by an internal "program." A dog searching for food and a homing torpedo would be examples. In contrast, a bullet, which may well reach its target, is governed completely by external physical forces. Bullets do not have internal programs; dogs and homing torpedoes do. Mayr argued against Nagel's analysis, and Nagel returned the favor, arguing against Mayr. It is clear that both Mayr and Nagel thought their accounts were mutually exclusive, that at most one of them was right. But in my view these arguments were ill-conceived. They failed to appreciate the complementarity of Mayr's and Nagel's projects. They also failed to understand the differences between the projects.

Nagel was engaged in conceptual analysis. He was trying to give a characterization of goal-directedness that would be useful in the scientific investigation of the phenomena. Another way of thinking about this is to say that Nagel was trying to characterize the behavior, or the phenomenon, of goal-directedness. Why? So that we would recognize goal-directedness when we see it, and so we would know what would be the proper domain of a theory of goal-directedness. (Should it account for the behavior of bullets? No. Should it account for the behavior of homing torpedoes? Yes.)

Mayr was not doing this. He was framing a hypothesis concerning the underlying mechanisms of goal-directed systems. What enables them to be goal-directed? They have internal programs. This is an empirical hypothesis. How could it be anything else? It is certainly not part of the phenomenology of goal-directedness that there is an internal program driving the system. The claim that such programs drive the system would, if true, be an explanatory hypothesis. Indeed, in retrospect it seems obvious that Mayr needs something like Nagel's analysis; otherwise the subject matter of his hypothesis would be unclear. There is a remaining fuzziness in Mayr's hypothesis, namely, the concept of a program. I will not try to resolve that here. But it is important to see that an empirical hypothesis, such as Mayr's, need not be completely unambiguous, nor need it be readily testable, to be empirical. In science, testable hypotheses are preferable, everything else being equal, to fuzzy untestable hypotheses. But, contrary to the logical empiricists, real testability is not the sole criterion of empirical meaningfulness.

Just as Mayr and Nagel were engaged in complementary projects, I believe that the accounts of modularity discussed above are different in nature, but complementary. In the second section I offered a conceptual analysis of evolutionary modularity. In short, evolutionary modularity is the conjunction of functional and developmental modularity. Is my analysis true? That would be the wrong question. Is it useful in the scientific investigation of evolution? I hope so, but that is the right question. Conceptual analyses in science are neither true nor false. They are attempts to set conventions—conventions that govern the way we think about and investigate natural phenomena. Such conventions are either useful or not—empirical truth is not the issue.

In contrast, the Wagner–Altenberg model discussed in the third section makes specific claims about the underlying structure of evolutionary modules. Questions of empirical truth or falsity are pertinent here. Is it true that the genes that control mammalian forelimb development are largely different from those that control hind limb development? No? Then the model needs to be modified. I'm a philosopher of biology, not a developmental biologist, but my view is that we have a substantial amount of evidence that contradicts the model. We know of a number of developmental regulatory genes that are utilized in the construction of a large number of functionally and morphologically different structures. For example, the infamous *Pax6* gene, implicated in eye development in mammals and fruit flies, is also utilized in the development of other sensory organs, such as the nose and olfactory bulbs (Walther and Gruss, 1991). It is also part of the development of glucagon-producing cells in the pancreas and intestinal endotherm (Hill et al., 1999). And it is involved in patterning the cerebellum (Engelkamp et al., 1999) and parts of the cerebral cortex (Gotz et al., 1998). Similar stories can be told about *Hox* genes and the *hedgehog* gene.[1]

Maybe these cases are exceptional; maybe they are the rule. Maybe they can be handled with a minor modification of the model, or maybe they require a major overhaul.[2] My point has simply been to show that the Wagner–Altenberg model is an empirical hypothesis to be criticized according to how well or how poorly it corresponds to the facts.

Mayr and Nagel never saw that they were doing different, and in fact complementary, things. I hope to have shown conclusively that my conceptual analysis of evolutionary modularity is different from, and potentially complementary to, the work of Wagner and Altenberg. Seeing this now may well facilitate future progress.

Notes

1. I owe these examples to James Balhoff. His showing me the empirical difficulties of the Wagner–Altenberg model was critical in my realization that their project is different in kind from mine.

2. My first attempt at a modification of the model would be to distinguish between general-purpose "housekeeping" genes and special-purpose genes. If this distinction is viable, then perhaps the Wagner–Altenberg model correctly describes the special-purpose genes. If all that were correct, it would raise new questions—for instance, how is modular evolution achieved when it involves some general-purpose genes? But still there would be much of explanatory value in this modified model.

References

Altenberg L (1995) Genome growth and the evolution of the genotype–phenotype map. In: Evolution and Biocomputation: Computational Models of Evolution (Banzhaf W, Eeckman FH, eds), 205–259. Berlin: Springer.

Bonner JT (1988) The Evolution of Complexity by Means of Natural Selection. Princeton, NJ: Princeton University Press.

Brandon RN (1990) Adaptation and Environment. Princeton, NJ: Princeton University Press.

Brandon RN (1999) The units of selection revisited: The modules of selection. Biol Philos 14: 167–180.

Engelkamp D, Rashbass P, Seawright A, van Heyningen V (1999) Role of Pax6 in development of cerebellar system. Development 126: 3585–3596.

Gotz M, Stoykova A, Gruss P (1998) Pax6 controls radial glia differentiation in the cerebral cortex. Neuron 21: 1031–1044.

Hempel CG (1965) Aspects of Scientific Explanation. New York: Free Press.

Hill ME, Asa SL, Drucker DJ (1999) Essential requirement for Pax6 in control of enteroendocrine proglucagon gene transcription. Mol Endocrin 13: 1474–1486.

Lewontin RC (1978) Adaptation. Sci Amer 239(3): 156–169.

Mayr E (1974) Teleological and teleonomic: A new analysis. Boston Stud Phil Sci 14: 91–117.

Nagel E (1977) Teleology revisited. J Philos 74: 261–301.

Wagner GP (1995) The biological role of homologues: A building block hypothesis. N Jb Geol Paläont Abh 195: 279–288.

Wagner GP (1996) Homologues, natural kinds, and the evolution of modularity. Amer Zool 36: 36–43.

Wagner GP, Altenberg L (1996) Perspective: Complex adaptations and the evolution of evolvability. Evolution 50(3): 967–976.

Walther C, Gruss P (1991) Pax-6, a murine paired box gene, is expressed in the developing CNS. Development 113: 1435–1449.

4 Evolutionary Developmental Biology Meets Levels of Selection: Modular Integration or Competition, or Both?

Rasmus G. Winther

It is a truism that wholes are composed of parts. In recent years a number of biologists, particularly developmental biologists, have started a new field of inquiry into the parts of biological wholes, called "modules" (see, e.g., Wagner, 1996, 2001; R. A. Raff, 1996; Hartwell et al., 1999; von Dassow and Munro, 1999; Bolker, 2000; Winther, 2001a). In multicellular organisms, modules include repeated and conserved structures such as arthropod segments and developmental units such as germ layers, morphogenetic fields, and cell lineages. In social insect colonies, such as those of ants, modules include ant organisms with particular structures and distinct behaviors, such as reproductive ants (gynes, i.e., reproductive females, which, when they have established a colony, are called queens, and males) and potentially morphologically differentiated ant workers. When the colony is considered to be an individual (i.e., a superorganism), then the behavior of the ant organism can be thought of as a part of colony physiology. I use the term "individual" in a broad sense to include, for example, multicellular organisms and social insect colonies. I will focus on these because they are well-integrated modular individuals at two distinct and compositionally related levels of biological organization.

Do modules mechanistically interact or selectively compete, or both? Two perspectives answer this question differently. Investigations in the *integration* perspective are concerned with the interactive mechanisms among modules and with the patterns of evolutionary change of mechanisms and modules. Mechanisms of interest to this perspective can be roughly divided into two categories: (1) developmental, those involved in causing the dynamical path taken during the production of an individual, and (2) physiological, those processes occurring along every step of this path (tables 4.1 and 4.2). For example, in multicellular organisms, this perspective investigates both the different embryonic regions interacting with each other during development and the specialized physiological processes that ensure organismal functionality.

Research under the *competition* perspective explores the selective processes acting among units at different levels of the genealogical hierarchy (e.g., gene, organelle, cell, organism, superorganism, species, and even clade). Typically biologists and philosophers distinguish between (1) replicators, of which copies are made, and (2) interactors, which interact as a whole with their environment (Dawkins, 1976; Hull, 1980; Brandon, 1982; for a review, see Lloyd, 2000). Interactors at multiple levels (modules and individuals) compete directly and thereby lead to the differential survival of the particular replicators, which produce, or are at least statistically correlated with, them. Copies are then made of these

Table 4.1
Integration and competition perspectives at the organism level

	Integration Perspective	Competition Perspective
	What are the mechanisms among parts (modules) and parts–wholes (modules–individuals)? Physiology vs. development	Which selective forces among interactor modules *change* gene (replicator module) frequencies? Interactor vs. replicator
Partitioning Whole can be understood as an aggregate sum of parts	• Human Genome Project	• Selfish gene theory and game theory: Dawkins and Maynard Smith in 1970s and subsequently • Single-level kin selection: Maynard Smith and Szathmáry, 1995
Articulation *Organism as a legitimate individual* 1. Relations, not properties, of parts are important 2. Whole and parts provide mutual meaning 3. Constant interaction with environment	• Integrative mechanisms at multiple levels 1. Physiological approach 2. Developmental approach 3. Structural approach Gerhart and Kirschner (1) R.A. Raff (2) Wagner (3) • *Integratively defined* organism	• Two mechanisms 1. Group selection for cheater-suppression mechanisms 2. Multilevel kin selection Buss (1) Michod (1,2) • Multilevel selection theory employed • *Competitively defined* organism

Table 4.2
Integration and competition perspectives at the superorganism level

	Integration Perspective	Competition Perspective
	What are the mechanisms among parts (modules) and parts–wholes (modules–individuals)? Physiology vs. development	Which selective forces among interactor modules *change* gene (replicator) frequencies? Interactor vs. replicator
Partitioning Whole can be understood as an aggregate sum of parts	• E. O. Wilson in 1960s and subsequently	• Hamilton's inclusive fitness of the 1960s • Selfish gene theory and game theory: Dawkins and Maynard Smith in 1970s and subsequently
Articulation *Superorganism as a legitimate individual* 1. Relations, not properties, of parts are important 2. Whole and parts provide mutual meaning 3. Constant interaction with environment	• Gordon in 1980s and subsequently • *Integratively defined* superorganism	• Hamilton and Price hierarchical covariance approach of the 1970s • Multilevel selection theory: Wade and D. S. Wilson in 1970s and subsequently • *Competitively defined* superorganism

replicators. The competition perspective is fundamentally interested in the patterns and processes of changes in replicator (e.g., gene) frequencies, across generations, in populations of interactor modules and individuals (tables 4.1 and 4.2). For example, in social insect colonies, selection occurs at both the hymenopteran organism and the hymenopteran colony level. Rather than focus on the physiological and behavioral relations (i.e., integrative mechanisms) of colonies, this perspective explores, for example, the conditions necessary for the fixation of alleles for cooperation.

In short, work within the integration perspective focuses on how modules interact to form an individual and on the patterns of evolutionary change of mechanisms and modules, whereas research in the competition perspective emphasizes the selective dynamics, often at multiple levels of modules and individuals, that lead to changes in replicator frequencies. Although the replicator/interactor distinction could be employed in the integration perspective and mechanistic interactions could be investigated in the competition perspective, these are not the concerns guiding research in each perspective. As we shall see, the term "individual" is defined differently in each perspective. In this chapter, I will not further explore the abstract meaning of individuality, nor will I discuss individuals at levels higher than the superorganism, such as species and clades (for such discussion see, e.g., Eldredge, 1985, 1989; Ghiselin, 1974, 1997; Gould, 1995; Gould and Lloyd, 1999; Hull, 1978, 1980; Vrba and Eldredge, 1984).

Modular cooperation and defection are understood differently in each perspective. For the *integration* perspective, *modular cooperation* is understood not as a cooperative act, on the part of modules, to ultimately increase their inclusive fitness or their reciprocal fitness benefits, or both, but as a developmental and physiological process that establishes a functional integration of the whole. *Modular defection* is denied as a meaningful phenomenon. It is interpreted as mechanistic dysfunction; selection is understood as occurring among higher-level modular individuals.

For the *competition* perspective, *modular cooperation* is explained as a strategy, on the part of modules, to maximize their inclusive fitness or their reciprocal fitness benefits, or both, in a group context. Alternatively, modular cooperation can be enforced by various higher-level control methods subject to higher-level selection. *Modular defection* is accepted. It happens when a module gains short-term fitness benefits at the expense of other particular modules or the whole collection of modules (i.e., the individual), or both. Modular defection can occur whenever genetic relatedness among modules becomes too low, or when reciprocal fitness benefits in mutualistic relationships become too low, or when the higher-level individual fails to control lower-level defector variant modules—or a combination of all three possibilities.

Addressing the issue of modular process—integration or competition—is distinct from the issue of whether the individual (whole) is greater than the sum of its modules (parts) (tables 4.1 and 4.2). This second issue is addressed by two distinct research strategies. The *partitioning* research strategy explains an individual as a relatively direct and linear aggregation of its modules (see Wimsatt, 1984, 1986; Gerson, 1998). Properties of modules, rather than interactions among modules, account for the dynamics and fitnesses of individuals. In contrast, the *articulation* research strategy emphasizes the dynamical and analytical relations among parts and wholes (tables 4.1 and 4.2).

I use the term "articulation" to stress the complex nonlinear interactions among parts—this is sometimes called "holism." Although my distinction is similar to the reductionism–holism distinction, I prefer my terms because I intend to classify rather than prescribe. I want to distance myself from the evaluative overtones of "reductionism" and "holism." With respect to articulation, biologists and philosophers make three classes of interrelated claims: (1) parts interact hierarchically—relations (i.e., interactions), and not context-independent additive properties, are of primary explanatory importance; (2) the whole provides meaning to the parts—parts can neither be defined nor be described independently of the whole; and (3) both parts and wholes interact with, and change in response to, their environment, which is hierarchical and also is defined by its interaction with the whole and its parts (e.g., Kauffman, 1971; Wimsatt, 1974, 1984, 1986; Levins and Lewontin, 1985; Lewontin and Levins, 1988; Gerson, 1995, and personal communication; Wade and Goodnight, 1998; Wade et al., 2001).

A research strategy consists of bets (commitments) that certain particular protocols and techniques will be more advantageous than others (see Gerson, 1998 on commitments). A perspective coordinates phenomena, methodologies (including research strategies), theories, and questions of interest into a general program of scientific investigative activity (e.g., Wimsatt, 1974; Griesemer, 2000a, and personal communication; Gerson, personal communication). Both research strategies are employed in both perspectives. For example, in studying the integration of ant colonies, Edward O. Wilson adopts a partitioning strategy when he argues that both chemical signals (pheromones) and ant organisms of distinct castes have context-independent properties that determine colony-level behavior in a linearly aggregative fashion. Deborah M. Gordon, who also adopts the integration perspective, adheres to an articulation strategy when she focuses on both context-dependent interaction rates as communication mechanisms and the context-dependent task flexibility of workers.

Determining whether a biologist or field of inquiry relies on partitioning or articulation strategies is not always easy. The conceptual contrast between the two extreme positions is clear, but much research is done somewhere in between them; a few researchers employ both of them, though they tend to use one more frequently. I will not attempt the difficult task of placing every discussed researcher somewhere along the continuum between the two extremes. It is also beyond the scope of this chapter to discuss the abstract question of the possibility of adopting both research strategies. However, for each of the four cases of perspective and biological level discussed, I will contrast two or more investigators who fall near either extreme (tables 4.1 and 4.2). The chapter first discusses the integration and competition perspectives, in that order, in multicellular organisms; it subsequently discusses these perspectives in social insect colonies.

This chapter sketches a map of the intellectual region where the material processes and scientific investigations of evolutionary developmental biology meet those of levels of selection. This area concerns both the evolutionary emergence of new levels of individuality and the evolutionary change of developmental patterns and processes of well-integrated individuals (see Winther, 2001a). Typically, proponents of the integration perspective, who are usually developmental biologists, are interested in describing the nature of, and evolutionary changes in, developmental mechanisms. Adherents of the competition perspective, who are usually evolutionary biologists, are often concerned with the selective processes affecting replicator frequencies or giving rise to new levels of individuality (e.g., multicellularity from unicellularity; see Buss, 1987). The correlation among perspectives, fields, and interests is high. I do not argue for the superiority of either perspective. Elsewhere, I have articulated how these perspectives relate with respect to the questions posed (Winther, 2001a).

In the conclusion, I will explore further the possible relations between the perspectives with respect to other components of a perspective (i.e., methodologies, which include research strategies, phenomena, and theories). I do not commit to any one interpretation of the relation between the integration and competition perspectives. Furthermore, although I think that the partitioning research strategy is more workable, the articulation research strategy is required in order to realistically describe the actual complexity of the world. Thus, each research strategy has strengths and weaknesses. My aim is diagnosis: I seek to make researchers in both perspectives aware of the other perspective and, furthermore, I invite them to consider the utility of the two research strategies of partitioning and articulation.

Integration Perspective on Multicellular Organisms

Kinds, Criteria, and Levels of Modules

Almost all the conceptual work on modularity stems from molecular and developmental biologists working on organisms and adhering to the integration perspective. Since discussion on modularity has focused on organisms and since the concept of modularity has been extensively reviewed elsewhere, here I will only briefly discuss some of the salient points (see, e.g., R. A. Raff, 1996; von Dassow and Munro, 1999; Bolker, 2000; Winther, 2001a). Because of the discussion available elsewhere, the first two sections, on multicellular organisms, will not be as detailed as the last two sections, on social insect colonies. I will, however, explore the concept of organismic modularity in the context of each perspective and will, in this first section on the integration perspective, relate it to other issues in the field of evolutionary development biology (sometimes called "evo-devo"). Evolutionary developmental biology is the study of how patterns and processes of development and heredity change during, and also influence, evolution (see R. A. Raff, 2000; Wagner et al., 2000).

There are various kinds of modules, and they differ among fields (Winther, 2001a). Systematics and comparative morphology study *structural* modules; developmental biology examines *developmental* modules; and physiology and functional morphology investigate *physiological* modules. Structural modules are the parts that compose an individual at a time slice of ontogeny; individualized vertebrate bones or arthropod segments are paradigmatic examples. Developmental modules can be of two subkinds—either parts that change over time or parts that induce other modules to change. Two examples of the latter kind are morphogenetic fields, which determine gradients that provide positional information for incipient structures (S. Gilbert et al., 1996; R. A. Raff, 1996; Wolpert, 1996), and *cis*-regulatory regions of the DNA, to which transcription factors bind (Arnone and Davidson, 1997). Physiological modules are individuated by their activity. For example, the production of insulin in the pancreatic islets of Langerhans uniquely distinguishes these clusters of cells as physiological modules. Any particular module may fulfill no, one, or multiple functional roles. It is important to distinguish physiological or developmental *processes* from abstract *functions*: these processes are activities, whereas functions are selective or analytic *reasons* for these processes. A process may not have a function (S. Gilbert and Bolker, 2001, p. 443; Winther, 2001a, pp. 117–118; on function, see Allen et al., 1998; Godfrey-Smith, 1993, 1996).

A number of criteria are employed to individuate structural, developmental, and physiological modules (R. A. Raff, 1996; Wagner and Altenberg, 1996; Gerhart and Kirschner, 1997; Bolker, 2000; Winther, 2001a). The more criteria a focal unit fulfills, the more justified we are in deeming it a module; thus, there are degrees of modularity. Here I discuss four criteria. First, modules have differential genetic specifications. There is genetic overlap (e.g., pleiotropy) across modules, but on average each module is unique both in its set of expressed genes and in the way these genes interact among themselves and with their environment. Modules consisting *of* genes are also clearly genetically different from one another. Second, modules are often repeated and conserved (a) within or across taxa, (b) at or across hierarchical levels within individuals (e.g., molecular, cellular, and histological), and (c) in different and similar contexts. Repetition and conservation occur because modules are more likely to arise from the duplication of preexisting modules, followed by their co-option in new functional contexts, rather than from the development of new modules. Third, there is strong connectivity within, and weak connectivity among, modules. Different modules are semiautonomous during both development and evolution. Perhaps they can be thought of as "individuals" in some respects (Wagner, 1989, pp. 1160–1163; Bolker, 2000, p. 773), but I prefer to use that term to describe an independently existing whole. Fourth, modules vary and change over ontogenetic and phylogenetic time. Two of the main research goals of evolutionary developmental biology are to precisely map modular *variation* among, and within, taxa, and to describe modular *change* over time. These four criteria are used to individuate modules (Winther, 2001a).

Modules exist at a variety of levels. At the molecular level, they exist both in genes and in proteins. Arnone and Davidson (1997) use "modules" to refer to segments of *cis*-regulatory regions of DNA. The term could also be applied to exons. Although they do not use the term, Gerhart and Kirschner describe the 16 basic kinds of intercellular signaling systems (Gerhart and Kirschner, 1997, chap. 3). Each system, which is composed of transmembrane proteins, has a particular structure and engages in the process of transducing one kind of extracellular signal to another kind of intracellular one. Thus, each signaling system is a particular structural and physiological "module-kind" (on module-kinds versus module-variants-of-a-kind, see Winther, 2001a, p. 120). At the organismic level, modules exist as physiological adult structures, such as cells with limited behavioral repertoires (Larsen and McLaughlin, 1987; Larsen, 1992), internal organs, and segments in segmented taxa. Modules are also developing parts such as germ layers or morphogenetic fields in the developing limb buds of vertebrates (R. A. Raff, 1996, chap. 10).

Three Approaches to Modules and Modular Interaction
Gerhart and Kirschner, R. A. Raff, and Wagner each investigate the roles of modules in development and evolution. Although they share assumptions about the hierarchy and the criteria of modularity, they approach their study systems differently.

Here I argue that their approaches are primarily physiological, developmental, and structural, respectively (table 4.1). However, in what follows, I do not intend to suggest in any way that Raff, for example, is not interested in structure. However, Raff *focuses* on developmental modules rather than on physiological or structural ones.

Gerhart and Kirschner are, respectively, a biochemist and a cellular biologist; their investigations focus on these levels. Two important themes in their book (Gerhart and Kirschner, 1997) are the conservation of basic modular processes within and across taxa, and the intensity of interaction among modules. They explain the conservation of protein function and structure in terms of processes of connectivity among multiple intracellular metabolic and regulatory systems. This connectivity is "contingent" in that molecular and cellular networks require multiple inputs for proper functioning. For example, protein kinases, which change the conformation of other proteins by adding a phosphate group from ATP to them, quickly activate or inhibit the other proteins (Gerhart and Kirschner, 1997, pp. 80ff.). This is an example of contingent connectivity in that catalytic proteins *depend* on other catalytic proteins for their state of activation or inhibition. Contingent connectivity also indicates the importance of complex networks of processes—networks of biochemical modules, in Gerhart and Kirschner's approach. In their approach, biochemical physiological modules are crucial in explaining development and evolution (see also Kirschner and Gerhart, 1998).

R. A. Raff and Wagner both focus on the organismic level, but Raff concentrates on developmental modules, whereas Wagner emphasizes structural modules. Raff and coworkers have studied the genetic and developmental differences between two sister species of sea urchins (e.g., R. A. Raff, 1996; R. A. Raff and Sly, 2000), *Heliocidaris erythrogramma* and *H. tuberculata*. The former species is a direct developer, whereas the latter is an indirect developer; it has a pluteus larval stage (R. A. Raff, 1996, 2000). In the direct developer, all of the early cells of the morula are the same size. In contrast, in the indirect developer, morula cells have significantly different sizes. Furthermore, in direct developers "cell types homologous to those of indirect developers have different precursors" (R. A. Raff, 1996, p. 231). The splitting of these lineages occurred only 4–5 million years ago (personal communication to Rudolf Raff by Kirk Zigler, who used calibration data of Lessios et al., 1999), which makes the evolution of their radical developmental differences remarkable. In addition to investigating differences in developmental patterns, Raff and coworkers successfully hybridized the two species to explore genetic and developmental patterns and processes in the hybrids (E. C. Raff et al., 1999; Nielsen et al., 2000). This research, as well as his theoretical discussion of morphogenetic fields (S. Gilbert et al., 1996; R. A. Raff, 1996, chap. 10), indicates Raff's emphasis on developmental modules.

Wagner studies morphology, developmental biology, systematics, and theoretical population genetics. For Wagner, modules and homologues are deeply connected: all modules of a particular kind are homologues and vice versa (Wagner, 1996, and personal communication). Unlike Raff and many others, Wagner prefers a structural rather than a phylogenetic definition of homology (Wagner, 1994, 1995, 1996). Modules are character complexes determined by unique sets of expressed genes (Wagner, 1996; Wagner and Altenberg, 1996; Mezey et al., 2000). Wagner also describes modules or "building blocks" as structures (homologues) that are stabilized and constrained during development and evolution (Wagner, 1994, 1995; see also Roth, 1994). In this sense, modules are structures that are conserved across taxa, not only because of descent from a common ancestor but also because of developmental constraints ("generative" and "morphostatic" constraints; Wagner, 1994).

Wagner is clearly interested in development as well as structure; this can be gleaned from his studies on bottom-dwelling blenny fish (Wagner, 1989, 1994). I consider his approach to modularity and homology to be primarily structural, however, because although he considers developmental mechanisms and constraints to be explanatory, what he seeks *to* explain is the structural identity and individualization of morphological sets of characters during development and evolution (Wagner, 1994, 1995).

These four investigators all explore mechanisms of modular integration at various levels. However, each approaches modules and modular interaction differently.

Partitioning and Articulation: How Powerful Are Genes?

Genes are often considered to be the agents of development. The four investigators discussed in the previous section, however, are aware of the complex hierarchical environment in which genes, proteins, and other molecules interact nonlinearly during development. They endorse an articulation strategy. Other adherents of the integration perspective employ a partitioning strategy by focusing on the context-independent power of genes (table 4.1).

Because the Human Genome Project (HGP), as described by key advocates, ultimately seeks to investigate genes as mechanistic prime movers of the development of morphology, physiology, behavior, and disease, it employs partitioning within the integration perspective. Proponents of the HGP claim that "the genetic messages encoded within our DNA molecules will provide the ultimate answers to the chemical underpinnings of human existence" (Watson, 1990, p. 44). Since we are "dictated by our genetic information," the HGP will allow us to "find sets of genes for such conditions as heart disease, susceptibility to cancer, or high blood pressure" (W. Gilbert, 1992, pp. 96, 94). Although the publication in *Science* presenting the human genome warns of the "fallacies of determinism and reductionism" (Venter et al., 2001, p. 1348), it is clear that the HGP adheres to a partitioning strategy which is premised on determinism and reductionism. A number of authors have discussed the weaknesses of this strategy (Lewontin, 1991; E. F. Keller, 1992; Griesemer, 1994; Lloyd, 1994; Oyama, 2000a, 2000b; Oyama et al., 2001).

The integration perspective on organisms focuses on the mechanisms of organismic development and physiology. This contrasts with the competition perspective, which I will now explore.

Competition Perspective on Multicellular Organisms

Selective Processes, Replicator Modules, and Interactor Modules

The competition perspective focuses on the competitive dynamics within multicellular organisms. In a typical selection scenario, units replicate *differentially* because they have heritable differences *and* also because they vie for a common pool of limited resources, such as food or mates, or are subject to differential predation or parasitism, or a combination of any of these factors. The competition perspective analyzes the origin and maintenance of multicellularity as a case of selection at, potentially, multiple levels (e.g., genes, cells, organisms, and even groups of organisms as well as species and clades; in this chapter I do not discuss species or clade selection).

The distinction between replicators and interactors highlights a *functional* difference of biological units (Griesemer, 2000b, in press). Some units, the interactors, interact with their environment and with other units. Interactors have a hierarchical structure. They are (1) parts of either higher-level parts or wholes, or (2) wholes themselves (i.e., interactors are either modules or individuals). Replicators are units of which copies are made, and they are statistically correlated with—or, more contentiously, cause—the development of the interactors (Dawkins, 1976; Hull, 1980). The higher the correlation, the more efficient the selection process—where efficiency is the realized heritability (Michael Wade, personal communication). Replicators also have a hierarchical structure (Brandon, 1982, 1990). Sometimes they are independent genes with context-independent, additive phenotypic or fitness effects, or both, but they can also be genes in linkage disequilibrium due to, for example, epistasis for fitness. Replicators can even be groups of organisms or species.

Here I will be concerned with genes, potentially in linkage disequilibrium, as replicators. Selection of interactors leads to the differential reproductive success of replicators involved in producing interactors (see Lloyd, 2000 for a review). Modules exist both in an interactor and in a replicator hierarchy. Since I will mainly analyze those selective processes among interactors that biologists have suggested for the origin and maintenance of multicellularity, the term "modules" will refer to inter-actor modules unless otherwise stated.

Suborganismic modules pertinent to the competition perspective include both genes that function as interactors (e.g., selfish transposons; Werren et al., 1988) and cell lineages. Because most work has been done on cell lineages, I will focus on them. The key question for the competition perspective is: Why did, and do, cell lineages in organisms cooperate rather than defect? Defection appears to be favored at the cell level; cell lineages leaving more cell offspring have a higher fitness, at that level, than those that do not. Cooperation seems to be disadvantageous at that level; somatic specialization and curtailed reproduction lower the immediate fitness of a particular cell. I divide my discussion into investigators who claim that cell-lineage defection is mainly absent in well-integrated organisms and those who claim that it is prevalent.

Cell-Lineage Defectors Are Mostly Absent in Well-Integrated Organisms

The Evolution of Individuality: Higher-Level Control Methods and Kin Selection
The locus classicus for discussion of the evolution of multicellularity is Buss's *The Evolution of Individuality* (1987; see also Buss, 1983, 1985, 1999). Buss notes that an organism is actually not a "genetically homogeneous unit" (1987, p. 19), although it does consist of "clonal lineages" (1987, p. 77). An organism is an environment "pop-ulated by normal and variant cells" (1987, p. 76). Cell developmental modules, which are interactors, compete within this "somatic ecology" (1987, p. 139). Variant (i.e., defector) cell lineages divide in an uncontrolled fashion and contribute little to *somatic* cell function. They are often detrimental to the whole organism. There is also strong cell-lineage selection for variant cell lineages to enter into the areas, or be part of the lineages, that fulfill the *reproductive* functions. Defector cell lineages can therefore disrupt both somatic and reproductive organismic functions, accord-ing to Buss.

Since organismal stability, early in the evolution of multicellularity, was threat-ened by defector cell lineages, methods of controlling defection were strongly favored at the organismic level. For example, (1) the evolution of a sequestered germ-line eliminated the possibility that a variant developmental module could be heritable *across* organismic generations. Furthermore, since germ cells undergo sig-nificantly fewer cell divisions than somatic cells, there is a smaller likelihood of mutation occurring in them, given an approximately constant mutation rate per cell division. Buss also considers (2) maternal control of early development and (3) inductive interactions as control methods. If maternally derived egg cytoplasmic mRNA and proteins determine division patterns and cell fate, then a cell-lineage variant, with its own genotypic and phenotypic properties, cannot arise until maternal control stops (1987, pp. 54ff.).

Similarly, inductive interactions between cell lineages "restrain [ ] or direct [ ] the activities of neighboring cells, . . . [thereby] enhanc[ing] their own replication and the survivorship of the individual harboring them" (1987, p. 78). This last control method is the only one in which the direction of selection is the *same* at both the cellular and the individual levels. Note that all these methods of control are considered higher-level adaptations at the organismic level.

Sometimes these methods of control fail, as when mammalian cancers occur (1987, p. 51). Less dramatic somatic mutations can also occur. Thus, Buss implies that defection by developmental somatic modules does occur in well-integrated organisms, despite numerous control methods. Germ-line mutational variants are, however, rare because there are "overlapping periods of maternal direction and germ-line sequestration" (1987, p. 116). The generation of germ-line variation is mainly a consequence of meiosis and recombination.

In his book's last chapter, aptly titled "The Evolution of Hierarchical Organization," Buss argues that "The history of life is a history of transitions between different units of selection" (1987, p. 171). This is because "Any given unit of selection, once established, can come to follow the same progression of elaboration of a yet higher organization, followed by stabilization of the novel organization" (1987, p. 172). Thus, stabilization of individuals through control methods occurs after transitions to *that* level of individuality. Buss's book is an exploration of the transition to, and stabilization of, the multicellular level. In his conclusion he notes that there are many other levels requiring investigation. I will consider the social insect level in the last two sections of this chapter.

Buss explored one explanation for the evolution of cellular cooperation—higher-level control methods. Another crucial explanation investigated subsequently to Buss is kin selection, which Hamilton first developed in a mathematically rigorous fashion. The basic idea of kin selection is that an allele that is correlated with, or causes, a behavior lowering the immediate fitness of the benefactor may actually increase in frequency (i.e., be selected) when the recipients of the behavior are close kin who have a high probability of carrying the same allele. Some proponents of kin selection propose that it operates at only a single level—the gene or the organism (Dawkins, 1976; Maynard Smith, 1976); others argue that kin selection has components at multiple levels (e.g., Price, 1970, 1995; Hamilton, 1975; Uyenoyama and

Feldman, 1980; Wade, 1980, 1985; Queller, 1992a, 1992b; Sober and Wilson, 1998). I will develop the logic of kin selection in more detail in the section on the competition perspective on social insects.

Michod and coworkers provide detailed models that employ both multilevel kin selection and higher-level control methods to explain the origin and maintenance of multicellularity (Michod and Roze, 1997; Michod, 1999a, 1999b). Michod has both cell-level and organism-level fitness parameters in his models (e.g., replication rate of defector versus cooperator cells; organism-level fitness as a function of cooperator cell frequency). If higher-level (i.e., organismic) selection is sufficiently strong, alleles for cooperation will increase in frequency in the population. However, a number of parameters need to be considered to determine whether cooperation can reach fixation[1]: total number of cell divisions in an individual, mutation rate, and relative benefit to a defector cell (which can be less than 1; in this case mutations are deleterious at both the cell and the organism levels).

In most of the parameter space Michod explores, alleles for cooperation do *not* reach fixation even though interactor cell modules are related by common descent (Michod, 1999a, chap. 5). Kin selection is not sufficient for the origin of organismic individuality; higher-level control methods such as germ-line sequestration and defection-policing mechanisms (e.g., immune systems) are necessary (Michod, 1999a, chap. 6). Questions about how kin selection, control methods, and mutualism—cooperation through expected long-term reciprocal benefits—are related still require investigation. As we will see in the section on the competition perspective on social insects, these questions have been addressed in more detail in social insects, where an important control method is parental manipulation and reciprocal mutualistic benefits among organisms are easily conceptualized.

Partitioning and Articulation: Interactors and Replicators By employing an explicitly multilevel selection framework and emphasizing the importance of higher-level control methods, Buss and Michod employ an articulation research strategy (table 4.1). They stress the relations among parts at, and across, hierarchical levels and among parts and wholes, as well as the importance of higher-level mechanisms. In their book on transitions, Maynard Smith and Szathmáry argue that control methods are not required for the stabilization, during evolution, of higher-level organisms (Maynard Smith and Szathmáry, 1995; see also Szathmáry and Maynard Smith, 1995). Genetic similarity among modules of an organism is sufficient to arrest any potential conflicts (Maynard Smith and Szathmáry, 1995, pp. 8, 244). Alleles for cooperation can reach fixation given such high degrees of relatedness ($r \approx 1$) between modules. With this argument, these authors ignore the potentiality, and reality, of mutation and they do not consider the full range of selective parameters (e.g., when defection benefit is high). If mutation occurs, and the selective differential is sufficiently high, cooperation will not reach fixation despite generally high relatedness (Michod, 1999a).

Regarding kin selection, Maynard Smith has argued that it should be understood as happening at the level of the individual—the organism (Maynard Smith, 1976, 1982). But when he and Szathmáry discuss cell (*sensu* individual) versus organism (*sensu* group) selection, it is unclear whether they argue that kin selection has only a cell-level component or whether it also has an organism-level component (see, e.g., their analogy between the "stochastic corrector model" and kin selection in Szathmáry and Maynard Smith, 1995, pp. 227–229). They do, however, state their allegiance to the "gene-centered approach" of Williams (1966) and Dawkins (1976), and they do not cite any of the literature on multilevel kin selection theory. Both of these actions imply that Maynard Smith and Szathmáry believe that cellular kin selection occurs only at the cell level. By denying the importance of higher-level control methods and by implying that kin selection operates at a single low level, Maynard Smith and Szathmáry reveal partitioning research strategies (table 4.1). Thus, with respect to interactors, Buss and Michod adopt an articulation strategy, whereas Maynard Smith and Szathmáry endorse a partitioning strategy.

The articulation versus partitioning research strategy distinction has thus far been applied to interactors. I will how briefly discuss this distinction in light of the replicator question (Lloyd, 2000). Dawkins argues that the replicator in evolution is the individual selfish gene, which has context-independent additive effects (Dawkins, 1976, 1982). Linkage disequilibrium (which Dawkins mentions only in passing) due to, for example, epistasis for fitness "*simply* increases the size of the chunk of the genome that we can usefully treat as a replicator" (Dawkins, 1982, p. 89; emphasis added). Dawkins's position is partly based on Williams's partitioning argument for the universal validity of calculating additive genetic effects by averaging the effect of a gene across all genetic backgrounds in a population; this is justified, according to Williams, "no matter how functionally dependent a gene may be, and no matter how complicated its interactions with other genes and environmental factors" (Williams, 1966, p. 57; for critical discussion that emphasizes considerations such as small natural population sizes and nonrandom distributions of genetic backgrounds, see Wimsatt, 1984; Lloyd, 1988, chaps. 5 and 7, and 2000; Wade, 1992; Wade and Goodnight, 1998; Wade et al., 2001). Dawkins's partitioning strategy can be clearly gleaned from the concluding sentences of *The Extended Phenotype*: "The integrated multicellular organism is a phenomenon which has emerged as a result of natural selection on primitively independent selfish replicators. It has paid replicators to behave gregariously. . . . In practice the organism has arisen as a partially bounded local concentration, a shared knot of replicator power" (Dawkins, 1982, p. 264).

On the other hand, views which utilize articulation strategies in investigating replicators include (1) those that emphasize the importance and ubiquity of interactions among genes, that is, epistasis which cannot be removed by averaging across genetic backgrounds (e.g., Avery and Wasserman, 1992; Wade and Goodnight, 1998; Wolf et al., 2000; Wade et al., 2001), as well as (2) the "reproducer" and "developmental systems" views developed, respectively, by Griesemer and Oyama (Griesemer, 2000a, 2000b, in press; Oyama, 2000a, 2000b; Oyama et al. 2001). These last two authors, however, are attempting to dismantle the replicator versus interactor distinction.

Cell-Lineage Defectors Are Common in Well-Integrated Organisms

A number of authors who endorse the competition perspective claim that cell-lineage selection occurs with substantial frequency in well-integrated organisms. Otto and coworkers have investigated germ-line cell-lineage selection in contemporary organisms (Otto and Orive, 1995; Otto and Hastings, 1998). They argue that the number of cell divisions from zygote to zygote is sufficiently large to consider their mutations and mutation rate evolutionarily important (e.g., 50 in corn, 25 for *Drosophila*, 25 for female mice, and 23 for human females, per generation; Otto and Hastings, 1998, p. 510). The models of Otto and coworkers indicate that, depending on the hierarchical (i.e., cell-level and individual-level) costs and benefits of mutations, intraorganismal selection can increase or reduce the mutation rate. Furthermore, selection between germ-line developmental modules can also decrease the mutation load in a population because deleterious mutations in such modules will tend to be eliminated as they compete.

These selective scenarios differ crucially from Buss's, Michod's, and Maynard Smith and Szathmáry's in that selection is interpreted as often acting in the *same* direction at both levels. Most loss-of-function mutations that are deleterious at the individual level are also deleterious at the cellular level. Furthermore, "mutations that improve the efficiency of metabolic pathways may often be beneficial at both levels" (Otto and Hastings, 1998, p. 520). Insofar as selection operates in the same direction at both levels, control methods are not necessary. However, if a fraction of mutations have beneficial cell-level effects, but deleterious individual-level effects, control methods will be necessary. An interesting research project investigating the relative frequency of this case of opposing directions of selection at the two levels, and the evolution of control methods as a function of its increased frequency, awaits exploration (Sarah Otto, personal communication).

Nunney also emphasizes the reality of modular competition, in the form of cancer, in well-integrated organisms (Nunney, 1999a, 1999b). Growth-controlling genes that suppress the defector consequences of mutations in other growth-controlling genes would be selectively advantageous at the organism level. Despite such controls, cancers are almost inevitable in modules with high replication and turnover rates (e.g., epithelial cells in the skin, hemopoietic cells in bone marrow and lymphatic tissue). Like Michod and Buss, Nunney emphasizes the case of multilevel selection operating in opposite directions. However, he differs from them in believing that modular defection is ubiquitous.

Investigations on multicellular organisms differ significantly, depending on the perspective employed. The integration perspective attempts to understand the integrative mechanisms occurring among modules within organisms. Adherents of this perspective do not deny selection, but they generally argue that it does not occur *within* organisms. Selection occurs as a consequence of the ecological context in which whole organisms are found. Conversely, proponents of the competition perspective agree that myriad kinds of mechanisms occur within organisms. But they are interested in the ones directly pertinent to fitness, particularly those concerning cell-lineage defection and those pertinent to the control of lower-level module defection. They are typically concerned with hierarchical selective dynamics. The two perspectives thus guide distinct kinds of research on multicellular organisms.

Integration Perspective on Social Insects

Modularity in Social Insects

The better-integrated a hymenopteran colony is, the more it can be interpreted as a superorganism (i.e., an individual) with component parts. These parts, the hymenopteran organisms, as well as various symbiotic organisms and structures such as the nest, can be usefully interpreted as modules. In particular, the hymenopteran organisms can be viewed as structural, developmental, and physiological modules. Hymenopteran organisms serve as the structural modules that compose the reproductive and somatic task force of the colony. They also develop over time, as developmental modules, through egg, larval, pupal, and adult stages. Depending on species and conditions, they can take distinct morphological courses during development. Hymenopteran organisms and labor groups (e.g., foragers) are also physiological modules—they engage in particular processes that maintain the colony. In the next two sections, I will explore the two perspectives with regard to the relations among hymenopteran modules.

Early Twentieth-Century Work on Superorganismic Integration

In addition to being the first to clearly and explicitly state that the ant colony was analogous to an organism, William M. Wheeler was also a pioneer in suggesting explanations for the origin of eusociality. He saw hymenopteran social organization as a special case of the "sociogenic" "tendencies of life" (1939 [1911], p. 26). In 1918 he suggested that "trophallaxis" (i.e., the sharing of nutrition among adults and larvae) caused potentially reproductive females, in the phylogenetic past, to stay in their mother's colony and help her rear more offspring, which were sisters to the worker ants. Although this explanation is not incompatible with Hamilton's subsequent kin selection explanation, Wheeler was focused on interorganismal physiological mechanisms, rather than on genetic selective dynamics, for both the origin and the maintenance of eusociality.

The termite expert Alfred Emerson and his collaborators further articulated a superorganismic perspective on hymenopteran and termite integrative mechanisms. They did this at the University of Chicago from the 1930s to the 1950s. In his thorough review of the superorganism concept, Emerson notes, "We find that the important ecological principle of natural selection acts upon the integrated organism, superorganism or population" (Emerson, 1939, p. 197). But his brief remark on hierarchical levels of selection is hidden among a swarm of examples and citations concerning hierarchical integrative mechanisms. Listing the section headings in his 1939 article provides a feeling for its colony-level developmental and physiological emphasis: "Division of Labor"; "Ontogenetic Coordination and Integration," which is divided into five sections—"Chromosomal Foundations of Integration," "Activity Gradients and Symmetry," "Chemical Integration," "Nervous Integration," and "Rhythmic Periodicity"; and "Superorganismic Phylogeny." Emerson also developed the idea of superorganismic homeostasis with negative feedback loops (Emerson, 1956).

Emerson's colleague at Chicago, the population geneticist Sewall Wright, was working on hierarchical selection in developing his Shifting Balance Theory. Thomas Park, Emerson's ecological colleague, viewed his own work on competition as a study in the physiology of populations (Michael Wade, personal communication). Furthermore, it was partly in reaction to Emerson's research program that George C. Williams, a figure instrumental in the history of the competition perspective, developed his criticisms of group-level integration and adaptation (Williams, 1966; see also Sober and Wilson, 1998, p. 36, which recounts Williams's critical response to a lecture by Emerson while Williams was a postdoctoral student at Chicago in the 1950s). The University of Chicago is thus a fascinating locus for investigating the theoretical and experimental cooperation and conflict between the two perspectives analyzed in this chapter. Historical research investigating this university would be useful for exploring the possibilities of synthesis between the two perspectives (see Mitman, 1992; Gerson, 1998).

Partitioning and Articulation in the Behavioral Ecology of Ants: Edward O. Wilson and Deborah M. Gordon

Two important researchers in the behavioral ecology of ants, Edward O. Wilson and Deborah M. Gordon, both study the behaviors and communication systems of ant organisms and ant colonies from an integration perspective. But whereas Wilson adopts the partitioning strategy, Gordon employs the articulation strategy (table 4.2).

The myrmecologist, and later sociobiologist, E. O. Wilson started working on the behavioral ecology, systematics, and communication systems of ants in the 1950s. One of his first conceptual pieces was a strong criticism of the superorganism concept (Wilson, 1967). He writes, "There is . . . a shared faith that characterizes the reductionist spirit in biology generally, that in time all the piecemeal analyses will permit the reconstruction of the full system *in vitro*. In this case an *in vitro* reconstruction would mean the full explanation of social behavior by means of integrative mechanisms experimentally demonstrated and the proof of that explanation by the artificial induction of the complete repertory of social responses on the *part of isolated members of insect colonies*" (p. 36; emphasis added). Note that he is interested in integrative mechanisms, for example, pheromonal communication signals. Note also that Wilson believes a comprehensive understanding of the whole can be achieved through the manipulation of the parts. Each part (i.e., module) has particular social responses, which are context-independent. The implicit idea is that social behavior is a linear extension of individual behavior. The employment of partitioning is further evidenced in his autobiography, written three decades later, in which he notes, "This reductionism [partitioning], as Lewontin expressed and rejected it, is precisely my view of how the world works" (Wilson, 1994, p. 346).

Wilson's adoption of partitioning can be best understood by analyzing two particular cases: caste membership as a determiner of tasks performed, and context-independent meaning of chemical signals. A minority of ant species have workers of different sizes and allometric proportions (the relative dimensions of their body parts are not scaled equally). For example, leaf cutter ants of the genus *Atta* often have four castes. Wilson and others argue that tasks performed are highly correlated with caste membership; ants of different castes specialize in different tasks (e.g., E. O. Wilson, 1968, 1971; Oster and Wilson, 1978; Hölldobler and Wilson, 1990).

Therefore, colony-level productivity is correlated with caste distribution, and there should be optimal caste distributions if selection can act on the colony, a premise Wilson accepted as early as a 1968 article. However, he does not explicitly argue for the utility of the superorganism concept, which includes the idea of colony-level selection, until the 1990 book written with Hölldobler (E. O. Wilson, 1968; Hölldobler and Wilson, 1990); in both Oster and Wilson (1978) and Wilson (1985), Wilson is implicitly sympathetic to some aspects of the superorganism concept. Wilson thus understands colony-level productivity as an aggregative function of caste membership, which is an ant organismal (i.e., modular) property.

Wilson also argues for the context-invariant meaning of chemical signals. In the chapter titled "Communication," Hölldobler and Wilson present a table in which they attribute context-invariant responses to particular chemical emissions from glands (1990, p. 228). In his autobiography, Wilson recounts his first discovery of an ant pheromone. He interprets the function of this ant pheromone as follows: "The pheromone in the gland is . . . both the command and the instruction during the search for food. The chemical was everything" (E. O. Wilson, 1994, p. 291). The meaning of the pheromone (i.e., the action it elicits) does not, for Wilson, depend on other pheromones simultaneously employed, the quantity of pheromone present, the state the perceiving ant is in, or any other factor. The meaning is context-independent. Furthermore, Wilson postulates that "mass communication" occurs through the aggregation of pheromones by workers walking along the same food trail (Wilson, 1994, p. 291; Wilson and Hölldobler, 1988). Thus, according to Wilson, communication among ants is a linearly aggregative function of context-invariant chemical signals. Wilson's partitioning strategy can be seen in how he (1) argues for the context-independent properties of both ant organisms and chemical signals, and (2) suggests that colony-level behavior is a linear aggregation of these properties.

The myrmecologist Deborah M. Gordon started her investigations on ant behavioral ecology and communication systems during the 1980s (Gordon, 1989, 1996, 1999). She investigates the ecology and the behavior of a species of desert seed harvester ant, *Pogonomyrmex barbatus*. This species, like most ant species, lacks morphologically distinct castes. Furthermore, particular ants switch between different tasks: foraging, patrolling, colony maintenance, midden work, and resting. Task-switching by ant organisms depends on circadian and seasonal rhythms, weather conditions, colony age, food availability, and presence of neighbors. Thus, the behavior of an individual ant is extremely sensitive to surrounding conditions. Gordon has not found any particular organismal property that may correlate with task proclivities. In collaboration with others, she has developed models that capture the flexibility of individual ants (Gordon et al., 1992). Gordon opposes Wilson's caste perspective and instead emphasizes the interactive context dependency of ant organism and ant colony behavior. She employs an articulation research strategy, which endorses—even requires—a superorganism view of ant colonies.

Gordon also emphasizes the role of interactions in ant communication. She distinguishes between two kinds of relations: interactions and interaction *rates* (see Winther, 2001b). Antennal contact between two ants is an example of an interaction. The number of different ants a particular ant interacts with in such a manner per unit time is an example of an interaction rate. Gordon's experiments and models indicate that ants change behavior in different contexts as a consequence of interaction rate, rather than due to any message carried in the interaction itself. Response to interaction rate may also occur in "brains, immune systems, or any place where the rate of flow of a certain type of unit, or the activity level of a certain type of unit, is related to the need for a change in the rate of flow" (Gordon, 1999, p. 169). Interaction rate, rather than interaction per se, explains temporal behavioral variance of complex dynamical systems: "the interaction pattern may be more important than the message" (Gordon, 1999, p. 156). Furthermore, information concerning ecological conditions surrounding the system can be transmitted through interaction rate. Interaction rates of pheromonal transmission and reception could also be modeled.

Note that focusing on either the message (e.g., pheromone) or the interaction rate is, strictly speaking, distinct from adhering to either context dependence or independence (i.e., pheromonal meanings could be context-dependent and a particular interaction rate could have a context-independent meaning). However, in this case, Wilson focuses on messages with context-independent meanings, whereas Gordon emphasizes the context-sensitive interaction rates of pheromonal, tactile, and nutritional communication as the ways that modules of an ant colony communicate.

The contrast between Wilson and Gordon with respect to their views on caste and communication in ants exemplifies how partitioning and articulation strategies are employed. Their common interest in understanding behavior as a consequence of physiological and ecological processes indicates their shared employment of an integration perspective.

Is an Evolutionary Developmental Biology of Social Insects Possible?

An evolutionary biologist once claimed that "evolution is the control of development by ecology" (van Valen, 1973). A significant amount of work has been done on the behavioral ecology of social insects (see West-Eberhard, 1987; Franks, 1989; Bourke and Franks, 1995; Gordon and Wilson references above). Thus, we have significant data on the ecological context in which social insect evolution occurs. The competition perspective has also provided voluminous information on evolutionary genetic aspects of insect societies. However, there is much less work on the development of insect societies considered as physiologically integrated wholes in an ecological context. In other words, we know relatively little about the patterns and processes of differential gene expression, and the developmental pathways, of social insect modules and colonies. An evolutionary developmental biology of social insects requires that we investigate this.

Schneirla (1971) performed an early set of investigations into development of army ant colonies. Army ant colonies have two discrete stages: nomadic and statary. The 15-day nomadic stage of *Eciton burchelli* starts when a cohort of adults has just emerged from their pupal case and a distinct cohort of eggs has just hatched into numerous hungry larvae (on the order of hundreds of thousands) requiring large amounts of food. The 20-day statary phase commences when these larvae pupate. Ten days into the statary phase, the queen starts laying eggs again. The timing of this 35-day cycle, with the synchronized timing of the development of the two generations, is the result of multiple reciprocal chemical, tactile, and nutritive interactions among queens, workers, and brood. Schneirla, following Wheeler, called these interactions "trophallaxis" (Schneirla, 1971; see Hölldobler and Wilson, 1990, pp. 577–579 for a note of skepticism).

More detailed research on developmental integration of social insect colonies has appeared subsequently (e.g., E. O. Wilson, 1985; D. E. Wheeler, 1986, 1991; Hölldobler and Wilson, 1990; Robinson et al., 1997; Hartfelder and Engels, 1998; Evans and Wheeler, 1999, 2001; Robinson, 1999). In her review of the different mechanisms involved in reproductive-somatic caste determination, Diana Wheeler discusses queen effects (parental manipulation) on (1) worker behavior, (2) larval development, and (3) egg production and/or quality (Wheeler, 1986). In bees and wasps, a queen pheromone affects the building of gyne wax cells; the pheromone usually suppresses its construction. Larvae in gyne wax cells receive more food from workers. As a consequence, they have higher levels of juvenile hormone (JH), which is necessary to develop into a reproductive female. The production of JH being contingent on nutrition quantity is called a "nutritional switch." Further elaboration of organism and colony physiology leading to reproductive caste differentiation in honeybees (*Apis mellifera*) can be found in Hartfelder and Engels (1998).

Similar mechanisms involving pheromones, nutrition, and JH are found in ants despite the absence of brood cells. A queen pheromone acts during a critical period of ant larval development to induce the loss of the capacity of larvae to develop as gynes. This inhibition occurs before the nutritional switch. Regarding egg production and quality, the ant queen can control how many eggs she lays—which is, of course, a function of how much nutrition she ingests, the temperature to which she is exposed, and other factors. More important, the queen can allocate different amounts of nutrition, mRNA, or hormones, or a combination of all three, to different eggs. In *Formica polycenta*, for example, large eggs with relatively large amounts of maternal mRNA develop into gynes (D. E. Wheeler, 1986). There are multiple strategies available to hymenopteran queens and workers for reproductive caste determination of the developing brood.

The hymenopteran colony can be interpreted as an individual with mechanisms of developmental differentiation and an internal physiology. When evolutionary developmental biology investigators study module differentiation in organisms, they study patterns and processes of differential gene expression. An evolutionary developmental biology of social insect superorganisms requires a search for such patterns and processes in hymenopterans. Evans and Wheeler (1999, 2001) found reliable differences in patterns of gene expression between honeybee workers and queens. They also found that "several genes with caste-biased expression in honey bees show sequence similarity to genes whose expression is affected by hormones in *Drosophila*" (Evans and Wheeler, 2001, p. 64). Thus, hormones such as JH may be involved in differentially activating genes correlated with morphological and physiological differences between workers and queens. Thus, the hymenopteran colony has an internal physiology that induces differential gene expression of its modules.

Behavior is a crucial factor in colony development and physiology. Although Evans and Wheeler do not ignore behavior, their focus is at the molecular and organism-physiological level. Gene Robinson and coworkers have explicitly called for the study of "the molecular genetics of social behaviour in ecologically relevant contexts" (Robinson, 1999, p. 204; see also Robinson et al., 1997). Their research program seeks to synthesize processes involving gene expression, hormones, pheromones, neurophysiology, behaviors, and ecology. They want to do this because "focusing on genes provides a common language and convergent research themes" (Robinson et al., 1997, p. 1099). Whether a synthetic theory of colony integration requires a genetic focus merits further discussion.

Since about 1990, several biologists have investigated the developmental and physiological mechanisms of social insect colonies. This application of the integration perspective has similarities to the evolutionary developmental biology synthesis that has been occurring at the organism level. Is an evolutionary developmental biology of social insect colonies possible? Clearly, differential gene expression of colony modules has been found. But this does not necessarily imply that we can consider social insect colonies as individuals when it comes to assessing module homologies, establishing ancestral and derived colony-level developmental patterns, describing the origin of colony-level innovations, and linking colony-level microevolutionary with colony-level macroevolutionary change (see R. A. Raff, 2000, p. 75; Wagner et al., 2000, p. 820).

Progress has been made, however, on some of these aspects of individuality as applied to social insect colonies. Colony-level properties such as nest morphology have been used in determining robust phylogenetic trees in wasps (Wenzel, 1993); some superorganismic modules can therefore be used for establishing homologies. Furthermore, Anderson and McShea (2001) argue that organs or "intermediate-scale structures," such as teams (workers adopting different subtasks in order to perform a task, e.g., carrying a prey item) and nests, exist in social insect colonies. These results, in combination with the work discussed in this section, indicate that an evolutionary developmental biology of social insect colonies is possible. We should embark on such a project, which would also involve an investigation of the individuality of social insect colonies.

Competition Perspective on Social Insects

Kin Selection, Inclusive Fitness, and Multilevel Selection, 1964–1975: The Work of Hamilton and Price

Why do some organisms, such as hymenopteran workers, become sterile? The origin and maintenance of cooperation, which is often called altruism, was a problem that Darwin wrestled with in his *Origin of Species* (Darwin, 1964 [1859], chap. 7). His prescient answer appealed to family-level selection of "fertile parents which produced most neuters with . . . profitable modification[s]" (Darwin, 1964 [1859], p. 239).

Hamilton developed the mathematics of this group selection argument more than 100 years later. Initially, however, he argued against group selection and felt that inclusive fitness made kin selection an extension of individual selection (Hamilton, 1963, 1964a, 1964b). Hamilton was interested in why organisms would reduce their fitness, for the benefit of other organisms, through behaviors such as defending the other organism or helping it to reproduce. Qualitatively, he argued that an allele that caused a behavior detrimental to a particular individual would increase in frequency when the recipients of the behavior were close kin who, with a high probability, carried the same allele (Hamilton, 1963, 1964a, 1964b). Quantitatively, he noted the conditions under which alleles for cooperation could increase in frequency: $(rB - C) > 0$, or $r > (C/B)$. In this equation, r is the coefficient of relatedness (e.g., in diplo-diploids such as mammals, organism-to-sibling $r = 0.5$; organism-to-first cousin $r = 0.125$); B is the fitness benefit the given behavior provides to the recipient; and C is the fitness cost to the benefactor (particular individual) of performing the behavior. This equation is known as Hamilton's rule.

What Hamilton noted was that in hymenopterans, females are more closely related to their sisters ($r = 0.75$) than to their offspring of either sex ($r = 0.5$), provided that the females have the same father. This high relatedness occurs because hymenopterans are haplo-diploid: males have only one chromosome of each pair of chromosomes, whereas females have both chromosomes of each pair. Thus, on relatedness grounds alone, a female should choose to help her mother rear offspring, which are her sisters: "Our principle tells us that even if this new adult had a nest ready constructed and vacant for her use she would prefer, other things being equal, returning to her mother's and provisioning a cell for the rearing of an extra sister to provisioning a cell for a daughter of her own" (Hamilton, 1996 [1964b], p. 58).

There are complications to this simple theory, however. Hamilton noted, as others subsequently have, that multiply mated queens produce female offspring with a relatedness coefficient smaller than 0.75. If the queen has mated with two males, and assuming equal contribution from the two males and no sperm competition, among-sibling relatedness is 0.5. If the queen has mated with more than two males, the relatedness coefficient is smaller than 0.5 and converges to 0.25 as the number of males gets very large, given the assumptions stated above (Hamilton, 1996 [1964b], p. 62; Hamilton did not explicitly mention sperm competition). Hamilton noted that despite this, cooperation would still be favored, given appropriate B and C parameters. Furthermore, in some genera (e.g., worker fire ants of the genus *Solenopsis*, which lack ovaries), reproduction is not a possibility. Thus, other parameters and conditions besides r, B, and C need to be considered.

Another complication that Hamilton discussed, but did not suggest an explanation for in his early articles, is that a worker is related by only 0.25 to her brothers, whereas she would be related by 0.5 to a son. Trivers and Hare (1976) subsequently suggested that workers would still prefer to raise sisters rather than offspring if they could skew the sex ratio of sibling reproductives toward a 3 : 1 gyne : male ratio (see also Crozier and Pamilo, 1996). Complications such as these have convinced investigators that Hamilton's rule is a shorthand for more complex quantitative and population genetic models.

Investigators in the competition perspective employ Hamilton's rule. Their focus is on the reproductive dynamics of social insect colonies leading to gene frequency change. Furthermore, they have tended only to estimate r (see Gadagkar, 1991; Bourke and Franks, 1995; Queller and Strassman, 1998). This is in part because it remains conceptually unclear how to estimate B and C, which both depend on ecological conditions. For example, which metric could we use to compare alternative worker strategies of egg-laying and foraging in estimating B and C? (Deborah Gordon, personal communication).

Hamilton initially emphasized that kin selection was an extension of individual selection; he used the term "inclusive fitness" (e.g., Hamilton, 1996 [1964a]; table 4.2). Price's covariance approach to selection radically changed Hamilton's view on kin selection. (Covariance is a probabilistic and statistical measure of the *correlation* between two variables.) Price sought to develop a "general selection theory" (Price, 1995, p. 389; Price, 1970). He realized that selection could be thought of as a covariance between the fitness of the units under study and their properties. These properties could be genotypic or phenotypic. Price, and others, have shown mathematically that this covariance can be decomposed into two components, each of which describes selection at one of two levels—one *within* the interactor unit and one *among* interactor units (Price, 1970, 1995; Wade, 1980, 1985; Frank, 1995). A hierarchical selection process causes gene frequency change.

In social insects, a nonzero first component representing within-colony selection could be caused by workers altering the colony sex ratio and thereby altering the normal Mendelian ratios (i.e., underrepresenting maternal genes by destroying males, overrepresenting paternal genes by not destroying gynes). This is analogous to meiotic drive in organisms (see Werren et al., 1988; Hurst et al., 1996). In social insects, a nonzero second component indicating among-colony selection would occur whenever some colonies left more offspring colonies than other colonies. Such selection is also part of some sex-ratio evolution models (Michael Wade, personal communication). Among-colony selection is analogous to organismal selection in organisms. Price's multilevel selection equation, which decomposes the causes of gene frequency change, can be further expanded to any number of levels so that we can have, for example, among-colony, among-organism (i.e., among-ant-organism module), and within-organism (i.e., within-ant-organism module, such as meiotic drive in queen ants) selection in social insects.

Hamilton employed Price's equation in an article in which he argued that kin selection was, indeed, a multilevel selection process (Hamilton, 1996 [1975]; table 4.2). Cooperation could evolve (i.e., alleles for cooperation increase in frequency) if among-colony selection for such alleles was stronger than within-colony selection against such alleles. One way to increase among-colony additive genetic fitness variance was precisely to have colonies with only one or a few queens mated with only one or a few males. This is analogous to unicellular bottleneck reproduction of sexual organisms (see Michod, 1999a). In these cases most additive genetic fitness variance would be among colonies rather then within them.

Kin selection, whether conceptualized as a single-level or a multilevel process, describes the selective dynamics among interactor modules, such as ant organisms, of social insect colonies. Furthermore, replicator modules, such as alleles for cooperative behavior, can increase in frequency as a consequence of these dynamics.

Other Mechanisms for the Origin and the Maintenance of Cooperation: Parental Manipulation and Mutualism

Two other explanations for the evolution of cooperation in social insects have been suggested. The first is a kind of control method, parental manipulation: offspring are inhibited during ontogeny so that they become sterile and help their mother (Michener and Brothers, 1974; Alexander, 1974). Parental manipulation can occur through pheromones, physical force, or limited nutrition. We encountered these mechanisms, as integrative mechanisms, in the section on the integration perspective on social insects. The second explanation is mutualism: social cooperative interactions between two organisms, each of which can reproduce, are beneficial to each organism in the long run, even if they are, on occasion, detrimental in the short run (Trivers, 1971; Axelrod and Hamilton, 1981; Maynard Smith, 1982, chap. 13, "The Evolution of Cooperation"). Evolutionary game theory has been used to model this mechanism. Both of these mechanisms can be interpreted as pertaining to modules of social insect colonies that have, respectively, (1) asymmetric power relationships or (2) long-term fitness benefits.

Some authors have presented the three mechanisms for the evolution of cooperation as *distinct* alternatives (Hölldobler and Wilson, 1990; Seger, 1991). Hölldobler and Wilson consider kin selection and parental manipulation to be distinct explanations for the origin of sterile castes (1990, p. 182). Other investigators, however, imply that it is meaningless to attribute relative importance to each of these mechanisms because they operate simultaneously and actually influence one another. Instead, these researchers have developed models that explicitly incorporate all of these mechanisms (e.g., L. Keller and Reeve, 1999; Crespi and Ragsdale, 2000; Timothy Linksvayer, personal communication).

Partitioning and Articulation: Kin Selection, Inclusive Fitness, and Multilevel Selection, 1976–2001

Subsequent to the development of Price's equation, multilevel selection theory was expanded by a number of investigators (Uyenoyama and Feldman, 1980; Wade, 1980, 1985, 1996; D. S. Wilson, 1980; D. S. Wilson and Colwell, 1981; Queller, 1992a, 1992b; for historical and philosophical reviews, see Lloyd, 1988, 2000; Sober and Wilson, 1998). Broadly stated, these investigators found that most cases of selection can be understood as hierarchical selection processes with hierarchical selective components. All cases of selection in populations with social interactions can be decomposed into at least two components: group and individual selection. Thus all cases of kin selection are hierarchical (see Wade, 1980). Not all cases of hierarchical selection need involve kin, however: consider selection on symbiotic relations such as lichens.

There are few cases in nature to which a hierarchical approach could not be applied. Those cases that approach the Fisherian idealization of extremely large, randomly mating, unstructured populations are candidates. The hierarchical selection approach, which implies an articulation research strategy (table 4.2), has been applied to social insects (e.g., Bourke and Franks, 1995). It can be used to understand the selective dynamics occurring among and within social insect colonies.

But the hierarchical approach has been met with resistance from investigators adopting a partitioning strategy regarding kin selection (table 4.2). A number of behavioral ecologists still interpret inclusive fitness and kin selection as an organism- (or gene-) level process or property, as Hamilton originally did in his articles from the 1960s (e.g., Dawkins, 1976; Grafen, 1984; Krebs and Davies, 1993). For these investigators, complete models can be built using inclusive fitness defined only at the single level of the organism or gene.

But perhaps the difference between articulation and partitioning research strategies is not significant. A number of modelers have cogently shown that single-level inclusive fitness is equivalent to hierarchical selection if the former is defined, modeled, and estimated correctly (Queller, 1992b; Dugatkin and Reeve, 1994; Bourke and Franks, 1995; Kerr and Godfrey-Smith, 2002). Two points should be made in response to this. First, hierarchical selection models still reveal articulation strategies in that they make the investigator aware of *all* the selection processes actually occurring in nature. A problem with individual-level inclusive fitness models is that they require that fitness parameters be averaged from the hierarchical selective parameters; a loss of theoretical and empirical information regarding selective dynamics occurs as a consequence of this averaging (Lloyd, 1988; Wade, 1992; Wade and Goodnight, 1998; Wade et al., 2001; Sober and Wilson, 1998). Second, claims about the equivalence of models should be assessed on a case-by-case basis.

In concluding this section I want to return to the superorganism, which I described in the section on the integration perspective on social insects. In the integration perspective the superorganism, as well as the multicellular organism, is defined in terms of developmental and physiological integration mechanisms. Although such mechanisms are not denied in the competition perspective, this perspective defines the superorganism, and the multicellular organism, in terms of the strength of selection at multiple levels (D. S. Wilson and Sober, 1989; Ratnieks and Reeve, 1992). Wilson and Sober state the definition succinctly: "When between-unit selection overwhelms within-unit [between-module] selection, the unit itself becomes an organism [individual] in the formal sense of the word" (1989, p. 343). According to the competition perspective, an increase in any of three factors will increase the among-unit component of total additive genetic fitness variance: relatedness, control methods, and the benefits from mutualism. If the among-unit component is greater than (i.e., "overwhelms") the within-unit component, an individual will have been "formally" identified. Each perspective employs its own definition of individuality. A further discussion of overlaps and tensions in these definitions is necessary for a synthesis of the two perspectives.

On Material Nature and Theoretical Perspectives: Modular Integration or Competition, or Both?

In this chapter, I have explored two perspectives on modular processes. Since I have been interested in exploring two distinct levels of individuality, I have focused on cells and social insect (particularly hymenopteran) organisms as parts of a whole. Under the integration view, these parts are intermediate-level modules involved mainly in mechanistic processes. Under the competition view, these parts are interactor modules engaged primarily in selective processes. But genes have also been important modules in my analysis. Genes are important mechanistic modules in the integration view and are, generally, replicator modules in the competition view. Partitioning research strategies in *both* perspectives tend to focus on the context-independent properties, and powers, of genes.

An apt distinction between gene-P (phenotypic gene) and gene-D (developmental gene) serves to contrast the two perspectives (Moss, 2001). Moss defines a gene-P in terms of "its relationship to a phenotype albeit with no requirements as regards specific molecular sequence nor with respect to the biology involved in producing the phenotype" (Moss, 2001, p. 87). A gene-D, in contrast, "*is defined by its molecular sequence. A Gene-D is a developmental resource (hence the "D") which in itself is indeterminate* with respect to phenotype" (Moss, 2001, p. 87). Whereas the integration perspective's goal is to unravel the mechanisms involving genes-D, the competition perspective is concerned with the change in frequencies of genes-P. The integration view seeks to unpack the molecular activities of genes-D. The competition view is not concerned with developmental mechanisms; statistical correlations, produced by genes-P, between genotype and phenotype are sufficient.

In either perspective, genes are undoubtedly crucial. However, in both perspectives, articulation research strategies go beyond the partitioning assumption of invariant monadic properties of genes by articulating the complex, nonlinear mechanistic and selective relationships that hierarchical modules, including genes, have (1) among themselves both at and across levels, (2) with respect to the whole in which they exist, and (3) with respect to the hierarchical environment in which they exist.

The historical origins of these two perspectives are worth noting. The field of evolutionary developmental biology, which is the primary territory of the integration perspective, first investigated organisms. This had started with Darwin's, Haeckel's, and Weismann's nineteenth-century syntheses of evolution, development, and heredity (e.g., Churchill, 1987; Gerson, 1998; Winther, 2000, 2001c). Investigators studying organisms also have developed the concept of modularity. Only very recently has the possibility of an evolutionary developmental biology research program of social insect colonies become real. Conversely, the field of levels of selection, which is the main territory of the competition perspective, first investigated groups of individual organisms, in particular social insect colonies. This started with Darwin's worries about the evolution of sterility in the Hymenoptera. It is not surprising that levels of selection theory started with the Hymenoptera; after all, it is here that the drastic reduction of immediate organism (or gene) fitness in favor of a higher-level group of related organisms is most obvious. Only recently has a levels of selection research program been applied to multicellular organisms and modularity.

In this chapter, I have presented the two perspectives without explicitly discussing how they are related or whether they are even different. A detailed analysis of the relationships between the perspectives would require a separate article. Therefore, in concluding this chapter I will merely sketch some of the relationships.

Perspectives coordinate scientific activity (Wimsatt, 1974; Griesemer, 2000a and personal communication; Gerson, personal communication). Phenomena, methodologies (including research strategies), theories, and questions of interest are all involved in scientific activity. Thus, perspectives coordinate relations among these aspects of science.

When considering these aspects it becomes clear that the integration and competition perspectives are distinct. For example, they are committed to different *methodologies*. Developing a detailed narrative of gene expression patterns or morphogenesis, or both, requires elaborate molecular techniques, whereas investigating gene frequency changes in hierarchically structured populations involves detailed quantitative genetic and population mathematical genetic models and simulations.

Work in the integration perspective could, and does, involve mathematical models, and simulations, whereas work in the competition perspective could, and does, involve molecular techniques. But the overall pattern of commitment to techniques, and the ways the techniques are used, are distinct in the two perspectives.

When considering the conclusions arrived at from abstract mathematical *models*, there can be a significant overlap between the two perspectives. A family of models constitutes a theory. Some modelers have used the Price multilevel selection equation to model the evolution of integration (Sahotra Sarkar, personal communication). But there is still a difference between the model types generally employed in the two perspectives: mechanistic narrative models in the integration perspective, as opposed to abstract mathematical models in the competition perspective. Furthermore, even when the integration perspective uses mathematical models, it tends to employ models that do not explicitly consider selection but, instead, describe interactions among modules and the effect of such processes on the whole (e.g., Gordon et al., 1992; von Dassow et al., 2000). When selection is considered in the mathematical models of the integration perspective, the models are different from the prevalent ones in quantitative and population genetics. Investigators in the integration perspective conceive of selection as acting on the whole individual; that is, modules do not compete (e.g., Oster and Wilson, 1978; Kauffman, 1993). Thus, by considering techniques and theories, which are families of models, it becomes clear that the integration and competition perspectives are distinct.

Elsewhere I have articulated how the two perspectives relate with respect to the *questions* posed: (1) the questions are different—this is the *different questions* interpretation (What are the mechanisms among parts and among parts and higher-level wholes? versus How do gene frequencies change as a consequence of competition among parts?); (2) the questions are the same, but concern different episodes in the history of life—integrative mechanisms work within stable individuals, whereas competitive dynamics are crucial during transitions to higher levels of individuality; this is the *nonoverlapping* interpretation (What developmental and evolutionary forces shape stable individuals? versus What developmental and evolutionary forces act during transitions?); (3) the questions are the same and explanations, integrative or competitive, are in conflict—this is the *irreconcilable* interpretation (What developmental and evolutionary forces shape stable individuals? The integration perspective would emphasize integrative mechanisms and would interpret apparent modular competition simply as mechanistic dysfunction; the competition perspective would emphasize competitive mechanisms and would tend to see the whole as a population with competing parts rather than as an integrated individual.) (Winther, 2001a). Only under the last interpretation are the perspectives in conflict.

Further work is required to express the multiple relationships between these two perspectives in terms of the phenomena, methodologies (including research strategies), theories, and questions of interest that they coordinate and employ. These perspectives guide the work in the scientific fields investigating the processes of evolutionary development biology and levels of selection. We need to understand the differences, similarities, conflicts, and complementarities between these perspectives in order to develop a complete contemporary synthesis of scientific fields pertinent to evolution, development, and genetics.

Acknowledgments

I thank Werner Callebaut and Diego Rasskin-Gutman for inviting me to the workshop at the Konrad Lorenz Institute near Vienna, Austria. I wrote this chapter for that conference. Numerous individuals provided encouragement, answered questions, read prose, and patiently discussed—verbally and via E-mail—myriad issues that were occasionally esoteric. I am grateful to the following for their invaluable help: Janice Alers-García, Lauren Ancel Meyers, Jessica Bolker, Mark Brown, Elihu Gerson, Deborah Gordon, James Griesemer, Abigail Harter, Benjamin Kerr, Sandhya Kilaru, Ellen Larsen, Timothy Linksvayer, Curtis Lively, Elisabeth Lloyd, Daniel McShea, Richard Michod, Narisara Murray, Amir Najmi, Sarah Otto, Susan Oyama, Rudolf Raff, Dick Repasky, Sahotra Sarkar, Gerhard Schlosser, Michael Wade, Günter Wagner, Aage and Grethe Winther, my sister, Rikke Winther, and Kritika Yegnashankaran.

Note

1. As any of these three parameters increases, the equilibrium frequency of cooperation diminishes (Richard Michod, personal communication). Note that by "fixation" I mean an extremely high frequency of alleles for cooperation: recurrent mutation ensures that no allele is ever fixed, strictly speaking.

References

Alexander RD (1974) The evolution of social behavior. Ann Rev Ecol Syst 5: 325–383.

Allen C, Bekoff M, Lauder GV (1998) Nature's Purposes: Analyses of Function and Design in Biology. Cambridge, MA: MIT Press.

Anderson C, McShea DW (2001) Individual versus social complexity, with particular reference to ant colonies. Biol Rev (Cambridge) 76: 211–237.

Arnone MI, Davidson EH (1997) The hardwiring of development: Organization and function of genomic regulatory systems. Development 124: 1851–1864.

Avery L, Wasserman S (1992) Ordering gene function: The interpretation of epistasis in regulatory hierarchies. Trends Genet 8: 312–316.

Axelrod R, Hamilton WD (1981) The evolution of cooperation. Science 211: 1390–1396.

Bolker JA (2000) Modularity in development and why it matters to Evo-Devo. Amer Zool 40: 770–776.

Bourke AFG, Franks NR (1995) Social Evolution in Ants. Princeton, NJ: Princeton University Press.

Brandon RN (1982) The levels of selection. In: PSA 1982, vol. 1 (Asquith PD, Nickles T, eds), 315–323. East Lansing, MI: Philosophy of Science Association.

Brandon RN (1990) Adaptation and Environment. Princeton, NJ: Princeton University Press.

Buss LW (1983) Somatic variation and evolution. Paleobiology 9: 12–16.

Buss LW (1985) The uniqueness of the individual revisited. In: Population Biology and Evolution of Clonal Organisms (Jackson JBC, Buss LW, Cook RE, eds), 467–505. New Haven, CT: Yale University Press.

Buss LW (1987) The Evolution of Individuality. Princeton, NJ: Princeton University Press.

Buss LW (1999) Slime molds, ascidians, and the utility of evolutionary theory. Proc Natl Acad Sci USA 96: 8801–8803.

Churchill FB (1987) From heredity theory to Vererbung: The transmission problem, 1850–1915. Isis 78: 337–364.

Crespi B, Ragsdale J (2000) A skew model for the evolution of sociality via manipulation: Why it is better to be feared than loved. Proc Roy Soc London B267: 821–828.

Crozier RH, Pamilo P (1996) Evolution of Social Insect Colonies: Sex Allocation and Kin Selection. Oxford: Oxford University Press.

Darwin CR (1964) [1859] On the Origin of Species by Means of Natural Selection, or the Preservation of Favoured Races in the Struggle for Life. Cambridge, MA: Harvard University Press.

Dawkins R (1976) The Selfish Gene. Oxford: Oxford University Press.

Dawkins R (1982) The Extended Phenotype. Oxford: Oxford University Press.

Dugatkin LA, Reeve HK (1994) Behavioral ecology and levels of selection: Dissolving the group selection controversy. Adv Stud Behav 23: 101–133.

Eldredge N (1985) Unfinished Synthesis: Biological Hierarchies and Modern Evolutionary Thought. New York: Oxford University Press.

Eldredge N (1989) Macroevolutionary Dynamics: Species, Niches, and Adaptive Peaks. New York: McGraw-Hill.

Emerson AE (1939) Social coordination and the superorganism. Amer Mid Nat 21: 182–206.

Emerson AE (1956) Regenerative behavior and social homeostasis of termites. Ecology 37: 248–258.

Evans JD, Wheeler DE (1999) Differential gene expression between developing queens and workers in the honey bee, Apis mellifera. Proc Natl Acad Sci USA 96: 5575–5580.

Evans JD, Wheeler DE (2001) Gene expression and the evolution of insect polyphenisms. BioEssays 23: 62–68.

Frank SA (1995) George Price's contributions to evolutionary genetics. J Theor Biol 175: 373–388.

Franks NR (1989) Army ants: A collective intelligence. Amer Sci 77: 138–145.

Gadagkar R (1991) On testing the role of genetic asymmetries created by haplodiploidy in the evolution of eusociality in the Hymenoptera. J Genet 70: 1–31.

Gerhart J, Kirschner M (1997) Cells, Embryos, and Evolution. Malden, MA: Blackwell Science.

Gerson EM (1995) Generalization in Natural History. Presentation at International Society for the History, Philosophy, and Social Studies of Biology meetings, Louvain, Belgium.

Gerson EM (1998) The American System of Research: Evolutionary Biology, 1890–1950. Sociology dissertation, University of Chicago.

Ghiselin M (1974) A radical solution to the species problem. Syst Zool 23: 536–544.

Ghiselin M (1997) Metaphysics and the Origin of Species. Albany: State University of New York Press.

Gilbert SF, Bolker JA (2001) Homologies of process and modular elements of embryonic construction. In: The Character Concept in Evolutionary Biology (Wagner GP, ed), 435–454. San Diego: Academic Press.

Gilbert SF, Opitz JM, Raff RA (1996) Resynthesizing evolutionary and developmental biology. Dev Biol 173: 357–372.

Gilbert W (1992) A vision of the grail. In: The Code of Codes: Scientific and Social Issues in the Human Genome Project (Kevles DJ, Hood L, eds), 83–97. Cambridge, MA: Harvard University Press.

Godfrey-Smith P (1993) Functions: Consensus without unity. Pac Phil Quart 74: 196–208.

Godfrey-Smith P (1996) Complexity and the Function of Mind in Nature. Cambridge: Cambridge University Press.

Gordon DM (1989) Caste and change in social insects. In: Oxford Surveys in Evolutionary Biology (Harvey PH, Partridge L, eds), 55–72. Oxford: Oxford University Press.

Gordon DM (1996) The organization of work in social insect colonies. Nature 380: 121–124.

Gordon DM (1999) Ants at Work: How an Insect Society Is Organized. New York: Free Press.

Gordon DM, Goodwin B, Trainor LEH (1992) A parallel distributed model of the behaviour of ant colonies. J Theor Biol 156: 293–307.

Gould SJ (1995) The Darwinian body. N Jb Geol Paläont Abh 195: 267–278.

Gould SJ, Lloyd EA (1999) Individuality and adaptation across levels of selection: How shall we name and generalize the unit of Darwinism? Proc Natl Acad Sci 96: 11904–11909.

Grafen A (1984) Natural selection, kin selection and group selection. In: Behavioural Ecology: An Evolutionary Approach (Krebs JR, Davies NB, eds), 62–84. Sunderland, MA: Sinauer.

Griesemer JR (1994) Tools for talking: Human nature, Weismannism, and the interpretation of genetic information. In: Are Genes Us? The Social Consequences of the New Genetics (Cranor CF, ed), 83–88. New Brunswick, NJ: Rutgers University Press.

Griesemer JR (2000a) Development, culture, and the units of inheritance. Phil Sci 67: S348–S368.

Griesemer JR (2000b) The units of evolutionary transition. Selection 1: 67–80.

Griesemer JR (in press) The informational gene and the substantial body: On the generalization of evolutionary theory by abstraction. In: Varieties of Idealization (Cartwright N, Jones M, eds). Amsterdam: Rodopi.

Hamilton WD (1963) The evolution of altruistic behavior. Amer Nat 97: 354–356. Reprinted in Hamilton 1996, 6–8.

Hamilton WD (1964a) The genetical evolution of social behaviour, I. J Theor Biol 7: 1–16. Reprinted in Hamilton 1996, 31–46.

Hamilton WD (1964b) The genetical evolution of social behaviour, II. J Theor Biol 7: 17–52. Reprinted in Hamilton 1996, 47–82.

Hamilton WD (1975) Innate social aptitudes of man: An approach from evolutionary genetics. In: Biosocial Anthropology (Fox E, ed), 133–153. London: Malaby Press. Reprinted in Hamilton 1996, 329–351.

Hamilton WD, ed (1996) Narrow Roads of Gene Land. Vol 1, Evolution of Social Behavior. New York: Freeman.

Hartfelder K, Engels W (1998) Social insect polymorphism: Hormonal regulation of plasticity in development and reproduction in the honeybee. Curr Top Dev Biol 40: 45–77.

Hartwell LH, Hopfield JJ, Leibler S, Murray AW (1999) From molecular to modular cell biology. Nature 402 Supp: C47–C52.

Hölldobler B, Wilson EO (1990) The Ants. Cambridge, MA: Harvard University Press.

Hull DL (1978) A matter of individuality. Phil Sci 45: 335–360.

Hull DL (1980) Individuality and selection. Ann Rev Ecol Syst 11: 311–332.

Hurst LD, Atlan A, Bengtsson BO (1996) Genetic conflicts. Quart Rev Biol 71: 317–363.

Kauffman SA (1971) Articulation of parts explanation in biology and the rational search for them. Boston Stud Phil Sci 8: 257–272.

Kauffman SA (1993) The Origins of Order: Self-Organization and Selection in Evolution. Oxford: Oxford University Press.

Keller EF (1992) Nature, nurture, and the Human Genome Project. In: The Code of Codes: Scientific and Social Issues in the Human Genome Project (Kevles DJ, Hood L, eds), 281–299. Cambridge, MA: Harvard University Press.

Keller L, Nonacs P (1993) The role of queen pheromones in social insects: Queen control of queen signal? Anim Behav 45: 787–794.

Keller L, Reeve HK (1999) Dynamics of conflicts within insect societies. In: Levels of Selection in Evolution (Keller L, ed), 153–175. Princeton, NJ: Princeton University Press.

Kerr B, Godfrey-Smith P (2002) On the relations between individualist and multi-level models of selection in structured populations. Biol Philos 17: 477–517.

Kirschner M, Gerhart J (1998) Evolvability. Proc Natl Acad Sci USA 95: 8420–8427.

Krebs JR, Davies NB (1993) An Introduction to Behavioural Ecology. 3rd ed. Oxford: Blackwell.

Larsen EW (1992) Tissue strategies as developmental constraints: Implications for animal evolution. Trends Ecol Evol 7: 414–417.

Larsen EW, McLaughlin HMG (1987) The morphogenetic alphabet: Lessons for simple-minded genes. BioEssays 7: 130–132.

Lessios HA, Kessing BD, Robertson DR, Paulay G (1999) Phylogeography of the Pantropical sea urchin *Eucidaris* in relation to land barriers and ocean currents. Evolution 53: 806–817.

Levins R, Lewontin RC (1985) The Dialectical Biologist. Cambridge, MA: Harvard University Press.

Lewontin RC (1991) Biology as Ideology: The Doctrine of DNA. New York: HarperCollins.

Lewontin RC, Levins R (1988) Aspects of wholes and parts in population biology. In: Evolution of Social Behavior and Integrative Levels (Greenberg G, Tobach E, eds), 31–52. Hillsdale, NJ: Erlbaum.

Lloyd EA (1988) The Structure and Confirmation of Evolutionary Theory. Princeton, NJ: Princeton University Press.

Lloyd EA (1994) Normality and variation. The Human Genome Project and the ideal human type. In: Are Genes Us? The Social Consequences of the New Genetics (Cranor CF, ed), 99–112. New Brunswick, NJ: Rutgers University Press.

Lloyd EA (2000) Units and levels of selection: An anatomy of the units of selection debates. In: Thinking About Evolution: Historical, Philosophical and Political Perspectives (Singh E, Krimbas C, Paul D, Beatty J, eds), 267–291. Cambridge: Cambridge University Press.

Maynard Smith J (1976) Group selection. Quart Rev Biol 51: 277–283.

Maynard Smith J (1982) Evolution and the Theory of Games. Cambridge: Cambridge University Press.

Maynard Smith J, Szathmáry E (1995) The Major Transitions in Evolution. Oxford: Freeman.

McShea DW, Venit EP (2001) What is a part? In: The Character Concept in Evolutionary Biology (Wagner GP, ed), 259–284. San Diego: Academic Press.

Mezey JG, Cheverud JM, Wagner GP (2000) Is the genotype–phenotype map modular? A statistical approach using mouse quantitative trait loci data. Genetics 156: 305–311.

Michener CD, Brothers DJ (1974) Were workers of eusocial Hymenoptera initially altruistic or oppressed? Proc Natl Acad Sci USA 71: 671–674.

Michod RE (1999a) Darwinian Dynamics: Evolutionary Transitions in Fitness and Individuality. Princeton, NJ: Princeton University Press.

Michod RE (1999b) Individuality, immortality, and sex. In: Levels of Selection in Evolution (Keller L, ed), 53–74. Princeton, NJ: Princeton University Press.

Michod RE, Roze D (1997) Transitions in individuality. Proc Roy Soc London B264: 853–857.

Mitman G (1992) The State of Nature: Ecology, Community, and American Social Thought, 1900–1950. Chicago: University of Chicago Press.

Moss L (2001) Deconstructing the gene and reconstructing molecular developmental systems. In: Cycles of Contingency: Developmental Systems and Evolution (Oyama S, Griffiths PE, Gray RD, eds), 85–97. Cambridge, MA: MIT Press.

Nielsen MG, Wilson KA, Raff EC, Raff RA (2000) Novel gene expression patterns in hybrid embryos between species with different modes of development. Evol Dev 2: 133–144.

Nunney L (1999a) Lineage selection and the evolution of multistage carcinogenesis. Proc Roy Soc London B266: 493–498.

Nunney L (1999b) Lineage selection: Natural selection for long-term benefit. In: Levels of Selection in Evolution (Keller L, ed), 238–252. Princeton, NJ: Princeton University Press.

Oster GF, Wilson EO (1978) Caste and Ecology in the Social Insects. Princeton, NJ: Princeton University Press.

Otto SP, Hastings IM (1998) Mutation and selection within an individual. Genetica 102/103: 507–524.

Otto SP, Orive ME (1995) Evolutionary consequences of mutation and selection within an individual. Genetics 141: 1173–1187.

Oyama S (2000a) The Ontogeny of Information: Developmental Systems and Evolution. 2nd ed. Durham, NC: Duke University Press.

Oyama S (2000b) Evolution's Eye: A Systems View of the Biology–Culture Divide. Durham, NC: Duke University Press.

Oyama S, Griffiths PE, Gray RD, eds. (2001) Cycles of Contingency: Developmental Systems and Evolution. Cambridge, MA: MIT Press.

Price GR (1970) Selection and covariance. Nature 227: 520–521.

Price GR (1995) The nature of selection. J Theor Biol 175: 389–396.

Queller DC (1992a) A general model for kin selection. Evolution 46: 376–380.

Queller DC (1992b) Quantitative genetics, inclusive fitness, and group selection. Amer Nat 139: 540–558.

Queller DC, Strassman JE (1998) Kin selection and social insects. BioScience 48: 165–175.

Raff EC, Popodi EM, Sly BJ, Turner FR, Villinski JT, Raff RA (1999) A novel ontogenetic pathway in hybrid embryos between species with different modes of development. Development 126: 1937–1945.

Raff RA (1996) The Shape of Life: Genes, Development, and the Evolution of Animal Form. Chicago: University of Chicago Press.

Raff RA (2000) Evo-Devo: The evolution of a new discipline. Nat Revs: Genet 1: 74–79.

Raff RA, Sly BJ (2000) Modularity and dissociation in the evolution of gene expression territories in development. Evol Dev 2: 102–113.

Ratnieks FL, Reeve HK (1992) Conflict in single-queen hymenopteran societies: The structure of conflict and processes that reduce conflict in advanced eusocial species. J Theor Biol 158: 33–65.

Robinson GE (1999) Integrative animal behaviour and sociogenomics. Trends Ecol Evol 14: 202–205.

Robinson GE, Fahrbach SE, Winston ML (1997) Insect societies and the molecular biology of social behavior. BioEssays 19: 1099–1108.

Roth VL (1994) Within and between organisms: Replicators, lineages, and homologues. In: Homology: The Hierarchical Basis of Comparative Biology (Hall BK, ed), 301–337. San Diego: Academic Press.

Schneirla TC (1971) Army Ants: A Study in Social Organization (Topoff HR, ed). San Francisco: Freeman.

Seger J (1991) Cooperation and conflict in social insects. In: Behavioural Ecology: An Evolutionary Approach (Krebs JR, Davies NB, eds), 338–373. Oxford: Blackwell.

Sober E, Wilson DS (1998) Unto Others: The Evolution and Psychology of Unselfish Behavior. Cambridge, MA: Harvard University Press.

Szathmáry E, Maynard Smith J (1995) The major evolutionary transitions. Nature 374: 227–232.

Trivers RL (1971) The evolution of reciprocal altruism. Quart Rev Biol 46: 35–57.

Trivers RL, Hare H (1976) Haplodiploidy and the evolution of social insects. Science 191: 249–263.

Uyenoyama M, Feldman MW (1980) Theories of kin and group selection: A population genetics perspective. Theor Pop Biol 17: 380–414.

Van Valen LM (1973) Festschrift. Science 180: 488.

Venter JC et al. 2001. The sequence of the human genome. Science 291: 1304–1351.

Von Dassow G, Meir E, Munro EM, Odell GM (2000) The segment polarity network is a robust developmental module. Nature 406: 188–192.

Von Dassow G, Munro E (1999) Modularity in animal development and evolution: Elements of a conceptual framework for EvoDevo. J Exp Zool (Mol Dev Evol) 285: 307–325.

Vrba ES, Eldredge N (1984) Individuals, hierarchies and processes: Towards a more complete evolutionary theory. Paleobiology 10: 146–171.

Wade MJ (1980) Kin selection: Its components. Science 210: 665–667.

Wade MJ (1985) Soft selection, hard selection, kin selection, and group selection. Amer Nat 125: 61–73.

Wade MJ (1992) Sewall Wright: Gene interaction and the Shifting Balance Theory. In: Oxford Surveys in Evolutionary Biology, vol. 8 (Futuyma D, Antonovics J, eds), 35–62. Oxford: Oxford University Press.

Wade MJ (1996) Adaptation in subdivided populations: Kin selection and interdemic selection. In: Adaptation (Rose MR, Lauder GV, eds), 381–405. San Diego: Academic Press.

Wade MJ, Goodnight CJ (1998) Perspective: The theories of Fisher and Wright in the context of metapopulations: When nature does many small experiments. Evolution 52: 1537–1553.

Wade MJ, Winther RG, Agrawal AF, Goodnight CJ (2001) Alternative definitions of epistasis: Dependence and interaction. Trends Ecol Evol 16: 498–504.

Wagner GP (1989) The origin of morphological characters and the biological basis of homology. Evolution 43: 1157–1171.

Wagner GP (1994) Homology and the mechanisms of development. In: Homology: The Hierarchical Basis of Comparative Biology (Hall BK, ed), 273–299. San Diego: Academic Press.

Wagner GP (1995) The biological role of homologues: A building block hypothesis. N Jb Geol Paläont Abh 195: 279–288.

Wagner GP (1996) Homologues, natural kinds, and the evolution of modularity. Amer Zool 36: 36–43.

Wagner GP, ed (2001) The Character Concept in Evolutionary Biology. San Diego: Academic Press.

Wagner GP, Altenberg L (1996) Perspective: Complex adaptations and the evolution of evolvability. Evolution 50(3): 967–976.

Wagner GP, Chiu C-H, Laubichler MD (2000) Developmental evolution as a mechanistic science: The inference from developmental mechanisms to evolutionary processes. Amer Zool 40: 819–831.

Watson JD (1990) The Human Genome Project: Past, present, and future. Science 248: 44–49.

Wenzel J (1993) Application of the biogenetic law to behavioral ontogeny: A test using nest architecture in paper wasps. J Evol Biol 6: 229–247.

Werren JH, Nur U, Wu C-I (1988) Selfish genetic elements. Trends Ecol Evol 3: 297–302.

West-Eberhard, MJ (1987) Flexible strategy and social evolution. In: Animal Societies: Theories and Facts (Itô Y, Brown JL, Kikkawa J, eds), 35–51. Tokyo: Japan Science Society Press.

Wheeler DE (1986) Developmental and physiological determinants of caste in social Hymenoptera: Evolutionary implications. Amer Nat 128: 13–34.

Wheeler DE (1991) The developmental basis of worker caste polymorphism in ants. Amer Nat 138: 1218–1238.

Wheeler WM (1911) The ant-colony as an organism. J Morphol 22: 307–325. Reprinted in Wheeler WM (1939) Essays in Philosophical Biology. Cambridge, MA: Harvard University Press.

Wheeler WM (1918) A study of some ant larvae, with a consideration of the origin and meaning of the social habit among insects. Proc Amer Phil Soc 57: 293–343.

Williams GC (1966) Adaptation and Natural Selection: A Critique of Some Current Evolutionary Thought. Princeton, NJ: Princeton University Press.

Wilson DS (1980) The Natural Selection of Populations and Communities. Menlo Park, CA: Benjamin/Cummings.

Wilson DS, Colwell RK (1981) Evolution of sex ratio in structured demes. Evolution 35: 882–897.

Wilson DS, Sober E (1989) Reviving the superorganism. J Theor Biol 136: 337–356.

Wilson EO (1967) The superorganism concept and beyond. In: L'effet de groupe chez les animaux. Colloques Internationaux Centre National de la Recherche Scientifique 173: 27–39.

Wilson EO (1968) The ergonomics of caste in the social insects. Amer Nat 102: 41–66.

Wilson EO (1971) The Insect Societies. Cambridge, MA: Harvard University Press.

Wilson EO (1985) The sociogenesis of insect colonies. Science 228: 1489–1495.

Wilson EO (1994) Naturalist. Washington, DC: Island Press.

Wilson EO, Hölldobler B (1988) Dense heterarchies and mass communication as the basis of organization in ant colonies. Trends Ecol Evol 3: 65–68.

Wimsatt WC (1974) Complexity and organization. In: PSA 1972 (Schaffner KF, Cohen RS, eds), 67–86. Dordrecht: Reidel.

Wimsatt WC (1980) Reductionist research strategies and their biases in the units of selection controversy. In: Scientific Discovery. Vol 2: Case Studies (Nickles T, ed), 213–259. Dordrecht: Reidel.

Wimsatt WC (1986) Forms of aggregativity. In: Human Nature and Natural Knowledge (Donogan A, Perovich N Jr, Wedin M, eds), 259–291. Dordrecht: Reidel.

Winther RG (2000) Darwin on variation and heredity. J Hist Biol 33: 425–455.

Winther RG (2001a) Varieties of modules: Kinds, levels, origins and behaviors. J Exp Zool (Mol Dev Evol) 291: 116–129.

Winther RG (2001b) Review of *Ants at Work. How an Insect Society Is Organized,* by Deborah Gordon. Phil Sci 68: 268–270.

Winther RG (2001c) August Weismann on germ-plasm variation. J Hist Biol 34: 517–555.

Wolf JB, Brodie ED, III, Wade MJ, eds (2000) Epistasis and the Evolutionary Process. New York: Oxford University Press.

Wolpert L (1996) One hundred years of positional information. Trends Genet 12: 359–364.

5 Modularity in Evolution: Some Low-Level Questions

Lee Altenberg

The Question of Multiplicity

A good deal of work in recent years has shown that the structure of the genotype–phenotype map is of fundamental importance to the process of evolution. The variational properties of the genotype–phenotype map—how genetic variation maps to phenotypic variation (Altenberg, 1994a, 1995; Wagner and Altenberg, 1996)—largely determine whether mutations and recombination can generate the sequence of phenotypes with increasing fitness that produce adaptation.

A most important property of the genotype–phenotype map is its *modularity*. The concepts of "modularity" and "module" are being employed now in novel contexts in the fields of genetics, behavior, and evolution. Their precise meaning has been fluid. "Modular" will be used in the current discussion to describe a genotype–phenotype map that can be decomposed (or nearly decomposed; Simon, 1962, 1969) into the product of independent genotype–phenotype maps of smaller dimension. The extreme example of modularity would be the idealized model of a genome in which each locus maps to one phenotypic trait. For the converse, the extreme example of nonmodularity would be a genotype–phenotype map with uniform "universal pleiotropy" (Wright, 1968), in which every gene has an effect on every phenotypic variable. Real organisms, one could argue, have genotype–phenotype maps that range somewhere in between these extremes.

It may seem intuitively obvious why modularity in the genotype–phenotype map should benefit evolution: if genetic changes tend to map to changes in a small number of phenotypic traits, then the genome can respond to selection on those traits alone, independently of the rest of the phenotype, with a minimum of deleterious pleiotropic side effects. Hence modularity would enhance the ability of the genetic system to generate adaptive variants, which one can refer to as its "evolvability" (Altenberg, 1994a, 1995).

In a genotype–phenotype map with low modularity, where genes have high pleiotropy, a genetic change that produces adaptation in one character may be confounded by maladaptive changes it causes in other characters. To produce adaptive changes, a patchwork of just the right mutations among modifier genes may be necessary to cancel out their overlapping negative pleiotropic effects. Therefore, the problem of pleiotropy points to a solution through polygeny.

Two kinds of constraints may prevent such solutions from being found. First, such patchwork may be be impossible to produce from any combination of genetic changes, so that only an approximation to the optimal phenotype can evolve. In other words, the phenotypes that are possible may span a subspace that does not include the optimum. I refer to this as a "subspace constraint."

Second, the kinetics of mutation, recombination, and selection may make optimal combinations of genetic changes unreachable by evolutionary processes. If coordinated mutations at a number of loci are required in order to produce a fitness advantage, and the single or double mutations along the way are deleterious or neutral, it becomes very improbable that such multiple mutations will ever appear (Riedl, 1975, 1977, 1978; Kauffman and Levin, 1987; Kauffman and Weinberger, 1991; Weinberger, 1991). This is a generic result, notwithstanding the complications of recombination and neutral networks (van Nimwegen et al., 1999). So in cases where adaptation requires the coordinated change of multiple loci, there may be no selective pathway to reach those changes, and the phenotype can remain stuck at a suboptimal genotype, resulting in a condition called "frustration" in statistical mechanics (McKay et al., 1982), or a "rugged fitness landscape" (Kauffman and Levin, 1987). I refer to this as a "kinetic constraint."

These two mechanisms—subspace and kinetic constraints—may prevent the simultaneous optimization of multiple phenotypic variables. A way to avoid these constraints would appear to be modularity, where genetic variation maps to small numbers of traits.

A Deconstruction of This Framework

While this explanation for the benefits of modularity may seem straightforward, a number of problems arise when we take a closer look. The advantage of modular genetic variation is seen to come from the small number of traits that are affected. By implication, this advantage is thus premised on the idea that selection tends to act on small numbers of traits alone.

What do we know about the nature of selection as it relates to numbers of traits? Here we find ourselves in a swamp, because the process by which traits are distinguished from one another is a human measurement process, dependent on the instruments and cognitive structures that we possess to parse the organism. How, for example, should we deal with a change in the size of an organism? Is this a change in all of the organism's measurements or, if allometric scaling relationships are maintained, in just one measurement? Is genetic variation modular if it causes just one part of an organism to change size, or if it causes the entire organism to change size? Suppose there were selection for sharper teeth. Would a genetic variant that made all teeth sharper be more or less modular than a genetic variant that made half of the teeth sharper, or just one pair of teeth, or just one tooth?

Consider another situation. Suppose that climate change has caused a simultaneous change in the optimal values of a number of organismal traits. Suppose further—just as a thought experiment—that genetic variation for some physiological variable happens to move many of these traits closer to their new optima. A gene with effects on these many traits would, under the common usage, be called pleiotropic and non–modular. And yet under this circumstance, such a gene, with the ability to move many phenotypic traits closer to their optima, would be an asset to the genome's evolvability.

Does the gene in our example "know" that it has several traits under directional selection rather than just one trait? Does the environment know that it has selected on several traits rather than just one trait? Where does the fact that several traits have been affected appear in the dynamics of this situation? The reality of what has occurred is that

1. There has been a change in climate

2. There is an allele that is now at a selective advantage under the new climate.

For "whom," then, are there multiple traits?

We see that when we try to apply our intuitive notion about the advantages of modularity, we run into the problem of how we parse the organism into traits. This is not a new problem—indeed, the problem of how to "carve nature at its joints" has been with us since Plato (c. 370 B.C.). Until this problem is resolved, we cannot say whether variation is modular or not.

In order to resolve the "question of multiplicity," there needs to be a way to get the human observer out of the way, and define modularity in terms of physical processes. I will offer two candidate ideas for this resolution:

1. The dimensionality of phenotypic variation

2. The causal screening off of phenotypic variables by other phenotypic variables.

Description and Degrees of Freedom

When we say that a gene affects multiple traits, we mean that it changes multiple features of the organism that are measured independently of one another. Each trait constitutes a variable that can take on a variety of values, distinct from the values other traits may take. To represent all the traits simultaneously therefore requires a multidimensional space, which will be the Cartesian product of the space of values each variable can take on. Thus, references to multiple traits are equivalent to references to multidimensional spaces of descriptive variables.

Thus, if S_1 is the space of possible values for trait x_1, and so forth for x_2 and the rest, then an organism with trait values $(x_1, x_2, \ldots, x_n)$ corresponds to a point x in the multidimensional space $S = S_1 \times S_2 \times \ldots \times S_n$.

While our multidimensional representation of the organism allots one degree of freedom for each trait, the critical question is whether these degrees of freedom have any physical reality as dimensions of variation in the organism, or dimensions of variation for selection. We can apply similar reasoning to the environment: a description of the environment can contain many variables, but we must ask whether these variables correspond to physical dimensions of variability in the environment.

Let us return to our thought experiment about climate change. Suppose the genetic underpinnings are the sort that Waddington (1942) considered for the evolution of canalization, where a physiological adaptive response, involving many phenotypic variables, is cued both by environmental signals—day length, temperature, and such—and by internal signals under genetic control. Genetic changes in how these adaptations are invoked may be capable of moving the whole complex response toward a more optimal match to a changed environment (such as time of flowering, moulting, hybernation, dormancy, budding, quantities of stored metabolites, etc.). While many traits would be observed to change under such genetic variation, there may in fact be only one degree of freedom if there is a single cueing mechanism that is being altered. In contrast to the apparent high dimensionality of the space of traits affected by the gene, the space of variation in this example may be a one-dimensional space merely embedded in the higher dimensions. This is illustrated in figure 5.1.

Embeddings and Dimension Reduction
To further illustrate the idea that low-dimensional variation may underlie what appears to be high-dimensional variation, I will draw attention to some recent work on dimension reduction. Dimension reduction has long been a part of morphometrics through the use of principal component analysis (PCA), but this technique assumes a linear form for the lower-dimensional subspaces. When the spaces of variation are nonlinear, other techniques are required to identify these spaces.

Two recent works provide algorithms that can take complex multidimensional data and discover when the variation is restricted to lower-dimensional manifolds, and can characterize these manifolds (Roweis and Saul, 2000; Tenenbaum et al., 2000).

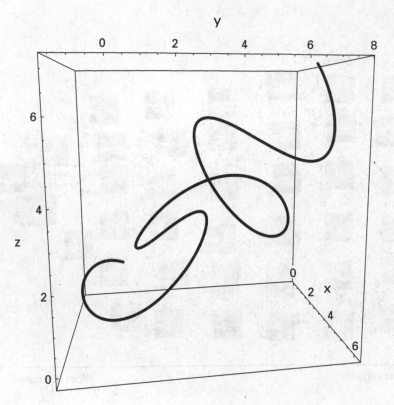

Figure 5.1
A one-dimensional space of variation embedded in three phenotypic dimensions.

Two illustrations from Tenenbaum et al. (2000) are reproduced here. Figure 5.2 shows a hand rotating at the wrist or opening its fingers. The hand is a complex object, here described by 64 × 64 pixel photographs, giving 4096 independent variables. All 4096 variables vary as a result of wrist rotation and finger extension. Yet, if each photograph is mapped to a single point in a 4096-dimensional space, the variation traced out by the set of photographs can be mapped to a two-dimensional manifold embedded in the 4096-dimensional space. This manifold is represented in the plane in figure 5.2.

The two-dimensional manifold of variation depicted here is produced by movement of the hand. Instead of a hand moving, we could just as well analyze a set of photographs of hands that represent the morphological variation in extant human populations. The structure and dimensionality of this space of phenotypic variation may very well be revealed by use of the "isomap" (Tenenbaum et al., 2000) or

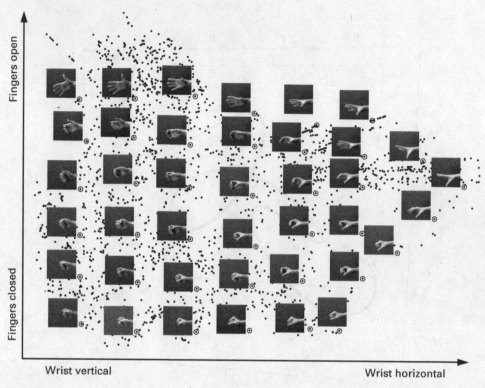

Fingers open — **Fingers closed** (vertical axis)

Wrist vertical — **Wrist horizontal** (horizontal axis)

Figure 5.2
An example of dimension reduction. While the hand can be described by many variables, in this ensemble of states there are really only two dimensions of variability: wrist rotation and finger extension. The two dimensions of variation are recovered from the 4096-dimensional image data by Tenenbaum, de Silva, and Langford using their Isomap algorithm. (Reprinted with permission from Tenenbaum et al., 2000. Copyright 2000, American Association for the Advancement of Science.)

"locally linear embedding" (Roweis and Saul, 2000) methods of nonlinear dimension reduction.

Figure 5.3 shows data from a two-dimensional manifold that is curled up in a 3-D embedding. Points that appear close in the 3-D embedding may actually be far apart in the manifold, as shown by the geodesic lines. Thus, naive interpretations of the dimensionality and "distances" represented by phenotypic variation may not reflect the real structure of the variation.

The wide application of the Isomap, locally linear embedding, and related nonlinear dimension-reduction methods (Verbeek et al., 2002; Agrafiotis and Xu, 2002) to morphometrics might prove fruitful at exposing unknown structures within phe-

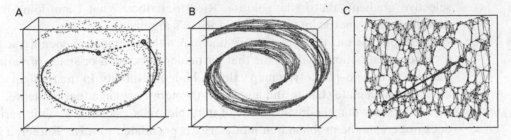

Figure 5.3
The nonlinear two-dimensional "Swiss Roll" manifold is recovered from its three-dimensional embedding by the Isomap algorithm of Tenenbaum, Silva, and Langford. Points that appear close together in the three-dimensional embedding may be far apart in the underlying manifold. (Reprinted with permission from Tenenbaum et al., 2000. Copyright 2000, American Association for the Advancement of Science.)

notypic variation. If low-dimensional manifolds are discovered amid the morphological variation found in different organisms, a new window may be opened on the question of "developmental constraints." The widespread characterizations of such manifolds across different taxa could provide the basis for a study of "morphomics," as it were (à la "genomics" and "proteomics"). To my knowledge, the application of these new methods to morphometrics has not yet been tried.

Evolvability and Alignment with Selection
I have argued that phenotypic variation which may appear to involve many variables may in fact represent the variation of very few parameters. Geometrical transformations of variables can change what appears to be high pleiotropy into low pleiotropy. We must ask, then, when does the geometry of variation make a difference to evolution? At this point we must consider how selection is involved.

In order for a subspace of phenotypic variation to allow a response to selection, it must pass through a selection gradient. Or, to be more precise, the space of variations must provide, with some reasonable probability, a sequence of genetic operations that produce monotonically increasing fitnesses. If the probability of such a sequence is too low, there is no evolvability. Two causes of such low probabilities are the phenotype being near its constrained optimum, and the situation where "frustration" prevails. Here we find ourselves back at the analysis of Riedl (1977).

Riedl proposes that the solution to the problem of adaptive frustration is the "systemization of the genome." By this he means the creation of new spaces of genetic variation that move the phenotype in directions that are under positive directional selection. Rather than modularity, it is the alignment of the space of variation with selective gradients that is the solution Riedl describes. What I am doing here is describing a geometric interpretation of Riedl's argument.

So, despite the fact that genetic variation may alter a number of phenotypic traits, if there is a selection gradient for that particular dimension or space of variation, then the genotype–phenotype map exhibits high evolvability. In the thought experiment about climate change and a gene that generates change that is adapted to it, I tried to show that the involvement of multiple traits, per se, in genetic variation does not create the problem that modularity is postulated to solve. Rather, it is the relationship of *selective gradients* to the *space* of *variation* that is the critical issue.

How does the earlier idea—that a nonmodular genotype–phenotype map produces frustration—hold up after this deconstruction? Frustration occurs because none of the spaces of phenotypic variability are able to provide, with suffcient probability, a sequence of genotypes that traverse the selective gradients that may happen to be present i.e. they are not aligned with selection gradients. Hence, the genome is unable to access regions of the phenotype that may be adaptive. Modularity, if it is to be a means to attain evolvability, must somehow imply an alignment between the spaces of phenotypic variation and the selection gradients. One can conclude either that

1. Modularity is one means to such an alignment or that
2. Modularity should be *defined* in terms of such alignments.

The Underlying Degrees of Freedom

I have talked about geometrical aspects of variation, and its relation to selection, without delving into the possible causes behind such properties. Here I will explore this issue a little further. I will propose that the notion of causal "screening-off" (Salmon, 1971, 1984; Brandon, 1984, 1990) can be used to describe the sort of modularity in the genotype–phenotype map that matters to evolvability.

The fundamental dimensions of variation in the genotype are determined by the spectrum of genetic changes that can occur. These include

• Point mutation, in which one nucleotide is replaced by another nucleotide

• Deletions and insertions

• Gene duplication, in which a sequence of nucleotides copied from an existing sequence is inserted in a chromosome

• Gene conversion

- Polyploidy, in which an entire genome is duplicated one or more times
- Translocation
- Transposition
- Recombination
- Segregation and syngamy
- Methylation change
- Horizontal genetic transmission (e.g., plasmid exchange)
- A variety of taxon-specific genetic mechanisms.

Each of these variation processes produces its own space of genetic variation (Stadler et al., 2002), distinguished not so much by the nature of the phenotypic changes that they produce as by the evolutionary paths they make possible between genotypes.

Let me be more concrete in describing the spaces of genetic variation. A genome of L nucleotides can be represented as a point in the genotype space $S = \{A, T, C, G\}^L$ (ignoring, for the sake of discussion, the metasequence properties such as methylation, chromosome structure, etc.). Under the action of point mutation, there exist L degrees of freedom for the genotype. The magnitude of L varies from being on the order of 10^6 for prokaryotes to 10^{11} for lungfish and trumpet lilies. A million to a hundred billion is clearly a vast number of degrees of freedom for point mutations, but each degree of freedom constitutes only a minuscule space—comprising only four points, in fact—the four nucleotide bases A, T, C, and G. It really makes no sense even to speak of directional selection or a selection "gradient" on a space of four discrete points; direction is undefined.

However, the process of gene expression groups these individual degrees of freedom into new spaces of variation with fewer degrees of freedom but many more elements. To begin with, DNA triplets in transcribed sequences map to the space of amino acids, $\{A, T, C, G\}^3 \mapsto \mathcal{P}$ (where $\mathcal{P}$ includes the 20 amino acids and the stop codons).

The dynamics of protein folding and molecular interactions in turn group the amino acids in a protein into a new set of variables that characterize the protein and its interactions. This is the first point in this chain of "decoding" where real-valued variables enter, such as the geometry of the protein fold, the kinetic rates for interaction with other molecules, binding energies, catalytic rates, thermal stability, hydrophobicity, and so on. Nontranscribed DNA has different mechanisms of expression, and real-valued variables can be seen to emerge immediately in characterizing its phenotypic effects, such as its affnities for binding with regulatory molecules, methylation enzymes, replication and transcription complexes, and so on.

Screening Off

While a great many real-valued variables are needed to describe, for example, a protein, it is typically the case that only a small subset of variables is needed to describe the causal effects of a gene on the organism—a catalytic rate, binding constants, levels of expression, the timing of expression, half-life, and so forth. Variation in a gene will not cause phenotypic variation except in how it varies these variables. In other words, there is some set of variables that *screen off* (Salmon, 1971, 1984; Brandon, 1982, 1984, 1990, 2002) the causal impact of genetic variation: if one knows the values of these variables, there is nothing more one needs to know about the gene in order to determine its effect on the organism. By "small number" I mean small relative to the typical number of nucleotides in a gene, which ranges from 10^3 to 10^5 in eukaryotes.

The processes of gene expression, ontogeny, and physiology convert the large number of essentially "digital" degrees of freedom in the genome into degrees of freedom of a smaller set of real-valued variables (similar, really, to what happens in electronic digital-to-analog conversion). The variables that screen off the properties of a gene may themselves be screened off by other variables that summarize their effects on other functions in the organism. For example, many factors contribute to levels of cortisol in vertebrates. But to the extent that they affect the organism through the action of cortisol, the cortisol level contains all the information about their effect.

In common usage, when people refer to the "function" of part of an organism, they may mean one of two things: "What does the part *do*?" or "What is the part *for*?" In both cases, nevertheless, the positing of a "what" implicitly uses the concept of screening-off. The "what" refers to a function that screens off the detailed characteristics of this organismal part, a variable that summarizes one of its causal consequences for the organism, or one of its purposes, respectively.

Regarding the latter notion—purpose—I will invoke the popular rejoinder "Let's not go there." Much has been written—indeed, books—on the notion of function as "what something is for." In particular I note the line of thought about "proper functions" developed by Ruth G. Millikan (1984). Throughout this chapter it will be "what something does" that I mean when I refer to "function."

The idea of phenotypic variables that screen off other variables can be expressed mathematically by saying that there is a set of functions, $\{F_i(g_k)\}$, that forms a complete description of the causal consequences for the organism due to variation in

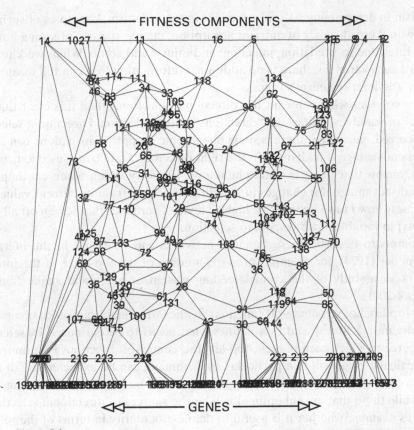

Figure 5.4
The "function network," showing the relationship between genes, variables that screen-off the gene's causal effects, and variables that screen-off these variables, and so on—all the way to fitness components.

the gene g_k. To be complete, and account for gene–environment interactions and epistasis, these functions will need to have other arguments that include genes and environmental variables, ϕ_j, and therefore be of the form $F_i(\{g_k\}, \{\phi_j\})$. These functions in turn may be screened off for their organismal effects by other sets of functions, $\{G_j(F_{i_1}, F_{i_2}, \ldots)\}$. Organisms have many chains of such dependence, which could be said to form a "function network." A simplistic illustration of the "function network" can be seen in figure 5.4.

Ultimately, one arrives at the variables that describe the rates of mortality and fertility of an organism as functions of its interactions with the physical and biotic environment. These variables screen off all other phenotypic properties of the organism in determining natural selection on the organism. Examples of such variables would be efficiency of nutrient absorption; mating success; offspring number; death rates due to predation, infection, and injury; and so on. When we know the value of such variables, there is no additional information that can tell us anything further about an organism's fitness.

If we consider what is meant by "directional selection," then it is clear that this last tier of variables defines the "directions" under selection. Directional selection, as conceived, means that there is some phenotypic property which alone can confer a fitness advantage if it changes in the right direction. Stabilizing selection, as conceived, means that there is some phenotypic property which alone can impose a fitness disadvantage if it changes in any direction away from its current value. This is precisely how I have defined the highest level of variables that screen off all other phenotypic variables with respect to selection.

Attempts to describe these function networks can be found in the literature. Dullemeijer (1974), for example, presents a graph of the top layer of the function network in a study of the cranial feeding system of a crotalid snake (cited in Schwenk 2001).

The top level of screening-off functions defines what matters to each component of an organism's fitness, and thus defines what needs to be optimized by selection. The degree to which each of these top-level screening-off functions is optimized, in any particular organism in a particular environment, can be expected to fall along a spectrum: the functions that are nearly optimal will be sources of stabilizing selection, while those that are suboptimal will be the sources of directional selection.

Let us examine whether it is useful to define modularity in terms of the relation between spaces of phenotypic variation and these top-level screening-off functions. The genotype–phenotype map is defined as modular if very few of these functions are affected throughout a space of phenotypic variation. When would such modularity enhance evolvability? First, it is of no use to evolvability for there to be a modular genotype–phenotype map when all the modules are under stabilizing selection, since there is no adaptive opportunity no matter how it is sliced (Wagner, 1996). Modularity among functions under stabilizing selection may nevertheless have other kinetic population genetic consequences, as found in the study by Waxman and Peck (1998).

Suppose, on the other hand, that the space of variation maps only to functions under *directional selection*. The interactions of these functions in determining fitness would define the selection gradient on the space of variation.

In order for modularity to enhance evolvability, it must parcellate the functions under stabilizing selection from those under directional selection. The importance of this cleavage between stabilizing and directional selection has been recognized for some time as an important element of evolvability, and motivated the development of the "corridor model" (Wagner, 1984, 1988; Bürger, 1986). Mechanisms that that produce the evolution of such a cleavage are discussed by Altenberg (1995). Evolvability is enhanced when there are spaces of phenotypic variation that fall narrowly within the functions under directional selection, and remain orthogonal to the functions under stabilizing selection. This is the kind of modularity that is implicit in the naive framework I described at the beginning of this chapter. Genes with that sort of modularity would look like the ones in figure 5.4 with direct connections to variables on the top level that are under directional selection, and few connections— direct or indirect—to the top level of variables under stabilizing selection.

Hence, these top-level variables provide a way of describing the sort of modularity that is important to evolvability. There is nothing to prevent one from defining pleiotropy and modularity in terms of the map between a gene and the variables at any level in the function network—or, for that matter, between a gene and any observer-defined phenotypic characters. But pleiotropy or modularity so defined will not say anything about whether the spaces of variation are aligned with selection gradients, and so will not be relevant to evolvability. The pleiotropy that is relevant to evolvability is that which applies to the map between the gene and the top-level screening-off variables.

Let us now return to my earlier hypothetical question about selection for sharper teeth. Would a genetic variant that made all teeth sharper be more or less modular than a genetic variant that made half of the teeth sharper, or just one pair of teeth, or just one tooth? We now have some machinery to answer this question.

What are the organismal functions that screen off the causal effects of tooth morphology with respect to selection? Such functions would include the rate of catching and killing of prey, the size of food particles sent to the stomach, the amount of flesh removed from a carcass, the success rate for defenses against attack, morbidity and mortality due to tooth and gum infections, mating success, and so on. We must ask which of these functions would be altered by the different spaces of tooth variation.

Suppose that there was directional selection for a stronger mouth grip on prey. A genetic variant that sharpened just the front half of the teeth would improve that quantity, and would leave alone the grinding function of the back teeth. A genetic variant that sharpened all the teeth might also improve the prey-grabbing function, but make it harder to grind food, and thus increase the particle size of food in the stomach, decrease nutrient absorption, and adversely affect fitness. It would affect two top-level functions instead of just one, involving both directional and stabilizing selection instead of just directional selection, and would thus be more pleiotropic and less modular than the mutation of just the front teeth, even though fewer characters (teeth) were altered.

Any number of variants of this example can be posed and analyzed in the same way. Whether genetic variation has a modular effect depends on how it affects the top-level screening-off functions.

Spaces of Environmental Variation

Can this conceptual framework for modularity give us any guidelines as to what we should expect from nature regarding modularity in the genotype–phenotype map? We have three principal features to consider: the top-level screening-off functions, the partitioning of these functions into those under stabilizing selection and those under directional selection, and the modularity of the genotype–phenotype map with respect to this partition.

The amount of modularity with respect to directional selection actually exhibited by an organism will depend on the particular set of functions that are under directional selection. If it happens to be a set for which the organism has a modular genotype–phenotype map, then it will show a modular relation to selection.

What determines which functions are under directional selection? Clearly, environmental change—biotic and abiotic—would be the principal cause of directional selection by dislodging the phenotypic optima. So the possession of a modular genotype–phenotype map—in the way that matters to evolvability—would appear to depend on the vagaries of environmental change. The study of modularity in the genotype–phenotype map of organisms as it pertains to evolvability would thus be somewhat of a haphazard subject.

However, there may be processes that give modularity a more systematic existence than the vagaries of environmental change would lead one to expect. I will later describe population genetic mechanisms that can lead to the evolution of modularity that enhances evolvability. From the foregoing discussion, we would expect that such modularity would evolve for functions that were under recurrent directional selection. This brings us to the spaces of variation in the environment.

The environment is analogous to the phenotype in that it takes vast numbers of variables to describe it, yet its degrees of freedom for variation are few in comparison. If we go forward with the idea that low-dimensional manifolds characterize the variation of the environment, then each of these spaces will be characterized by different fluctuation statistics. The ones that are highly variable will induce recurring directional selection on those screening-off functions of the organism that are sensitive to these environmental variables. Modularity for these variables will enhance the organism's ability to respond evolutionarily to this recurring directional selection. Therefore, if evolvability-enhancing modularity can evolve as a response to directional selection, it will be most well developed for those functions that are under recurring directional selection (Altenberg, 1995; Wagner, 1996).

The upshot is that the spaces of variation in the organism may come to mirror the spaces of variation in the environment. This idea is really only a technical revision of the idea originally proposed by Riedl, that "the epigenetic system copies the functional interdependencies of the phene system" (Riedl, 1978, p. 93).

How great a degree of mirroring we can expect depends on the quantitative details of the processes that would produce the evolution of modularity. Such details are left for another day, and here are merely proposed as a possibility that merits investigation.

Mutational Kinetics, Modularity, and Evolvability

To pursue the foregoing discussion with a specific example, I utilize the "B-matrix" model of Wagner (1989). This model contains all the ingredients discussed thus far:

• Genes control multiple phenotypic variables, creating the dimensions of variation for the phenotype.

• Phenotypic variables are controlled by multiple genes.

• A fitness function is defined on the phenotypic variables.

In this model, the "function-network" has only this one, top, level of phenotypic variables; there are no other phenotypic variables that are screened off by these fitness-defining variables.

Because the B-matrix model is simple and well-defined, we can answer the question of how the alignment between the dimensions of variation and selection gradients affect evolvability, and derive a means to define modularity as a property intrinsic to the model, not imposed by subjective parsing of the phenotype. We shall discover that a critical feature for defining modularity turns out to be the magnitude of mutation effects.

Wagner's B-Matrix Model

In the B-matrix model of Wagner (1989), selection is optimizing, acting on multiple traits controlled additively by multiple loci. There are three spaces in this model: genotype, phenotype, and fitness. Genetic variables are mapped to phenotypic variables, and these in turn are mapped to fitness. Each phenotypic variable has an optimal value, and fitness is defined as a Gaussian function of the departure of the traits from the optimum. I will describe each of these mappings.

The Phenotype–Fitness Map   The optimal value of each phenotypic variable is set to 0 for simplicity. Letting x represent the vector of phenotypic variables, the fitness w is defined to be

$$w = \exp\left(-\frac{1}{2}x^{\top}Mx\right),$$

where M is a positive definite matrix (positive definiteness assures that fitness decreases as one departs from the optimum). Here, the function $\delta(x, M) = x^{\top}Mx$ is the sole top-level screening-off function in this system, since if we know δ, the fitness is $w = \exp(-\delta/2)$, and there is no additional information that x gives about fitness.

The Genotype–Phenotype Map   The vector of phenotypic variables x is itself a linear function of the underlying genetic variables y:

$$x = By.$$

The scalar value y_i can be interpreted as the lowest-level screening-off function for gene i, which summarizes the entire causal effect that gene i has on the organism.

    The fitness function, expressed in terms of y, is

$$w(y) = \exp\left(-\frac{1}{2}y^{\top}B^{\top}MBy\right). \tag{1}$$

    In this model, the spaces of phenotypic variation are simple straight lines defined by the columns of B,

$$b_k = \begin{bmatrix} B_{1k} \\ B_{2k} \\ \vdots \\ B_{Lk} \end{bmatrix},$$

where L is the number of genetic variables (loci). Thus the space of phenotypic variation produced by variation at locus k is the line $S_k = \{y_k b_k : y_k \in \Re\}$.

Finite and Infinitesimal Models for Mutational Kinetics

In order for a genotype with these dimensions of variation to respond to selection, the earlier discussion claims that there must be selection gradients along the spaces of variation. When we wish to analyze the evolutionary dynamics, we see immediately that we must know something more about the magnitude of variation produced by mutation of the genotypic variables. In the quantitative genetic literature, we find two main kinetic models (Bürger, 2000) for the production of variation:

- The "random-walk" mutation model (Crow and Kimura, 1964)
- The "house–of–cards" mutation model (Kingman, 1977, 1978).

The random-walk model embodies the assumption that mutation perturbs the genetic variable away from its current value by a random variable ε, giving

$$x_i \rightarrow x_i + \varepsilon.$$

Typically, ε is distributed symmetrically around 0, having a Gaussian, exponential, or Γ, distribution. The transition probability (or density) is

$$T(x_i \leftarrow x_j) = u(x_i - x_j).$$

The "house–of–cards" model assumes that mutation "topples the house of cards" that adaptation has built up, producing a new phenotype that is independent of the old, with a value that is sampled from the same distribution regardless of the original value, giving

$$x_i \rightarrow \varepsilon.$$

The transition probability (or density) is

$$T(x_i \leftarrow x_j) = u(x_i).$$

A key difference between the models becomes apparent when they are adapted to the multivariate context. In the random-walk model, the perturbation caused by each individual mutation is taken to be small, and thus nearly neutral. Finite perturbations are taken to be the result of multiple small mutations. Under this process, multiple infinitesimal mutations can accumulate before selection can differentiate them, giving for the random-walk model:

$$x \rightarrow x + \sum_i \varepsilon_{\kappa_i} \mathbf{1}_{\kappa_i},$$

where $\kappa_i \in \{1, 2, 3, \ldots, L\}$ are independent random variables designating the index of the locus to be mutated, and $\mathbf{1}_{\kappa_i}$ is a vector for the ith mutation that has a single non–zero entry:

$$\mathbf{1}_k = \begin{bmatrix} 0 \\ 0 \\ \vdots \\ 1 \\ \vdots \\ 0 \end{bmatrix} \leftarrow k\text{th entry,}$$

so $\mathbf{1}_k$ has all 0 entries except for the kth entry, which is 1.

By the law of large numbers, $\Sigma_i \varepsilon_{\kappa_i} \mathbf{1}_{\kappa_i}$ approaches a multivariate Gaussian random variable $\boldsymbol{\varepsilon}$, so the mutation process gives

$$x \rightarrow x + \sum_i \varepsilon_{\kappa_i} \mathbf{1}_{\kappa_i} \approx x + \boldsymbol{\varepsilon}.$$

The mutation process will diffuse away from any "wild-type" genotypes and produce a cloud of genotypes surrounding it.

In the "house–of–cards" model, on the other hand, mutations are not infinitesimal in size, since the mutant genotype value is sampled from a fixed distribution independent of its current value. In the mutant genotype, a random locus κ is mutated, which replaces element y_κ with ε_κ in the vector y', leaving the other positions alone. A way to express the mutant genotype, y', is

$$y' = y \circ (\mathbf{1} - \mathbf{1}_\kappa) + \varepsilon_\kappa \mathbf{1}_\kappa = y + (\varepsilon_\kappa - y_\kappa) \mathbf{1}_\kappa, \tag{2}$$

where ε_κ is the random variable for the new genotype value, sampled from the distribution $u(x_\kappa)$.

As selection moves the population toward fitter genetic values in the "house–of–cards" model, a smaller and smaller fraction of the fixed distributions $u(x_i)$ is closer to the optimum; hence the probability of generating fitter mutants trails off with increasing adaptation. Contrary to the random-walk model, since mutations are not infinitesimal, they will not be nearly neutral, so selection will start to fix or purge mutations as soon as they occur. Thus it will not be possible to build up small mutations at multiple loci before selection acts. This has a significant impact, because the mutation process will no longer produce a multivariate Gaussian perturbation. Instead, the frequencies of single-, double-, and triple-locus mutants (and so on) will decrease exponentially (at a rate which is a function of the mutation rate and selection magnitude).

The "house–of–cards" and random-walk models are illustrated in figure 5.5. The points in figure 5.5A show the spectrum of phenotypes that are accessible under the

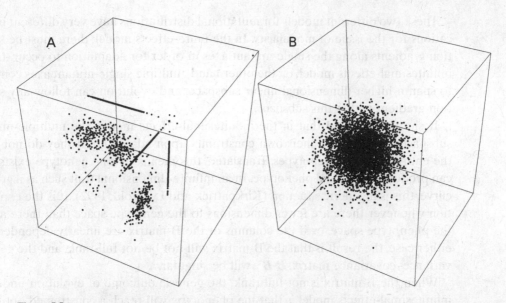

Figure 5.5
(A) Distribution of mutational effects under the "house of cards" assumptions. (B) Distribution of mutational effects under the "random-walk" assumptions. 5000 points are sampled.

"house–of–cards" assumptions. This graph is produced with the assumption that loci mutate independently, so single, double, and triple mutations occur with frequencies proportional to powers of the mutation rate. What matters, however, is how the frequency spectrum of multiple mutations affects the accessibility of the space. Most variants are single mutants, which fall along the axes of variation produced by each gene. Rarer double mutants fall along the planes defined by each pair of single-mutant axes. Even rarer triple mutants fall in the interior. The "wild type" is at the intersection of the single-mutant axes. The points in figure 5.5B show the spectrum of phenotypes that are accessible under the random-walk model, where multiple infinitesimal mutations allow access to the entire space around the wild-type phenotype.

I have presented the "house–of–cards" model as a paradigm for a mutational kinetics that generates variation along low-dimensional spaces, as depicted in figure 5.5A. However, what is critical to this result is not the "house–of–cards" assumption itself. Rather, it is that mutation is finite rather than infinitesimal in effect. Any mutational distribution that is dominated by finite effects will result in a population distribution similar to figure 5.5A, in which variation falls along the single-mutant axes. Therefore, I will refer to the two distinct paradigms for mutational kinetics as "finite" versus "infinitesimal" models.

These two different models for mutational distributions have very different implications for the issue of modularity. In the finite-effects model, there must be selection gradients along the single-mutant axes in order for adaptation to occur. In the infinitesimal-effects model, on the other hand, multiple single-mutant axes combine to span a higher-dimensional linear subspace, and evolution can follow any selection gradient within this subspace.

It should be noted that in the infinitesimal-effects model, the multiple-mutant subspaces may impose their own constraints upon adaptation if they do not span the entire space of phenotypes. Translated, this means that no genotype exists that can produce an optimal phenotype. In an infinite-dimensional trait such as a growth curve, this is a generic situation (Kirkpatrick and Lofsvold, 1992). It is the expectation whenever there are fewer dimensions to the genotype space than there are to the phenotype space, or if the columns of the B-matrix are linearly dependent. In either case, the result is that the B-matrix will not be not full-rank, and the genetic variance–covariance matrix, $B B^T$, will be singular.

When the B-matrix is not full rank, the generic outcome of evolution under an infinitesimal-effects model is that the phenotype will reach a constrained optimum within the space spanned by the B-matrix, at some distance from the global optimum (Kirkpatrick and Lofsvold, 1992; Altenberg, 1995). At this constrained optimum, additive genetic variation may exist for each phenotypic variable individually, but the reduced dimensionality of their joint variation will prevent any response to selection. There will remain a "latent" directional selection orthogonal to the space of variation (Altenberg, 1995).

The finite-effects model makes possible a form of constraint—frustration—above and beyond the constraint caused by a non-full-rank B-matrix. Frustration may prevent even the constrained optimum from being reached.

Frustration is a *kinetic constraint*, in that genotypes with the optimal phenotype may be possible, but the probability of generating them is minute because it requires multiple simultaneous mutations away from the wild type.

Riedl (1977) delves into the issue of finite versus infinitesimal effects in his discussion of alternative theories for the evolution of complex phenotypes. One which he calls the "storage theory" proposes that in cases where multiple mutations are needed to produce a particular adaptation, these mutations can be stored in the gene pool until they are brought together by recombination or hybridization. But this requires that the mutations, not valuable individually, be nearly neutral so as not to be expunged by selection. In order to be nearly neutral, they must be of extremely small effect. The storage theory, then, is an infinitesimal-effects model for mutational kinetics.

Alignment with Selection Gradients

With this distinction between these two models for mutational kinetics now spelled out, let us return to the thesis described at the beginning of this chapter about the advantage of modularity for evolvability. As should now be obvious, these ideas have as a core assumption that mutation follows a finite-effects kinetics.

Recalling that the fitness function in the B-matrix model is

$$w(y) = \exp\left(-\frac{1}{2}y^\top B^\top M B y\right),$$

then in the "house–of–cards model," with locus κ mutated, the fitness of the mutant genotype $y' = y + (\varepsilon_\kappa - y_\kappa)\mathbf{1}_\kappa$ (from equation 2) is

$$w(y') = \exp\left(-\frac{1}{2}y'^\top B^\top M B y'\right)$$

$$= \exp\left(-\frac{1}{2}\left[y^\top B^\top M B y + 2(\varepsilon_\kappa - y_\kappa)y^\top B^\top M b_\kappa + (\varepsilon_\kappa - y_\kappa)^2 b_\kappa^\top M b_\kappa\right]\right)$$

$$= w(y)\exp\left[(y_\kappa - \varepsilon_\kappa)y^\top B^\top M b_\kappa - \frac{1}{2}(y_\kappa - \varepsilon_\kappa)^2 b_\kappa^\top M b_\kappa\right].$$

So we see that whether the mutation is adaptive or not depends on the relationship of the column vectors b_κ with the matrix M and the current genotype y.

One could exactly quantify the magnitude of evolvability for this model by specifying the sampling distribution of ε, and deriving the probability that $w(y') > w(y)$. This, however, would go beyond the purpose of this chapter, which is merely to delineate the relationship between the different factors described here: modularity, spaces of variation, selection gradients, and evolvability.

I have claimed that the natural notion of modularity—a genotype–phenotype map that is decomposable into the product of lower-dimensional genotype–phenotype maps—is no more than a means (nor the only means) to enhance evolvability by making it easier to align the spaces of variation with selection gradients. In the B-matrix model, a modular genotype–phenotype map corresponds, in the extreme degree, to B being a diagonal matrix. Under this condition, we have

$$y^\top B^\top M b_\kappa = \sum_j y_j B_{jj} M_{j\kappa} B_{\kappa\kappa}$$

and

$$b_\kappa^\top M b_\kappa = B_{\kappa\kappa}^2 M_{\kappa\kappa};$$

hence

$$w(y') = w(y)\exp\left[(y_\kappa - \varepsilon_\kappa)\sum_j y_j B_{jj} M_{j\kappa} B_{\kappa\kappa} - \frac{1}{2}(y_\kappa - \varepsilon_\kappa)^2 B_{\kappa\kappa}^2 M_{\kappa\kappa}\right].$$

We notice that despite the modularity of the genotype–phenotype map, there are interaction terms $y_j B_{jj} M_{j\kappa} B_{\kappa\kappa}$ that signify epistasis between loci—that is, whether mutation at locus κ can generate adaptation depends on the state of the other loci, y_j. In fact, the situation with a modular genotype–phenotype map is really no different from the situation with a non–modular genotype–phenotype map, because we can write

$$w(y) = \exp\left(-\frac{1}{2}y^\top M'y\right),$$

where $M' = B^\top M B$ is a positive definite matrix, which is the same form as if B were the identity matrix.

Therefore, a modular genotype–phenotype map is not sufficient to ensure any special evolutionary capabilities of the variation-generating system. What is further required is that the M matrix itself be a diagonal. In that case, we obtain

$$w(y') = w(y)\exp\left[(y_\kappa - \varepsilon_\kappa)B_{\kappa\kappa}^2 M_{\kappa\kappa}\left[y_\kappa - \frac{1}{2}(y_\kappa - \varepsilon_\kappa)\right]\right]$$

$$= w(y)\exp\left[\frac{1}{2}(y_\kappa^2 - \varepsilon_\kappa^2)B_{\kappa\kappa}^2 M_{\kappa\kappa}\right].$$

Here, we see that the ability of a new mutation to produce a fitness increase depends solely on whether the new genotypic value, ε_κ, is closer to the optimum than the old genotypic value, y_κ. No other loci are involved. But we see that "modularity" here cannot be defined solely in terms of the genotype–phenotype map; it must also involve the matrix M, which describes how phenotypes map to selection. So again, what is more fundamental to evolvability than modularity in the genotype–phenotype map is the relationship between the spaces of genetic variation and the selection gradient.

Discussion

In this chapter I have tried to focus on some of the low-level issues that arise when trying to approach the issue of modularity in evolution. I have not delved at all into the question of how evolutionary dynamics may affect modularity and the alignment of spaces of variation with selective gradients. I will offer some comments on the evolutionary dynamics affecting modularity.

Constructional Selection

The role of gene origin in sculpting the modularity of the genotype–phenotype map is explored in Riedl's work (Riedl, 1975, 1977, 1978) and in several of my own papers (Altenberg and Brutlag, 1986; Altenberg, 1994b, 1995).

The central idea of this work is that we expect the dimensions of variation in the genome to be enriched with spaces that are in alignment with selective gradients. This enrichment process is a systematic outcome of the dynamics of genome growth. New genes that happen to change the phenotype along a positive selection gradient are much more likely to be preserved by selection than genes which produce variation that randomly perturbs the phenotype and is thus likely to be detrimental. Thus, the degrees of freedom in the genome should grow in the direction of greater evolvability. My shorthand term for this process is "constructional selection" because it pertains to the construction of the genome.

Modularity is one means, though not the only means, to achieve the correct alignment of the space of variation with respect to selection. So modularity is one feature that we expect to be enriched by the process of genome growth. Clear examples of this sort of modularity are the separation of regulatory function from coding function in eukaryotic genomes. Such separation is not a functional necessity, as seen in nonmodular genes where sequences carry both coding and regulatory function. But separation of these functions permits one of the dimensions of genetic variation—sequence duplication—to explore combinatorial spaces which preserve the regulatory and coding functions of the gene fragments. By maintaining these functions, but bringing them together in a new combination, such modular genetic elements have a greater likelihood to produce a selective advantage, and thus be kept by the genome. Therefore, the genome should become more enriched for such elements as it grows. This same process would also apply to elements within regulatory regions or within coding regions. And we find that many proteins are mosaics of function recombined from other genes (Hegyi and Bork, 1997).

What is important to remember is that the modularity that can result from selective genome growth is defined *in terms of the genetic operators* producing the genetic variation, in this case the processes of sequence duplication. So, for example, if sequence duplication happened to be restricted to a certain range of sequence lengths, it would be on that length scale that genome growth would select for modularity. And modularity is selected only with respect to its ability to increase the likelihood that the sequence duplication event is beneficial. All structural features that we would call "modular" are defined in terms of this probability rather than any a priori structural definitions that we might impose.

Failure to appreciate this essential point is a source of confusion when discussing the issue of modularity of exons (Logsdon, 1998; de Souza et al., 1998). Modularity with respect to exon shuffling can be achieved when protein domain boundaries correspond to exon boundaries. But a lack of correspondence is not in itself evidence against modularity. If functional properties of an exon are maintained after exon shuffling, then this exon exhibits modularity. It may not be necessary for domain and exon boundaries to correspond in order for the functional properties to withstand exon shuffling—other properties of the sequence can stabilize the functional elements. This distinction is subject to empirical testing because modifications of splice sites in exons with evolved modularity would be expected to decrease their modularity, whether or not the splice sites fall between protein domains.

Other Sources of Modularity in the Genotype–Phenotype Map

In addition to genome growth processes, there may be other sources that produce modularity in the genotype–phenotype map. These deserve some mention here.

Modularity "For Free"   There may be generic features of biology, chemistry, or physics that provide modularity in the genotype–phenotype map "for free"—to borrow the phrase from Kauffman (1995, chap. 4). Kauffman speaks of "order for free," that is, order in living organisms that arises not from Darwinian selection (order at a cost) but as a generic outcome of physical self-organizing processes. Similarly, there may be examples of "modularity for free" in the genotype–phenotype map that have a similar origin. In other words, there may be circumstances when we expect modularity to be a generic property of organisms that does not require natural selection to establish or maintain.

One obvious candidate source for modularity without natural selection is the branching structure of the cellular genealogy in multicellular organisms. The single-celled organism is the epitome of a module, and this is the ancestral state for the cells of multicellular organisms. While many single-celled organisms can have aggregate properties (e.g., production of biofilms), their tendency to separate, disperse, and become independent after replication is a generic property that makes unicellular organisms modules. Multicellular organisms have adaptations that counteract this independence after replication, and maintain proximity and interaction to varying degrees. However, a certain amount of separation and independence is inescapable among the cells in multicellular organisms. This would be a fundamental source of "modularity for free" in multicellular organisms.

Vascular plants maintain a close parallel between their physical structure and their genealogical structure, because their cells have less mobility than cells in animals. So cells which are genealogically distant also tend to be physically distant. This physical distance makes modularity in the genotype–phenotype map more easily realized, because phenotypic alterations in the structure of, say, a flower may have fewer physical interactions with, for example, a root.

In complex animals, there is less isolation between genealogically distant cells because of cellular mobility and physiological integration. Multiple tissue lineages participate in the construction of integrated organs. Hormonal and neuronal communication integrates genealogically distant cells in their function. Therefore, in animals one would expect to find significantly less "modularity for free" from the cellular genealogy.

Modularity "Included" It is possible that modularity in the genotype–phenotype map can "hitchhike" (Maynard Smith, 1974) along with traits under natural selection. This is what I mean by "modularity included"—it doesn't come free, but is included as a side effect of natural selection for traits under selection. A paradigmatic example of "modularity included" is the work on selection for robustness in RNA structures by Ancel and Fontana (chapter 6 in this volume). They find that as greater stability evolves in their molecular structures, most of the molecular sites become structurally neutral, while structural sensitivity to mutation concentrates in a tightly integrated core of sites.

It is possible that there is a physical explanation for this phenomenon, which may make it a generic property of molecular interactions. Structural stability depends on strong molecular bonding, and strong bonding requires physical proximity of bonding sites. Such physical proximity, however, can be shared by only a limited number of sites. Therefore, the strongest bonding interactions are expected to be limited to a selected set of sites, screening other sites from these high-energy bonds. Thus selection for strong bonding can have the side effect that these high bond energies become concentrated among a small number of sites.

This correlation between the strength of interaction and the specificity of interaction may be a generic feature of a wide class of molecules, especially ones where the interaction is specified by shape, such as proteins, nucleotides, receptors, and enzymatic reactions. There are obvious exceptions, such as peroxides, that achieve strong interaction with little specificity. But many biological molecules, especially proteins and nucleotides, may receive "modularity included" in selection for structural stability because of this correlation.

This mechanism for "modularity included" would also apply to spatial compartmentalization (Weng et al., 1999). Compartmentalization of reacting molecules increases the strength of interaction simply by increasing concentrations, but because of the conservation of matter, it decreases concentrations elsewhere, and thereby increases the specificity of interactions. Selection for high concentration of molecules may thereby bring along modularity as a side effect.

Direct Selection for Modularity   Specificity of interaction may be a side effect of selection for strong interaction, but it may also be a target of selection in its own right. Coordination of activities from the scale of the chromosome to the entire organism, or even an entire population, requires precise specificity between signals and receptors. Specificity is needed so that the control of different processes in the organism has the degrees of freedom needed to optimize their coordination. This specificity of interaction can translate directly into specificity for the phenotypic effects of genetic variation, also known as modularity.

Subfunctionalization

Where are we to place the phenomenon of subfunctionalization (Force et al., 1999; Lynch and Force, 2000) within this categorization scheme? Subfunctionalization is a process in which duplicate genes make themselves necessary to the organism by losing, rather than gaining, function. In the classical thinking about the fate of gene duplications, the duplicates had to gain new functions in order to avoid being redundant and eventually silenced by mutation (Ohno, 1970). However, if genes carry out multiple functions, and these functions can be silenced independently of one another, then a different set of functions can be silenced in each gene, and the remaining functions of each gene can be preserved by selection. In essence, after subfunctionalization, there is still only one gene functioning, but it is split up into two different loci and involves two different transcripts with complementary function. The complementation must therefore be *trans*-acting.

It should be immediately clear that subfunctionalization is not a means to produce modularity, but rather the reverse: it requires that functions of the gene already be modular, in that the gene has independent degrees of freedom for the loss of each function. The process of gene duplication and subfunctionalization will exhaust itself when the modules inherent in the original gene have been completely parceled out among the duplicate genes. A further gene duplication will not be able to simultaneously lose part of its function and complement the losses in other genes. It will be either redundant or necessary as a whole.

Subfunctionalization thus faces a finite limit on the process, which distinguishes it from constructional selection. In constructional selection, the amplification of modular elements in the genome is limited only by the selective opportunity for new combinations of modules. Subfunctionalization, on the other hand, is effectively conservative for module number—spreading out modules among multiple loci but not creating them. Therefore it cannot explain module origin, and thus is a consequence, rather than a source, of module-creating processes, such as constructional selection, genetic modification, and selection for properties that have "modularity included."

Conclusion

I have endeavored in this chapter to delve into some of the low-level conceptual issues associated with the idea of modularity in the genotype–phenotype map. My main proposal is that the evolutionary advantages that have been attributed to modularity do not derive from modularity per se. Rather, they require that there be an "alignment" between the spaces of phenotypic variation and the selection gradients that are available to the organism. Modularity in the genotype–phenotype map may make such an alignment more readily attained, but it is not sufficient; the appropriate phenotype–fitness map in conjunction with the genotype–phenotype map is also necessary for evolvability.

Acknowledgments

My greatest thanks to the Konrad Lorenz Institute for inviting me to participate in the workshop and creating a very stimulating scientific environment; to the editors, Diego Rasskin-Gutman and Werner Callebaut, for their forbearance; to the Department of Information and Computer Sciences at the University of Hawaii at Manoa; and to my family and friends on general principle.

References

Agrafiotis DK, Xu H (2002) A self-organizing principle for learning nonlinear manifolds. Proc Natl Acad Sci USA 99: 15869–15872.

Altenberg L (1994a) The evolution of evolvability in genetic programming. In: Advances in Genetic Programming (Kinnear JKE, ed), 47–74. Cambridge, MA: MIT Press.

Altenberg L (1994b) Evolving better representations through selective genome growth. In: Proceedings of the 1st IEEE Conference on Evolutionary Computation, part 1, 182–187. Piscataway, NJ: IEEE.

Altenberg L (1995) Genome growth and the evolution of the genotype–phenotype map. In: Evolution and Biocomputation (Banzhaf W, Eeckman FH, eds), 205–259. Berlin: Springer-Verlag.

Altenberg L, Brutlag DL (1986) Selection for modularity in the genome. Unpublished. Cited in Doolittle (1987), Tomita et al. (1996).

Ancel LW, Fontana W (2000) Plasticity, evolvability and modularity in RNA. J Exp Zool (Mol Dev Evol) 288: 242–283.

Brandon RN (1982) The levels of selection. In: PSA 1982, vol. 1 (Asquith P, Nickles T, eds), 315–323. East Lansing, MI: Philosophy of Science Association.

Brandon RN (1984) The levels of selection. In: Genes, Organisms, Populations: Controversies over the Units of Selection (Brandon R, Burian R, eds), 133–141. Cambridge, MA: MIT Press.

Brandon RN (1990). Adaptation and Environment. Princeton, NJ: Princeton University Press.

Brandon RN (2002) Philosophy of selection: Units and levels. Hum Nat Rev 2: 373. http://www.spee.ch/archive/evolution_philoso–phy_of_selection_units_and_levels.htm.

Bürger R (1986) Constraints for the evolution of functionally coupled characters: A nonlinear analysis of a phenotpyic model. Evolution 40: 182–193.

Bürger R (2000). The Mathematical Theory of Selection, Recombination, and Mutation. Chichester, UK: John Wiley.

Crow JF, Kimura M (1964) The theory of genetic loads. In: Proceedings of the XIth International Congress of Genetics, vol 2, 495–505. Oxford: Pergamon Press.

De Souza SJ, Long M, Klein RJ, Roy S, Lin S, Gilbert W (1998) Toward a resolution of the introns early/late debate: Only phase zero introns are correlated with the structure of ancient proteins. Proc Nat Acad Sci USA 95: 5094–5099.

Doolittle WF (1987) The origin and function of intervening sequences in DNA: A review. Amer Nat 130: 915–928.

Dullemeijer P (1974) Concepts and Approaches in Animal Morphology. Assen, Netherlands: Van Gorcum.

Force A, Lynch M, Pickett FB, Amores A, Yan YL, Postlethwait J (1999) Preservation of duplicate genes by complementary, degenerative mutations. Genetics 151: 1531–1545.

Hegyi H, Bork P (1997) On the classification of evolution of protein modules. J Prot Chem 16: 545–551.

Kauffman SA (1995) At Home in the Universe: The Search for the Laws of Self-Organization and Complexity. Oxford, UK: Oxford University Press.

Kauffman SA, Levin S (1987) Towards a general theory of adaptive walks on rugged landscapes. J Theor Biol 128: 11–45.

Kauffman SA, Weinberger ED (1991) The NK model of rugged fitness landscapes and its application to maturation of the immune response. In: Molecular Evolution on Rugged Landscapes: Proteins, RNA and the Immune System (Perelson AS, Kauffman SA, eds), 135–175. Redwood City, CA: Addison-Wesley.

Kingman JFC (1977) On the properties of bilinear models for the balance between genetic mutation and selection. Math Proc Cam Philos Soc 81: 443–453.

Kingman JFC (1978) A simple model for the balance between selection and mutation. J Appl Prob 15: 1–12.

Kirkpatrick M, Lofsvold D (1992) Measuring selection and constraint in the evolution of growth. Evolution 46(4): 954–971.

Logsdon JM (1998) The recent origins of spliceosomal introns revisited. Curr Opin Genet Devel 8: 637–648.

Lynch M, Force A (2000) The probability of duplicate gene preservation by subfunctionalization. Genetics 154: 459–473.

Maynard Smith J (1974). The hitch-hiking effect. Heredity 33: 130.

McKay SR, Berker AN, Kirkpatrick S (1982) Spin-glass behavior in frustrated ising models with chaotic renormalization-group trajectories. Phys Rev Lett 48: 767.

Millikan RG (1984) Language, Thought, and Other Biological Categories: New Foundation for Realism. Cambridge, MA: MIT Press.

Ohno S (1970) Evolution by Gene Duplication. Berlin: Springer-Verlag.

Plato In: Plato's Statesman: A Translation of the Politicus of Plato with Introductory Essays and Footnotes (Skemp JB, trans.) (1952) London: Routledge and Kegan Paul.

Riedl RJ (1975) Die Ordnung des Lebendigen: Systembedingungen der Evolution. Hamburg and Berlin: Parey.

Riedl RJ (1977) A systems-analytical approach to macroevolutionary phenomena. Quart Rev Biol 52: 351–370.

Riedl RJ (1978) Order in Living Organisms: A Systems Analysis of Evolution (Jefferies RPS, trans). Chichester, UK: John Wiley.

Roweis ST, Saul LK (2000) Nonlinear dimensionality reduction by locally linear embedding. Science 290: 2323–2367.

Salmon WC (1971) Statistical Explanation and Statistical Relevance. Pittsburgh, PA: University of Pittsburgh Press.

Salmon WC (1984) Scientific Explanation and the Causal Structure of the World. Princeton, NJ: Princeton University Press.

Schwenk K (2001) Functional units and their evolution. In: The Character Concept in Evolutionary Biology (Wagner GP, ed), 165–198. San Diego: Academic Press.

Simon HA (1962). The architecture of complexity. Proc Amer Phil Soc 106: 467–482. Reprinted in Simon, 1969.

Simon HA (1969) The Sciences of the Artificial. Cambridge, MA: MIT Press.

Stadler BMR, Stadler PF, Shpak M, Wagner GP (2002) Recombination spaces, metrics, and pretopologies. Zeit Physital Chem 216: 217–234.

Tenenbaum JB, de Silva V, Langford JC (2000) A global geometric framework for nonlinear dimensionality reduction. Science 290: 2319–2323.

Tomita M, Shimizu N, Brutlag DL, (1996) Introns and reading frames: Correlations between splice sites and their codon positions. Molec Biol Evol 13(9): 1219–1223.

Van Nimwegen EJ, Crutchfield JP, Huynen M (1999) Metastable evolutionary dynamics: Crossing fitness barriers or escaping via neutral paths? Bull Math Biol 62: 799–848.

Verbeek JJ, Vlassis N, Kröse B (2002) Fast nonlinear dimensionality reduction with topology preserving networks. In: Proceedings of the 10th European Symposium on Artificial Neural Networks (Verleysen M, ed), 193–198. Evere, Belgium: D-side Publications.

Waddington CH (1942) Canalization of development and the inheritance of acquired characters. Nature 150: 563–565.

Wagner GP (1984) Coevolution of functionally constrained characters: Prerequisites of adaptive versatility. BioSystems 17: 51–55.

Wagner GP (1988) The influence of variation and of developmental constraints on the rate of multivariate phenotypic evolution. J Evol Biol 1: 45–66.

Wagner GP (1989) Multivariate mutation-selection balance with constrained pleiotropic effects. Genetics 122: 223–234.

Wagner GP (1996) Homologues, natural kinds, and the evolution of modularity. Amer Zool 36: 36–43.

Wagner GP ed (2001) The Character Concept in Evolutionary Biology. San Diego: Academic Press.

Wagner GP, Altenberg L (1996) Perspective: Complex adaptations and the evolution of evolvability. Evolution 50(3): 967–976.

Waxman D, Peck J (1998) Pleiotropy and the preservation of perfection. Science 279: 1210–1213.

Weinberger ED (1991) Local properties of Kauffman's N-k model, a tuneably rugged energy landscape. Phys Rev A44(10): 6399–6413.

Weng G, Bhalla US, Iyengar R (1999) Complexity in biological signaling systems. Science 284: 92–96.

Wright S (1968) Evolution and the Genetics of Populations. Chicago: University of Chicago Press.

6 Evolutionary Lock-In and the Origin of Modularity in RNA Structure

Lauren Ancel Meyers and Walter Fontana

Modularity is a hallmark of biological organization and an important source of evolutionary novelty (Bonner, 1988; Wagner and Altenberg, 1996; Hartwell et al., 1999). Yet, the origin of modules remains a problem for evolutionary biology, even in the case of the most basic protein or RNA domains (Westhof et al., 1996). Biological modularity has been defined on many levels, including genetic, morphological, and developmental. The task of creating a general theory for the origins, ubiquity, and function of modularity requires the synthesis of these perspectives.

The first step in understanding the causes and consequences of modularity is to define modularity. The second step is to formulate methods of detecting it. The third is to build models and design experiments to study its origins and evolutionary implications.

In this chapter we offer a rigorous definition of modularity in RNA. It is the partitioning of molecules into subunits that are simultaneously independent with respect to their thermodynamic environment, genetic context, and folding kinetics. On the variational level, the module consists of a stretch (or stretches) of contiguous ribonucleotides held together by a covalent backbone. On the functional level, the relevant interactions include the covalent bonds of adjacent ribonucleotides and hydrogen bonds between nonadjacent bases. These elements of secondary structure provide the scaffold for tertiary structure, which underlies the functionality of the molecule.

We furthermore offer a practical tool for identification of modules in this semi-empirical framework. The melting profile of a molecule is the set of minimum free energy shapes attained as the temperature increases from 0° to 100° C. Modules are exactly those subunits that dissolve discretely without perturbing the remaining structure as temperature increases. A modular molecule is one made up entirely of such subunits.

Finally, we present a theory about the origins of modularity. Modularity may facilitate the evolution of more complex organisms through the combination of modules. If we try to explain the origins of modularity in terms of this evolutionary benefit, we run into a chicken-and-egg paradox. Modularity cannot produce a more sophisticated syntax of variation until it already exists.

Here we offer a more agnostic explanation for the evolution of modularity. Using a model of RNA folding, we show that modularity arises as a by-product when natural selection acts to reduce the plasticity of molecules by stabilizing their shapes. This is mediated by a statistical property of the RNA folding map: the more thermodynamically well-defined a shape, the more localized and less disruptive the effects of point mutations. We report two consequences of this relationship. First, under selection for thermodynamic stability, evolution grinds to a halt because of insufficient phenotypic variation. Second, the shapes trapped in this exploration catastrophe are highly modular. They consist of structural units that have become thermophysically, kinetically, and genetically independent.

Plasticity in RNA Secondary Structure

Under natural conditions, RNA molecules do not freeze in their minimum free-energy shape (henceforth ground state), but exhibit a form of structural plasticity. Thermal fluctuations (corresponding to environmental noise) cause molecules to equilibrate among alternative low-energy shapes. We model the genotype–phenotype map from an RNA sequence to its repertoire of alternative secondary structures (henceforth shapes), using an extension (Wuchty et al., 1999) of standard algorithms (Nussinov and Jacobson, 1980; Waterman, 1978; Zuker and Stiegler, 1981) which assist in the prediction of RNA secondary structure. For a given sequence, we compute all possible shapes having free energy within 3 kcal/mol of the ground state (at 37° C). We call this set of shapes the "plastic repertoire" of an RNA sequence (see figure 6.1A for an illustration). The partition function (McCaskill, 1990) of a sequence is $Z = \Sigma_\alpha \exp(-\Delta G_\alpha / kT)$, where ΔG_α is the free

Figure 6.1
Loss of plasticity and plastogenetic congruence. (*A*) On the right: all six shapes in the plastic repertoire of the most frequent sequence after 10^7 replications in the plastic simulation depicted in Figure 6.2A. The few shapes in the plastic repertoire of the evolved sequence are structurally similar to each other and the ground state. On the left: a subset of the 1208 shapes in the plastic repertoire of a randomly chosen sequence with the same ground state. Dots stand for many shapes not displayed. The number to the left of a shape is its equilibrium probability. The vertical lines on the right measure 3 kcal/mol, and each shape points to a height proportional to its energetic distance from the minimum free energy. (*B*) The graph illustrates the mechanisms underlying plastogenetic congruence in a small sequence. Aside from a graphical depiction of shapes, we also use a string representation in which a dot stands for an unpaired position, and a pair of matching parentheses indicates positions that pair with one another. A highly plastic sequence (i) is shown in its ground state shape α together with a list of all shapes in its plastic repertoire (numbers indicate equilibrium probabilities). Single-point mutations readily tip the energy balance in favor of another shape. For example, a point mutation from U to C in the loop of (i) makes its suboptimal shape β the new ground state in (ii). Generally, single-point mutations tip a highly plastic sequence in favor of shapes already present in its plastic repertoire. Single-point mutations can also act to reinforce the ground state. For example, a mutation from U to C (iii) generates a better stacking pair in the helix of α, and dramatically reduces the plasticity. As a consequence, the mutation that altered the ground state shape previously, (i) → (ii), no longer has a phenotypic effect, (iii) → (iv). Thermodynamic stabilization dramatically decreases variability—access to new structures through mutation.

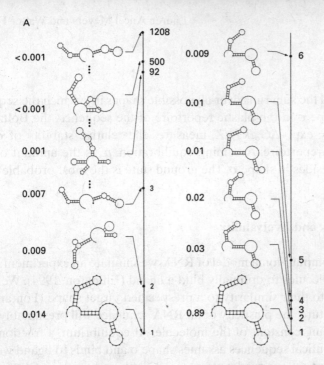

A

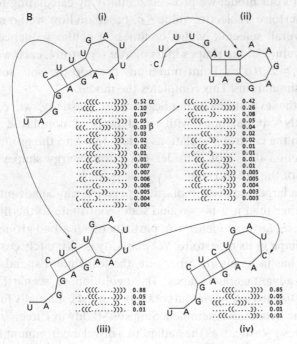

B

(i)   (ii)

(iii)   (iv)

energy of shape α and the sum runs over all possible shapes into which the sequence can fold. For any shape σ in the plastic repertoire of the sequence, the Boltzmann probability of σ, $p_\sigma = \exp(-\Delta G_\sigma/kT)/Z$, measures the relative stability of σ with respect to the entire repertoire. Assuming equilibration, p_σ is the amount of time the RNA molecule resides in shape σ. The ground state is the most probable shape for a molecule.

The Loss of Plasticity and Evolvability

Equipped with this computational model of RNA, we simulate an experimental protocol that evolves molecules to optimally bind a ligand (Ellington, 1994). We select sequences according to their similarity to a prespecified target shape (Fontana and Schuster, 1998). In nature, the plasticity of an RNA sequence will presumably influence the overall binding constant of the molecule. At equilibrium, a fraction p_σ of a large number of identical sequences assumes shape σ and binds to ligand with the corresponding constant. In our model we proceed similarly by calculating for each shape σ in the plastic repertoire a selective value $f(\sigma)$ based on how well σ matches the target shape. The overall selective value, or fitness, r of the sequence is the average of the selective values of the shapes in its plastic repertoire, each weighted by its occupancy time, $r = \Sigma_\alpha f(\sigma)p_\sigma$. Point mutations provide the sole source of genetic variation in our simulations. This completes the model.

In order to identify the evolutionary implications of plasticity, we compare simulations of plastic RNA populations with simulations of nonplastic control populations. Sequences in the control populations rigidly fold into the ground state only. Consequently, selection does not consider other low-energy shapes or the thermodynamic stability of the ground state.

High plasticity, that is, a large and diverse plastic repertoire, can be advantageous since multiple shapes, rather than just the ground state, contribute to the fitness of a sequence. In particular, a plastic sequence can partially offset a bad ground state with a good alternative shape in its repertoire. Yet plasticity is ultimately costly. The more shapes a molecule has in its plastic repertoire, the less time it spends in any one of them, including advantageous shapes. The evolutionary scenario under consideration—selection toward a constant target shape—must eventually favor the reduction of plasticity. The dynamics resemble a Simpson–Baldwin Effect, in which organisms gain and then lose plasticity as they adapt to a novel environment (Ancel, 1999; Baldwin, 1896). Figure 6.1A illustrates the plastic repertoire of a typical sequence present at the end of a plastic simulation (right). A comparison with a randomly chosen sequence folding into the same ground state (left) reveals the staggering reduction in plasticity. The number of shapes in the plastic repertoire decreases 200-fold, the fraction·of time spent in the ground state increases from 1.4% to 89%, and structural diversity within the 3 kcal band is nearly eliminated. Sequences found at the end of such simulations have well-defined ground states that are extremely resilient to thermal fluctuations.

This reduction in plasticity produces a remarkable effect evident in the evolutionary trajectories of typical plastic and control populations (figure 6.2A). Surprisingly, the population of plastic sequences evolves more slowly than the non-plastic control population, and quickly reaches an evolutionary dead end. Both evolve in a stepwise fashion with periods of phenotypic stasis punctuated by change toward the target shape (Ancel and Fontana, 2000; Fontana and Schuster, 1998).

Plastogenetic Congruence

The loss of evolvability stems from a statistical correlation between the thermodynamic plasticity of an RNA molecule and the mutability of its ground state structure through point mutations (Simpson, 1953). In particular, thermodynamic robustness (or lack of plasticity) is positively correlated to mutational robustness. Mutations which stabilize the ground state of a molecule also serve to buffer the molecule against structural changes due to mutations. Figure 6.1B illustrates the mechanism underlying this correlation. A highly plastic molecule wiggles among multiple alternative shapes that are energetically close to one another. A point mutation can easily tip the energy landscape of the molecule in favor of an alternative shape (without destroying sequence compatibility with the original ground state). This occurs in the transition from (i) to (ii), where the alternative shape β becomes the new ground state. At the same time, a plastic molecule also offers opportunities to stabilize a ground state, as in the transition from (i) to (iii). Note, however, that along with this reinforcement comes the immunity of the ground state to a point mutation which affected it in the previous sequence context, as in (iii) to (iv) versus (i) to (ii).

This capability of genes (or, in this case, ribonucleotides) to buffer other genes against the effects of mutations is known as epistasis. More generally, epistasis is the nonindependence of loci. When the fitness consequence of a mutation at one site depends on the nucleotides present at other sites, then there is epistasis between the sites. In RNA we observe the evolution of epistatic buffering, where critical base pairs render the minimum free-energy structure robust to most point mutations.

In summary, the plastic repertoire of an RNA molecule indicates how much and in which ways its ground state can be altered by mutation. We call this statistical

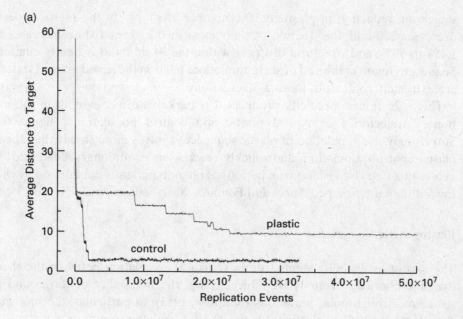

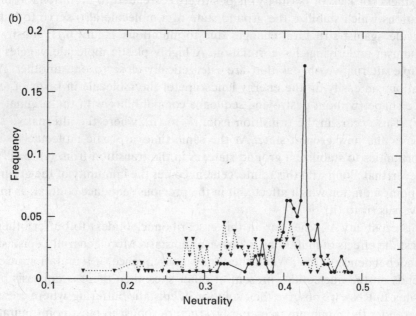

alignment between the thermodynamic sensitivity of the ground state and its genetic mutability *plastogenetic congruence*. We emphasize that this is not an assumption of our model, but a hitherto unknown statistical property of RNA folding algorithms that were developed independently of our evolutionary study. A similar correlation between the thermodynamic stability of the native conformation and its mutational robustness has been found in models of protein folding (Bornberg-Bauer and Chan, 1999; Vendruscolo et al., 1997; Bussemaker et al., 1997).

What are the consequences of plastogenetic congruence for our evolutionary model? Natural selection produces sequences with low plasticity (figure 6.1A). Low plasticity sequences are, by virtue of plastogenetic congruence, highly buffered against the effects of mutation. A large fraction of all possible single-point mutations on such sequences will preserve the ground state (figure 6.2B). Furthermore, the low-plasticity sequences not only reside mostly in their ground state, but spend the rest of their time in shapes that are structurally akin to it (figure 6.1A). Again, by plastogenetic congruence, the rare mutations that alter the ground state are likely to cause only slight structural changes. Thus, the phenotypic variability of the population, that is, the potential variation accessible through mutation, is dramatically curtailed. Plastogenetic congruence has steered the population into an evolutionary dead end which we call *neutral confinement*. This does not occur in the control populations (figure 6.2A).

Figure 6.2

Evolutionary dynamics and neutral confinement. (*A*) A simulated population of RNA sequences undergoes mutation, replication, and selection in a chemical flow reactor constrained to fluctuate around 1000 individuals. This process models a stochastic continuous-time chemical reaction system (Fontana and Schuster, 1998; Huynen et al., 1996). We graph the average population distance from the target shape with respect to replication events rather than external time. The selective value $f(\sigma)$ of a shape σ is defined in terms of the Hamming distance $d(\sigma,\tau)$ between the string representations (see caption for figure 6.1B) of σ and the target shape τ: $f(\sigma) = \dfrac{1}{0.01 + d(\sigma,t)/n}$ where n is the sequence length (here $n = 76$). The fitness (or replication constant) of a plastic sequence is given as $r = \sum_{\sigma_i} f(\sigma_i) \dfrac{\exp(-\Delta G_{\sigma_i}/kT)}{z}$, where the sum runs over all shapes σ_i with free energies ΔG_{σ_i} within 3 kcal/mol from the minimum free energy. The control population comprises sequences that are mapped simply to their minimum free-energy shapes σ_0, and hence have fitness $r = f(\sigma_0)$. Replication accuracy per position is 0.999. (*B*) For each sequence species present in a given population, we compute its neutrality, that is, the fraction of single-point mutants that preserve the ground state. The graph compares the distributions of neutralities for the plastic population (solid line) and the nonplastic control (dotted line) after 10^7 replications.

Origins of Modularity

Low plasticity is achieved through increasing the thermodynamic independence of
any one structural component from the remaining structure. At the same time, the
effects (if any) of mutations become increasingly limited to local shape features.
Wagner and Altenberg (1996) argued that modularity evolves in organisms by such
a decrease of pleiotropy. This process underlies the extreme modularity of ground
states that we observe at neutral confinement. We distinguish here between modules
defined in purely morphological (syntactical) terms and modules defined on the
basis of thermophysical, kinetic, and genetic autonomy. The former is a trivial notion
in RNA, since it is implicit in the very definition of a secondary structure as a combination of loops and helices. We address the latter. (In the case of RNA secondary
structures, these two notions are conveniently consistent with each other: the
thermodynamic stabilization of the ground state as a whole can easily proceed
through the stabilization of its component helices and loops.)

From a thermophysical perspective, low-plasticity shapes are composed of structural components that remain intact over large temperature regimes and that melt
in distinct phase transitions as discrete units (figure 6.3A). In particular, the melting
of one unit does not disturb the other units (figure 6.3B). This is in sharp contrast
to the melting behavior of high-plasticity sequences with the same ground state
(figures 6.3A and 6.3C). From a kinetic perspective, these same units fold independently, as suggested by a single folding funnel which dominates the energy landscape of a low-plasticity sequence (figure 6.3D). A conformational energy landscape
so organized prevents the occurrence of energetically trapped intermediates by
guiding the folding events reliably and quickly toward the ground state. Again, this
is in sharp contrast to the high-plasticity sequence, whose energy landscape provides
no guidance to the folding process (figure 6.3E).

Genetic autonomy is seen when sequence segments underlying these structural units are transposed from their original context into random contexts. Low-
plasticity segments maintain their original shape with a much higher likelihood than
the fragments of random sequences with the same shape. For example, the sequence
segments underlying the shape features labeled A and B in figure 6.3B maintain
their original shape with probabilities 0.83 and 0.94, respectively, when flanked by
random segments of half their size. These are much larger than the probabilities
0.017 and 0.015, respectively, for a random sequence with the same shape at 37°C
(figure 6.3C). There is computational evidence for such transposability in natural
sequences (Wagner and Stadler, 1999).

While it is unclear how natural selection could generate modularity directly, there are many scenarios in which natural selection favors the reduction of plasticity. In our model, modularity arises as a necessary by-product of that reduction. It is the nature of modules to resist change; hence the process that produces modularity simultaneously leads populations into evolutionary dead ends. By enabling variation at a new syntactical level, however, modularity may provide an escape from the evolutionary trap that produced it in the first place.

In the future, we will study RNA structural evolution in a fluctuating environment. Evolutionary theory teaches us that phenotypic plasticity is favored under sufficiently heterogeneous conditions. When RNA molecules evolve under macroscopic fluctuations, for example, in the presence of multiple binding targets or changing temperatures, natural selection may produce plastic RNA molecules. We will evaluate this hypothesis, and ask whether fluctuating environments consequently favor evoluability and preclude the evolution of modularity.

Acknowledgments

We thank Werner Callebaut and Diego Rasskin-Gutman for inviting our contributions to this volume and to the workshop on modularity at the Konrad Lorenz Institute. Thanks also to Ivo Hofacker, Peter Stadler, and Stefan Wuchty for their work on the Vienna RNA package. We are grateful to Leo Buss, Marc Feldman, Christoph

Figure 6.3
Thermophysical and kinetic modularity. (*A*) The calculated melting behavior (specific heat versus temperature) as it would appear in a differential scanning calorimetry experiment. The modular (evolved) RNA molecule (solid line) melts in two sharp phase transitions. The dotted line depicts the melting behavior of a high-plasticity sequence. (*B*) The ground states of the modular (evolved) RNA molecule as a function of temperature ($°$ C). The 37$°$ C features have extended thermal stability, and melt individually at distinct temperatures while leaving other parts of the shape unaffected. In other words, the melting behavior is discrete. (*C*) The succession of ground states with rising temperature of a high-plasticity sequence with the same shape at 37$°$ C as the evolved molecule in (*B*). The 37$°$ C shape of the sequence is unstable upon temperature perturbations in both directions, and undergoes global rearrangements as the 37$°$ C features destabilize with rising temperature. (*D*) Given a kinetic model of the folding process, the low-energy portion of a molecule's conformational landscape can be represented as a tree (Flamm et al., 2000). The ordinate measures free energy, and the abscissa has no meaning. A leaf corresponds to a shape at a local energy minimum, and the height of a branch point corresponds to the energy barrier between two local minima. (*D*) shows that the folding landscape of the evolved sequence is organized as a funnel leading directly to the ground state shape. Modules fold independently. (*E*) The energy landscape of the random sequence provides little or no guidance to the folding process, and results in frequent deadlocks at local minima.

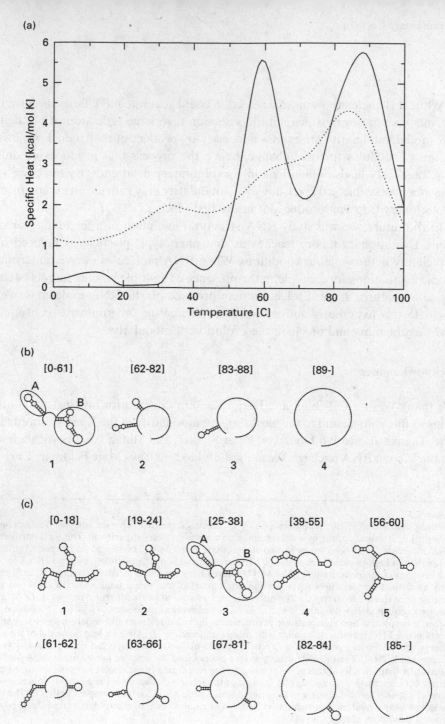

(a)

(b)

[0-61]   [62-82]   [83-88]   [89-]

A   B

1   2   3   4

(c)

[0-18]   [19-24]   [25-38]   [39-55]   [56-60]

A   B

1   2   3   4   5

[61-62]   [63-66]   [67-81]   [82-84]   [85-]

6   7   8   9   10

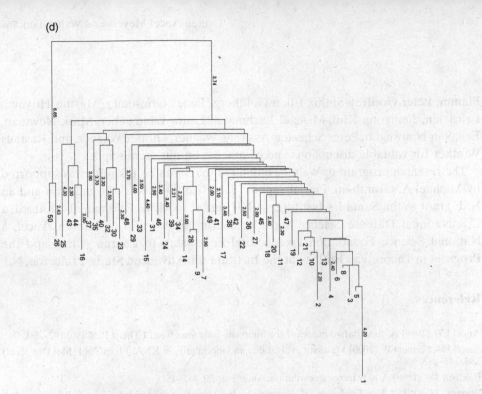

(d)

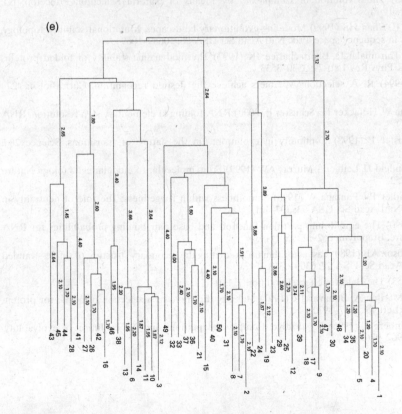

(e)

Flamm, Peter Godfrey-Smith, Ellen Goldberg, James Griesemer, Martijn Huynen, Erica Jen, Junhyong Kim, Michael Lachmann, Laura Landweber, Mark Newman, Erik van Nimwegen, Peter Schuster, Andreas Wagner, Günter Wagner, and Rasmus Winther for valuable discussions and comments on our manuscipts.

The research program of Walter Fontana at the Santa Fe Institute was supported by Michael A. Grantham. Partial support came from the Keck Foundation and an NSF grant to the Santa Fe Institute, an NIH grant to M. W. Feldman at Stanford, a U.S. National Defense Science and Engineering Fellowship to Lauren Ancel, a National Science Foundation Postdoctoral Fellowship to Lauren Ancel, and the Program in Theoretical Biology at the Institute for Advanced Study, Princeton, N.J.

References

Ancel LW (1999) A quantitative model of the Simpson–Baldwin effect. J Theor Biol 196: 197–209.

Ancel LW, Fontana W (2000) Plasticity, evolvability and modularity in RNA. J Exp Zool (Mol Dev Evol) 288: 242–283.

Baldwin JM (1896) A new factor in evolution. Amer Nat 30: 441–451.

Bonner JT (1988) The Evolution of Complexity by Means of Natural Selection. Princeton, NJ: Princeton University Press.

Bornberg-Bauer E, Chan HS (1999) Modeling evolutionary landscapes: Mutational stability, topology, and superfunnels in sequence space. Proc Natl Acad Sci USA 96: 10689–10694.

Bussemaker HJ, Thirumalai D, Bhattacharjee JK (1997) Thermodynamic stability of folded proteins against mutations. Phys Rev Lett 79: 3530–3533.

Ellington AD (1994) RNA selection: Aptamers achieve the desired recognition. Curr Biol 4: 427–429.

Flamm C, Fontana W, Hofacker IL, Schuster P (2000) RNA folding at elementary step resolution. RNA 6: 325–338.

Fontana W, Schuster P (1998) Continuity in evolution: On the nature of transitions. Science 280: 1451–1455.

Hartwell LH, Hopfield JJ, Leibler S, Murray AW (1999) From molecular to modular cell biology. Nature 402 supp.: C47–C52.

Huynen MA, Stadler PF, Fontana W (1996) Smoothness within ruggedness: The role of neutrality in adaptation. Proc Natl Acad Sci USA 93: 397–401.

McCaskill JS (1990) The equilibrium partition function and base pair binding probabilities for RNA secondary structure. Biopolymers 29: 1105–1119.

Nussinov R, Jacobson AB (1980) Fast algorithm for predicting the secondary structure of single-stranded RNA. Proc Natl Acad Sci USA 77: 6309–6313.

Simpson GG (1953) The Baldwin effect. Evolution 7: 110–117.

Vendruscolo M, Maritan A, Banavar JR (1997) Stability threshold as a selection principle for protein design. Phys Rev Lett 78: 3967–3970.

Wagner GP, Altenberg L (1996) Perspective: Complex adaptations and the evolution of evolvability. Evolution 50(3): 967–976.

Wagner A, Stadler PF (1999) Viral RNA and evolved mutational robustness. J Exp Zool (Mol Dev Evol) 285: 119–127.

Waterman MS (1978) Secondary Structure of Single-stranded Nucleic Acids. New York: Academic Press.

Westhof E, Masquida B, Jaeger L (1996) RNA tectonics: Towards RNA design. Folding and Design 1: R78–R88.

Wuchty S, Fontana W, Hofacker IL, Schuster P (1999) Complete suboptimal folding of RNA and the stability of secondary structures. Biopolymers 49: 145–165.

Zuker M, Stiegler P (1981) Optimal computer folding of larger RNA sequences using thermodynamics and auxiliary information. Nucleic Acids Research 9: 133–148.

7 Amphibian Variations: The Role of Modules in Mosaic Evolution

Gerhard Schlosser

Modules as Units of Evolution: A Testable Hypothesis?

Evolutionary changes of characters do not occur in a haphazard way. Some characters tend to change simultaneously and coordinatedly, whereas others tend to be modified largely independent from one another. Therefore the question arises, which factors determine whether some characters belong to the same or to different units of evolutionary changes (units of evolution). An answer is suggested by the observation that the organization of organisms is often modular (i.e., it consists of different character complexes, each of which develops and/or functions in an interdependent fashion but largely autonomous from other character complexes. Such modules are promising candidates for units of evolution, because characters that are strongly coupled, developmentally or functionally exclusively among themselves are likely to strongly constrain each other's evolution. In this chapter I will first characterize the concepts of "unit of evolution" and "module" more precisely, and suggest an approach to test the hypothesis that certain types of modules act as units of evolution. This approach will then be illustrated with several examples from amphibians.

Repeated Dissociated Coevolution and the Units of Evolution

It has long been known that evolution proceeds in a mosaic fashion, modifying some characters but not others in a certain lineage (Gould, 1977; Raff, 1996; Shubin, 1998). Moreover, mosaic evolution often exhibits trends where several characters tend to coevolve repeatedly while being easily dissociated from their context. "Dissociation" will here be used in a general sense to indicate evolutionary changes of one suite of characters relative to another (Needham, 1933; Raff and Wray, 1989; Raff et al., 1990; Raff, 1996). Dissociation comprises various kinds of evolutionary changes, such as: (1) loss of one suite of characters but not the other; (2) shift in timing of development of one suite versus the other (heterochrony); (3) shift in location of one suite versus the other (heterotopy); (4) redeployment of one suite in additional contexts; and (5) other kinds of correlated character changes (in size, shape etc.) in one suite versus the other. Such patterns of dissociation may affect all phases of development, possibly leading to the rearrangement of developmental events or "ontogenetic repatterning" (Roth and Wake, 1985; Wake and Roth, 1989), and may result from several different kinds of processes (for reviews see de Beer, 1958; Gould, 1977; McKinney and McNamara, 1991; Hall, 1992; Raff, 1996; Arthur, 1997).

Phylogenetic trends in the coevolution of dissociated characters can occasionally be due to the recurrence of combinations of environmental selection pressures acting on several characters independently (Maynard Smith et al., 1985; D. B. Wake, 1991). However, repeated dissociated coevolution may also occur without correlated environmental changes, suggesting that adaptation to a similar environment can be ruled out (e.g., Alberch, 1980, 1983; Maynard Smith et al., 1985; D. B. Wake, 1991; Shubin and Wake, 1996; Wagner and Schwenk, 2000). Such patterns are then more likely explained either by the fact that characters are not able to vary independently from each other or by the fact that characters reciprocally influence their respective fitness values (i.e., show epistasis for fitness). In either case, the interdependent suite of characters will act as a "building block" (Wagner, 1995; Wagner and Altenberg, 1996) or "unit of evolution" (Schlosser, 2002),[1] for which a high degree of both coordinated evolution and context dissociation is to be expected (i.e., which has a high coevolution probability). It should be noted that a character complex can act as a unit of evolution, in the sense intended here, even when it never acts as a unit of selection in actual selection processes.[2]

Modules May Act as Units of Evolution

What is the basis for the nonindependence of variability and the interdependence of fitness values underlying the recurrent evolution of several characters as a unit? In general, one might predict that one important determinant of the constraints that characters impose on each other's evolution should be their degree of developmental and functional coupling. Characters are functionally coupled when they are collectively required to perform a function (e.g., perform a vital behavior, develop an organ, etc.), whereas they are developmentally coupled when their development involves a common causal factor, such as a common inducer (see, e.g., Roth and Wake, 1989; Schlosser, 1998).

Developmental coupling often prevents characters from varying independently from each other. This imposes certain limits on the generation of possible phenotypes, often referred to as "developmental constraints" (see, e.g., Gould and Lewontin, 1979; Alberch, 1980, 1982; Maynard Smith et al., 1985; D. B. Wake and Larson, 1987; Schwenk, 1994; Shubin and Wake, 1996; Wagner and Schwenk, 2000). Functional couplings, on the other hand, imply the reciprocal fitness dependence (fitness epistasis) between the characters, resulting in "functional constraints." These constraints may lead to "internal selection," a phenomenon in which the most important selection pressures are due to the other characters of the complex rather than to the external environment (reviewed in Schwenk, 1994; Arthur, 1997; Wagner and Schwenk, 2000). While developmental and functional couplings may be clearly distinguished, there will often be (at least in theory) reciprocal interactions among characters of organisms, which will therefore be connected developmentally as well as functionally.[3]

More and more evidence is accumulating that the network of developmental and functional couplings in organisms is organized in a modular fashion (e.g., Wagner, 1995, 1996; Wagner and Altenberg, 1996; García-Bellido, 1996; Raff, 1996; Gilbert et al., 1996; Gerhart and Kirschner, 1997; Hartwell et al., 1999; von Dassow and Munro, 1999; Gilbert, 1998, 2000; Schlosser and Thieffry, 2000; Schlosser, 2004). Modules can be characterized as subnetworks (or subprocesses), which are highly integrated developmentally and/or functionally but which develop and function relatively independently from other subnetworks of the organism (Schlosser, 2002, 2004). In other words, modules are integrated and relatively context-insensitive units of development and/or function (figure 7.1).

For instance, limb buds continue to develop normally even after transplantation to an ectopic site (Harrison, 1918; Hinchliffe and Johnson, 1980). This indicates that the development of limb buds proceeds independently from their surrounding tissues (for a limited time period). Many gene regulatory networks or signal transduction pathways also seem to operate in an integrated and relatively autonomous fashion, judging by their repeated deployment during development. Interactions between the receptor Notch, its ligand Delta and several other proteins, for example, are involved in regulating cell fate decisions in a wide variety of structures of vertebrates, including the central nervous system, the ear, somites, and pronephros, to name only a few (reviewed in Gerhart and Kirschner, 1997; Gilbert, 2000).

Thus, it is tempting to speculate that the modular organization of organisms will be reflected in the coevolution probabilities of their characters and that modules not only constitute units of development and/or function but also act as units of evolution (Wagner, 1995, 1996; Wagner and Altenberg, 1996; Brandon, 1999; Schlosser, 2002, 2004). This hypothesis is supported by an increasing body of evidence, largely from developmental genetics, that indicates that modules, such as limbs (Shubin et al., 1997; Tabin et al., 1999; Ng et al., 1999) and other complex organ primordia; signaling cascades initiated by Wnt-, hedgehog-, TGF-β-, or Delta-Notch proteins (reviewed in Gerhart and Kirschner, 1997; Gerhart, 1999; Gilbert, 2000); and regulatory networks involving Hox- and Pax-transcription factors (e.g., Slack et al., 1993; Noll, 1993; Gellon and McGinnis, 1998; Heanue et al., 1999; Relaix and Buckingham, 1999) have been conserved in phylogeny in a broad range of metazoan taxa, although they often have acquired new roles and may operate in different developmental contexts in different lineages.

However, it is important to point out that modules as defined here are likely to act as units of evolution only when certain additional conditions are met (Schlosser, 2002, 2004). For example, explantation of the vertebrate limb bud allows relatively normal limb development and has little effect on trunk development. In this case, the module is not only context-insensitive but also largely dispensable for the development and/or function of its normal surroundings (as will often be the case in systems with an overall modular organization). This will, however, not always be true. For example, the vertebrate dorsal blastopore lip ("organizer") is able to develop normally into the notochord even at ectopic sites (context insensitivity), but its explantation disrupts many inductive processes required for normal development of surrounding tissues (indispensability). In such cases, there will be additional constraints on evolutionary change transcending module boundaries. Therefore, modules are promising candidates for units of evolution only when they are relatively insensitive to and also dispensable for the context (figure 7.1).

Testing Claims About Modules as Units of Evolution

A rigorous test of the hypothesis that certain types of modules (those that are both insensitive to and dispensable for their surroundings) tend to act as units of evolution for a given suite of characters in a certain lineage, however, requires a detailed analysis in two steps (see also Müller, 1991). First, it has to firmly establish that the suite of characters acts as a unit of evolution, and second, it has to be shown that it indeed constitutes such a module.

In the first step (figure 7.1A), comparative phylogenetic analysis is required to show that all characters of the suite repeatedly evolve in a coordinated fashion, but dissociated from other characters, and that they do so more frequently than they are disrupted (disruption would occur in case of dissociated coevolution of only some characters of the suite with characters not belonging to the suite). Only a *recurrent* pattern of dissociated coevolution—reflected in phylogenetic homoplasy, in particular parallelism—can establish that certain characters indeed are *likely* to coevolve (see also Alberch, 1980; Alberch and Gale, 1985; D. B. Wake and Larson, 1987; D. B. Wake 1991; Wray and Bely, 1994; Shubin and Wake, 1996), while only the recurrence of *coevolution* and *dissociation* events can establish which characters do and which do not belong to the unit of evolution. Neither overall evolutionary stasis nor unique dissociation events allow the identification of units of evolution because they are compatible with many different evolutionary scenarios. An "evolutionarily stable configuration" of characters in the sense of Wagner and Schwenk (2000), for instance, will be identifiable as a unit of evolution only if it was evolutionarily stable

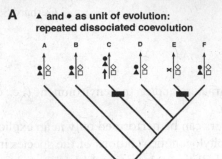

A ▲ and ● as unit of evolution:
repeated dissociated coevolution

B ▲ and ● as module:
(1) integration and (2) context insensitivity;
in some cases also (3) dispensability

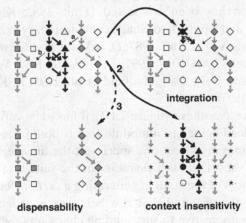

Figure 7.1
Hypotheses that a suite of characters (e.g., triangle and circle) acts as a module as well as a unit of evolution in a certain lineage are testable. (*A*) Mapping sequences of development of characters (symbols next to long arrows) on a phylogenetic tree of species A–F may reveal frequent dissociated coevolution of characters (i.e., a higher frequency of dissociated coevolution than disruption). In the case depicted, out-group comparison allows inference of two independent events (black bars) of dissociated coevolution of circle and triangle: a coordinated heterochronic shift (short arrow) relative to other characters in C, and a coordinated loss (cross) in E, while there have been no events disrupting the unit of circle and triangle (e.g., no events of dissociated coevolution of circle and square). (*B*) In order to establish that circle and triangle form a module, it has to be established that they generate a certain behavior (input–output relation) in (1) an integrated and (2) a relatively context-insensitive way (e.g., because they are relatively strongly coupled developmentally or functionally among each other [black arrows], but only relatively weakly influenced [dashed arrow a] by the context in which they act [squares, rhombs]. The integration (1) of circle and triangle is supported when experimental manipulations of one character (e.g., circle) affect the behavior of the suite via coperturbation (or modification of the effect of independent perturbations) of other characters of the suite (triangle). The context independence (2) of the interactions between circle and triangle is supported when they behave normally (i.e., their interactions produce the same input–output relation) after perturbation of the context (e.g., after transplantation of the suite of characters into a new context [asterisks]). Such experimental manipulation may also reveal that circle and triangle are largely dispensable (3) for their original context (e.g., due to weak influences on the latter; dashed arrow b), because the latter behaves normally after their removal. While dispensability is not part of the definition of a module, it is important for making modules likely candidates for units of evolution. The sequence of rows of symbols in (*B*) represents the same characters at subsequent steps in time; filled symbols indicate states of characters that are necessary for certain causal transitions (arrows).

despite repeated changes in other characters and/or the environment (i.e., despite repeated dissociation events).

The comparative analysis of characters can be performed only in an explicit phylogenetic framework. Only when the phylogenetic relations of the species in question are already established by prior cladistic analysis (see Eldredge and Cracraft, 1980; Wiley, 1981) can patterns of character coevolution be investigated by mapping character states on a given phylogeny. Dissociated coevolution of different suites of characters can then be detected by a variety of comparative methods (e.g., Felsenstein, 1985; Maddison, 1990; Harvey and Pagel, 1991; Pagel, 1999). In particular, out-group comparison has been advocated (Fink, 1982; Kluge and Strauss, 1985; Northcutt, 1990) and applied to analyze evolutionary changes in development of a wide range of taxa (e.g., Raff, 1987; D. B. Wake and Larson, 1987; Jeffery and Swalla, 1990; Northcutt, 1990, 1992, 1997; D. B. Wake, 1991; Wray and Raff, 1991; Mabee, 1993; Wray and Bely, 1994; Hadfield et al., 1995; K. K. Smith, 1996, 1997; Schlosser and Roth, 1997b).

In the second step of testing the hypothesis (figure 7.1B), it has to be verified that all characters of the suite do indeed belong to a module that is both insensitive to and dispensable for its surroundings in at least one species of the lineage. In order to support this, it is necessary to show that each character of the suite is (1) indeed integrated with the other characters of the suite in generating a certain "behavior" (i.e., a particular input–output relation), and that this behavior is at least for a limited time period (2) relatively insensitive to surrounding characters not belonging to the suite; and (3) largely dispensable for the behavior of the latter.

The first requirement can be met by showing that at least some perturbations of one character of the suite affect the behavior of the suite via coperturbation (or modification of the effect of independent perturbations) of other characters of the suite. The second and third requirements can be met by demonstrating experimentally that the suite of characters displays a relatively normal behavior in isolation, ectopically (e.g., after transplantations to different tissues) or after various kinds of perturbations of its normal surroundings (context insensitivity), while the development and functional performance of the original context is relatively unaffected by their removal (dispensability). Alternatively, the context insensitivity of interactions can be supported by showing that a similar pattern of interaction among characters is instantiated repeatedly and in different developmental or functional contexts in the species analyzed, for instance, by cluster analysis of coexpressed genes or similar methods (Wen et al., 1998; Eisen et al., 1998; Niehrs and Pollet, 1999).

In the following sections, the evolution of amphibian development will serve as a paradigm to illustrate this two-step approach. Amphibians are particularly well suited for such a study for two reasons. First, their development is particularly well studied because they have long served as model organisms for developmental biology. Second, they exhibit a variety of life history modes, sometimes with some dramatic differences in development.

Amphibian Life History Evolution

The phylogenetic relations among extant amphibians are depicted in figure 7.2. Most analyses suggest that all living amphibians belong to a monophyletic taxon, the Lissamphibia, and that caecilians, urodeles, and anurans each form a monophyletic taxon within Lissamphibia (Duellman and Trueb, 1986; Hillis, 1991; Bolt, 1991; Trueb and Cloutier, 1991; Cannatella and Hillis, 1993; Hay et al., 1995), but the relation among the three groups is unclear. The phylogeny of Ford and Cannatella (1993) for anurans and of Larson and Dimmick (1993) for salamanders is adopted here. While frog phylogeny in particular remains controversial at present, alternative hypotheses about anuran relations (e.g., Duellman and Trueb, 1986; Hillis et al., 1993; Hay et al., 1995) will not affect the main arguments of this chapter.

From phylogenetic analyses it can be inferred that ancestral amphibians had a biphasic life history, with free-living larval and adult stages separated by some sort of metamorphosis, which was probably mediated by thyroid hormones (Szarski, 1957; Fritzsch, 1990; Brown, 1997). In anurans, metamorphosis has become much more pronounced with increasing specialization of both the tadpole and the adult (Orton, 1953; Wassersug and Hoff, 1982; Alberch, 1987, 1989; Fritzsch 1990). The ancestral biphasic condition has been repeatedly modified in all amphibian lineages (for reviews see Noble, 1927; Lutz, 1947; Orton, 1951; Lynn, 1961; Dent, 1968; Salthe and Mecham, 1974; Duellman and Trueb, 1986; M. H. Wake, 1989; Hanken, 1989a, 1992; Duellman, 1992; Wakahara, 1996a; D. B. Wake and Hanken, 1996; Callery et al., 2001). For instance, many salamanders never undergo metamorphosis and retain most larval characters when they mature sexually, a phenomenon known as neoteny (see below).[4] On the other hand, a free-living larval stage has been frequently abolished in taxa with direct development, which has evolved repeatedly in all three orders. Reversals from direct development to a biphasic life history have probably also occurred (Duellman et al., 1988; Titus and Larson, 1996).

While neoteny is thought to have evolved in response to unfavorable terrestrial environments (Wilbur and Collins, 1973; Duellman and Trueb, 1986), direct development was presumably favored under the reverse condition, where aquatic environments were hostile—for instance, due to predation pressure or desiccation (Lutz,

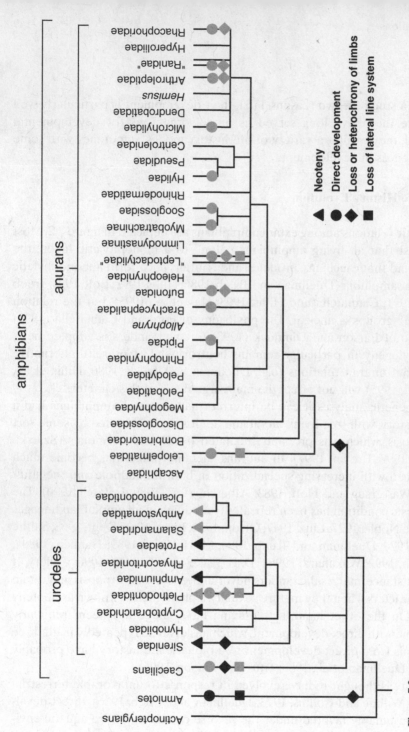

Figure 7.2

Phylogeny of amphibians based on Ford and Cannatella (1993) for anurans and Larson and Dimmick (1993) for urodeles. Symbols indicate alterations of life history (direct development and neoteny) as well as loss of limbs (caecilians, Sirenidae), shifts in the timing of limb development, and complete loss of the lateral line system. Black symbols indicate that the character change is typical for all species of a taxon, while gray symbols indicate character changes in only some of the species of a taxon. Direct development and neoteny evolved multiple times, but the number of independent events is difficult to determine for neoteny. There are insufficient data on limb and lateral line development for most taxa, so the numbers of independent changes depicted have to be regarded as minimum estimates. Only large heterochronic shifts of limb development are included; it is assumed that the ancestral tetrapod condition involved development of the hindlimbs well after forelimbs. This condition is retained in urodeles, but is altered in amniotes (simultaneous, early embryonic development of forelimbs and hindlimbs) and anurans (almost simultaneous, but postembryonic development of forelimbs and hindlimbs). In urodeles, simultaneous development of forelimbs and hindlimbs evolved secondarily in some plethodontids. In anurans, limb development during embryonic stages (prior to differentiation of cranial cartilages) evolved several times from the ancestral anuran condition. Data are based on the following sources: *direct development:* Duellman and Trueb (1986), M. H. Wake (1989), Duellman (1992), D. B. Wake and Hanken (1996); *neoteny:* Dent (1968), Wakahara (1996a), Shaffer and Voss (1996); *limb loss and heterochrony:* Warren (1922), de Villiers (1929), Orton (1949), Townsend and Stewart (1985), Duellman and Trueb (1986), Callery

1948; Lynn, 1961; Wilbur and Collins, 1973; Wassersug, 1974; Heyer et al., 1975; Callery et al., 2001). Sufficient nutrients for a protracted nonfeeding period appear to have been a necessary precondition for the evolution of direct development (Lutz, 1948; Salthe and Duellman, 1973; Duellman and Trueb, 1986; Elinson, 1987; D. B. Wake and Hanken, 1996; Callery et al., 2001), since direct-developing species are found only in taxa that have relatively large and yolky eggs or nurse the developing eggs in special skin pouches on their back (the reverse, however, is not true: many taxa with large eggs develop biphasically).

The diversity of amphibian life histories and their well-studied development provides many opportunities to investigate the relation between modules and units of evolution. The following paragraphs will discuss several examples, ranging from entire life history stages to organ primordia to cell types in order to show that such units can be identified at many different levels. In a first step, candidates for units of evolution will be identified by their repeated dissociated coevolution in amphibian phylogeny, and in a second step it will be explored whether these units of evolution do, in fact, correspond to known modules of development and/or function in amphibians. As a note of caution, I should add that while the following examples are well suited to illustrate the general approach, in most cases we still lack sufficient comparative or experimental evidence, so that modules or units of evolution can often be identified only tentatively.

Life History Stages as Modules and Units of Evolution

Metamorphosis as Module and Unit of Evolution in Urodeles

Neoteny in salamanders has evolved many times independently (figure 7.2); exactly how often is at present difficult to determine due to the repeated evolution of neoteny even within single genera such as *Ambystoma*. Elegant crossing experiments have shown that neoteny in the axolotl evolved mainly due to alterations of few genetic loci, suggesting that neoteny is easily generated during evolution. However, the genetic basis may differ among populations (see Voss, 1995; Shaffer and Voss, 1996; Voss and Shaffer, 1997, 2000). Moreover, not all cases of neoteny are likely to involve exactly the same kind of genetic change, because different types of neoteny appear to be due to alterations of different developmental processes. Whereas some taxa, such as several species of *Ambystoma* and *Triturus*, are neotenic only under certain environmental conditions (facultative neoteny), others show obligatory neoteny. Among the latter, metamorphosis is inducible by thyroid hormone treatments in some species (e.g., *Ambystoma mexicanum*), but not in others (e.g., *Siren*, *Proteus*, *Necturus*) (for reviews see Lynn, 1961; Dent, 1968; Wakahara, 1996a; Shaffer and Voss, 1996; Rosenkilde and Ussing, 1996). While failure of thyroid hormone receptor autoinduction may underlie neoteny in *Necturus*, a noninducible obligate neotenic salamander, inducible and facultatively neotenic species seem to be rather deficient in thyroid hormone production or action (Yaoita and Brown, 1990; Tata et al., 1993; Shaffer and Voss, 1996; Safi et al., 1997; Rosenkilde and Ussing, 1996).

Despite this variability in underlying mechanism, all cases of neoteny appear to involve some kind of disruption of the thyroid axis and the concurrent loss of thyroid hormone (TH)-dependent metamorphosis. Therefore, they present an exquisite example for the recurrent coordinated loss of an entire suite of characters, in this case all or many developmental events dependent on TH for metamorphic reorganization (reviewed, e.g., in Shi et al., 1996; Tata, 1996, 1998; Su et al., 1999). These include a wide variety of processes, such as cell death and development of new cell types in gut and epidermis; remodeling of cranial cartilages, muscles, and parts of the nervous system; and changes in globin types and liver enzymes (reviewed in Dodd and Dodd, 1976; Fox, 1981; Duellman and Trueb, 1986).

Only a few metamorphic changes, for example, globin transition, have been reported to occur in neotenic axolotls (Ducibella, 1974) as well as in *Hynobius retardatus* after blocking thyroid function (Wakahara and Yamaguchi, 1996), presumably because they require only low doses of TH. In the neotenic Cryptobranchidae and Amphiumidae, on the other hand, failure to metamorphose appears to be restricted to a few tissues. This indicates that changes in TH levels, besides promoting the dissociation of TH-dependent events from TH-independent events, may also lead to the dissociation among various suites of metamorphic events, because these may differ in their sensitivity or response characteristics to certain hormone levels (see also Hanken and Summers, 1988; Hanken et al., 1989; Schmidt and Böger, 1993; Rose, 1995c, 1996; Shaffer and Voss, 1996).

The repeated dissociated coevolution of various TH-dependent metamorphic events in neotenic urodeles indicates that they jointly constitute a unit of evolution in the sense defined above. But can TH-dependent metamorphosis also be regarded as a module of development or function? This suggestion may seem unusual at first, because modules are often thought of as well-circumscribed spatial units. However, modules are defined only as integrated and autonomous patterns of interaction, which do not need to be locally confined. Though metamorphic events occur in different locations throughout the body, they are all developmentally integrated by their joint dependence on thyroid hormones. Gonadal maturation is notably excluded from this suite of TH-dependent characters in urodeles and can proceed even when thyroid hormones are absent and TH-dependent metamorphic changes do not take place (e.g., Dodd and Dodd, 1976; Wakahara, 1996a; Hayes, 1997). In contrast, gonadal maturation (and other aspects of sex differentiation) in anurans is TH-dependent, which may explain the absence of neoteny in anurans (Hayes, 1997).

Metamorphic changes can be precociously initiated by TH injections at larval stages of urodeles and anurans, or delayed if not prevented by blocking endogenous TH production (reviewed in Dodd and Dodd, 1976; White and Nicoll, 1981; for urodeles see also Rose, 1995b, 1995c, 1996; Schmidt and Roth, 1996; Wakahara and Yamaguchi, 1996). This suggests that metamorphosis is relatively independent from (and dispensable for) the particular cellular environment in which it usually occurs, and does not depend on an exact temporal coordination with other developmental events (including gonadal maturation in urodeles). Increasing knowledge about the downstream targets of TH (Shi, 1996; Tata, 1998; Su et al., 1999) will allow us to assess whether the effects of TH during metamorphosis involve the same downstream genes and similar cellular processes as TH-induced changes later in development. Such a repeated employment of TH-mediated events during development would further support their relatively context-independent regulation.

Do Amphibian Larvae Act as Modules or Units of Evolution?

Whereas neotenic urodeles have simplified the ancestrally biphasic life history by retaining only the larva stage, direct-developing amphibians have secondarily lost a free-living larva. Direct development has evolved independently many times in amphibians: at least once in caecilians (M. H. Wake, 1989); one to five times in plethodontid salamanders (D. B. Wake and Hanken, 1996; Titus and Larson, 1996); and probably 12 times in frogs (Duellman and Trueb, 1986; Duellman, 1992). Do these multiple losses of a larva point to the entire larval stage as a unit of evolution, in analogy to the case of neotenic urodeles, where multiple losses of TH-dependent metamorphosis suggested its role as unit of evolution? Upon closer scrutiny, the answer is likely to be negative in this case. Although none of the direct-developing taxa have a free-living larva, most taxa recapitulate at least some larval structures within the egg. Moreover, the sparse data available suggest that the degree to which larval structures are developed varies greatly among different taxa. Among anurans, for example, direct developing species of *Pipa* (Pipidae) and *Nectophrynoides* (Bufonidae) pass through largely normal larval stages, whereas species of *Leiopelma* (Leiopelmatidae) and *Philautus* (Rhacophoridae) have lost many larva-typical structures (Noble, 1927; Stephenson, 1951; Orton, 1951; Lynn, 1961; Dent, 1968; M. H. Wake, 1980; Patil and Kanamadi, 1997).

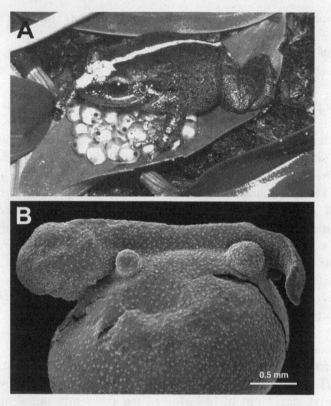

Figure 7.3
In the direct-developing frog *Eleutherodactylus coqui*, development is dramatically abbreviated and repatterned. (*A*) *E. coqui* male guarding a clutch of large, yolk-rich eggs. (*B*) Early *E. coqui* embryo (stage 4 of Townsend and Stewart, 1985) on top of the huge yolk sac: many structures typical for tadpoles (cement gland, mouthparts, etc.) are absent, but limb buds have already formed.

The most dramatic reduction of development, however, is found in the very speciose genus *Eleutherodactylus* (Leptodactylidae) (figure 7.3). Though a few larval characters such as Rohon Beard sensory neurons are transitorily present in *Eleutherodactylus*, most larval specializations of the integument and the circulatory, digestive, musculoskeletal, and nervous systems never develop, while limbs and other characters typical of postmetamorphic frogs develop precociously and sometimes in a very different temporal order than in most biphasically developing frogs (Lynn, 1942; Gitlin, 1944; Lynn and Lutz, 1946; Goin, 1947; Hughes, 1959a, 1959b; 1962, 1965a, 1965b; Adamson et al., 1960; Chibon, 1960; Townsend and Stewart, 1985; Elinson, 1990, 1994; Hanken et al., 1992; Hanken, Jennings, et al., 1997; Hanken,

Klymkowsky, et al., 1997; Callery and Elinson, 1996, 2000a, 2000b; Fang and Elinson, 1996, 1999; Richardson et al., 1998; Schlosser and Roth, 1997b, 1997c; Jennings and Hanken, 1998; Schlosser et al., 1999; Schlosser and Kintner, 1999; Callery et al., 2001; Schlosser, 2001). Likewise, the degree of larval losses and "ontogenetic repatterning" varies among different direct-developing salamanders, with bolitoglossine plethodontids showing the most drastic alterations (reviewed in D. B. Wake and Hanken, 1996).

The variability of ontogenies in direct-developing amphibians provides little evidence that the larva as a whole acted as a unit of amphibian evolution. Similarly, there is little to support that the larva as a whole acts as a module of development or function, although the "compartmentalization" of larval and adult cell populations in some amphibians might at a first glance invite such a view. Alberch (1987, 1989) has argued that with the evolution of increasing divergence between larva and adult in frogs and several salamanders, a compartmentalization into larval and adult cell populations was inevitable, because only minor metamorphic remodeling could be accomplished without it. Indeed, there is evidence for compartmentalization in many amphibian tissues, such as the epidermis, the digestive tract, and parts of the skeleton and muscles, where larval cell populations degenerate at metamorphosis and are replaced by a newly differentiating population of adult cells (e.g., de Jongh, 1968; Alberch and Gale, 1986; Alberch 1987, 1989; Alley, 1989, 1990; Yoshizato, 1992; Nishikawa and Hayashi, 1994, 1995; Rose, 1995a; Hourdry et al., 1996; Ishizuya-Oka, 1996; Schlosser and Roth, 1997a; Shi et al., 1998; Su et al., 1999). In addition, some larval structures completely degenerate at metamorphosis, such as the tail, the lateral line system (in anurans), and larva-specific muscles and neurons (e.g., Wahnschaffe et al., 1987; Lamborghini, 1987; Schlosser and Roth, 1997a; Berry, Schwartzman, et al., 1998), whereas several structures typical for adults, including many components of the middle ear, teeth (in anurans), and ipsilateral retinotectal projections begin to differentiate only at metamorphosis (e.g., Hoskins and Grobstein, 1985; Lumsden, 1987; Hetherington, 1988).

But despite this abundance of compartmentalization, the amphibian larva does not seem to act as a module for two reasons. First, compartmentalization is not complete and not all differentiated larval cells are replaced by different cell populations during metamorphosis. Many tissues that exhibit distinct larval specializations, including large parts of the skeleton and the nervous system, are merely remodeled during metamorphosis (e.g., de Jongh, 1968; Kollros, 1981; Alley and Barnes, 1983; Barnes and Alley, 1983; Alberch, 1987; Rose 1995a; Omerza and Alley, 1992; Alley and Omerza, 1998; Berry, Rose, et al., 1998). Second, although subpopulations of larval cells are definitely coupled among themselves developmentally or

functionally, there seems to be no particular developmental or functional integration encompassing all exclusively larval cells to the exclusion of cells that survive
into adulthood.

The larva, therefore, appears to be a good candidate neither for a unit of amphibian evolution nor for a module in amphibian development. However, the repeated
occurrence of direct development might be explained via the precocious activation
of TH-dependent metamorphosis, whose role as both a unit of evolution and a
module of development has already been substantiated.

Can Direct Development be Explained by Precocious Metamorphosis?

This has in fact been proposed by several authors (Lynn, 1942, 1961; Matsuda, 1987;
Rose, 1996; Hanken, Jennings, et al., 1997; Jennings and Hanken, 1998; Callery et al.,
2001). It is an attractive hypothesis because experimental elevations of TH levels
during early larval stages are known to cause both the precocious loss of many larval
characters and heterochronic shifts in the development of some adult characters relative to others (Moser, 1950; Hanken and Hall, 1988; Hanken and Summers, 1988;
Hanken et al., 1989; Rose, 1995b, 1995c, 1996; Schmidt and Böger, 1993; Schmidt and
Roth, 1996; Shi et al., 1996). Consequently, a simple alteration in hormone levels
may possibly account for many or most developmental alterations observed in direct
developers such as *Eleutherodactylus* or bolitoglossine salamanders.

In support of this hypothesis, early differentiation of the thyroid axis, including
precocious maturation of the thyroid gland and elevation of TH levels, has been
reported for *Eleutherodactylus* (Lynn, 1936; Jennings, 1997; Hanken, Jennings, et al.,
1997; Jennings and Hanken, 1998) and thyroid hormone receptors are already
present from early embryonic stages on (from stage 4 of Townsend and Stewart,
1985) (Jennings, 1997; Callery and Elinson, 2000a). Moreover, there is evidence for
TH dependence of developmental events in *Eleutherodactylus coqui* from around
stage 12 on, immediately following thyroid maturation (Lynn, 1948; Lynn and
Peadon, 1955; Hughes, 1966; Elinson, 1994; Callery and Elinson, 1996; 2000a). Taken
together, this suggests that precocious thyroid maturation may account for the
fact that several events associated with metamorphosis in biphasically developing
anurans, such as the remodeling of cranial skeleton and muscles (de Jongh, 1968;
Hanken et al., 1992; Hanken, Klymkowsky, et al., 1997; Schlosser and Roth, 1997a),
occur relatively earlier in *E. coqui* than in most biphasic frogs.

However, this model cannot explain many other modifications of development in
E. coqui that are already apparent much earlier in development, including the loss
of many larval structures such as the cement gland, lateral line placodes; larval
mouthparts; many larval cartilages, muscles, and nerves; early remodeling events of

the jaw skeleton and muscles (starting at stage 9); and the precocious development of limbs. Furthermore, early TH treatment of biphasically developing frogs or salamanders mimicks only some ontogenetic modifications of direct-developing species, as the model would predict, but not others (for all the examples above, see Lynn, 1942; Townsend and Stewart, 1985; Hanken and Hall, 1988; Elinson, 1990, 1994; Rose, 1995b; Schmidt and Roth, 1996; Fang and Elinson, 1996, 1999; Richardson et al., 1998; Hanken et al., 1989, 1992; Hanken, Klymkowsky, et al., 1997; Schlosser and Roth, 1997b; Schlosser et al., 1999).

While precocious thyroid maturation alone is clearly insufficient to explain the "excision of the larva" (Elinson, 1990; Callery and Elinson, 2000b; Callery et al., 2001) in *Eleutherodactylus*, it would be premature to conclude that many of its developmental events have been freed from thyroid control, as previously suggested (Lynn and Peadon, 1955; Lynn 1961; Dent, 1968; Hughes, 1966). The presence of maternal TH in the yolk (Jennings, 1997), together with the early onset of TH receptor expression in *E. coqui* (Jennings, 1997; Callery and Elinson, 2000a), is compatible with the possibility that altered patterns of early embryonic development in *E. coqui* may be due to precocious TH receptor expression. This hypothesis remains to be tested and needs to be reconciled with some conflicting observations, such as the apparent lack of TH responsiveness in early limb development of *Eleutherodactylus* (Lynn, 1948; Lynn and Peadon, 1955; Elinson, 1994).

In conclusion, while the repeated evolution of neoteny in urodeles can be explained by the fact that TH-dependent metamorphosis itself could easily be lost as an entire unit in evolution due to its modular nature, there is at present little evidence that the repeated evolution of direct development can be similarly explained. Neither the loss of the entire larva nor the precocious activation of TH-dependent metamorphosis seems to be able to fully account for the mosaicism of developmental alterations and its phylogenetic diversity in direct-developing amphibians. The next sections will show, however, that the dramatic patterns of dissociations observed in some direct developers are nonetheless very useful to investigate the role of modules (albeit at the smaller scale of organ primordia) as units of evolution.

Organ Primordia as Modules and Units of Evolution

Modules of Limb Development as Units of Evolution
Limbs have been reduced to varying degrees many times during vertebrate evolution, for instance, in whales, in snakes, and in other serpentiform reptiles (Lande,

1978; Raynaud, 1985). In many cases, limb buds form, but later regress or develop only into rudimentary structures. Several taxa, however, have completely lost fore-limbs and/or hindlimbs and never develop limb buds. These include snakes (no fore-limb buds; Raynaud, 1985; Cohn and Tickle, 1999), caecilians (neither forelimb nor hindlimb buds; Dünker et al., 2000) and sirenid salamanders (no hindlimbs, unclear if limb buds form; Duellman and Trueb, 1986).

Even more frequent than loss of the entire limb are heterochronic shifts of limb bud development in tetrapods (Richardson, 1995) and particularly in amphibians (Salthe and Mecham, 1974; Collazo and Marks, 1994; Elinson, 1994; Richardson et al., 1998). The most dramatic evidence for such heterochronic shifts comes from the direct-developing frog *E. coqui* (Townsend and Stewart, 1985; Elinson, 1990, 1994; Schlosser and Roth, 1997b; Carl et al., 1997; Richardson et al., 1998; Schlosser and Kintner, 1999). Late onset of limb development during larval stages and after com-pletion of the embryonic period of cranial morphogenesis and differentiation is typical for frogs and represents the ancestral anuran condition. In contrast, in *E. coqui*, limbs develop relatively much earlier and prior to the differentiation of many cranial structures, reminiscent of amniotes (e.g., Hamburger and Hamilton, 1951; Kaufman, 1992). Similar predisplacements of limb development have been described for other direct-developing frogs (figure 7.2) (Warren, 1922; Noble, 1927; Orton, 1949; de Villiers, 1929; Patil and Kanamadi, 1997). Whereas in frogs usually both limbs develop more or less simultaneously, in urodeles heterochronic shifts of limb development repeatedly have led to the temporal dissociation of forelimb and hindlimb development (Salthe and Mecham, 1974; Collazo and Marks, 1994).

The high degree of variability in the timing of limb development is compatible with several evolutionary scenarios, all of which involve multiple changes of the timing of limb development relative to other structures. Out-group comparison with the Australian lungfish *Neoceratodus* (Kemp, 1982, 1999) suggests that an onset of forelimb development at the end of embryogenesis, more or less simultaneous with the onset of pharyngeal cartilage differentiation, as observed in many urodeles (e.g., Bordzilovskaya et al., 1989; Northcutt and Brändle, 1995), may be closest to the ancestral tetrapod condition. Under this assumption, initiation of forelimb develop-ment may have been slightly predisplaced to earlier embryonic stages (well before pharyngeal cartilage differentiation) during the evolution of amniotes, but was greatly delayed into larval stages during the evolution of anurans (Schlosser, 2001). However, in several direct-developing frogs, forelimb development was shifted from larval to early embryonic stages, resembling the amniote condition (figure 7.2).

The retardation of hindlimb development relative to forelimb development in *Neoceratodus* (Kemp, 1982) suggests a slightly different scenario for the evolution of hindlimb development. Here, most urodeles and anurans may retain the ancestral condition of postembryonic hindlimb development (subsequent to pharyngeal cartilage differentiation), whereas amniotes and some direct-developing frogs and salamanders may have predisplaced hindlimb development secondarily.

The repeated losses and heterochronic shifts of limb development in evolution affect the entire limb but have little consequences for the development of other characters, indicating that the limb as a whole acts as a unit of evolution. This is further supported by the fact that the complex patterning system employed in the development of pectoral fins (forelimbs) has been redeployed for the development of pelvic fins (hindlimbs) at some early point in vertebrate evolution, resulting in the well-known serial homology of both types of paired appendages (Shubin et al., 1997; Coates and Cohn, 1998, 1999).

Besides acting as units of evolution, limbs are *the* classic example for a module, which not only has its distinct function but also develops in an integrated and autonomous fashion. Complex interactions between several patterning mechanisms intrinsic to the limb bud ensure the proper positioning of cartilages and other tissues along its proximodistal, anteroposterior and dorsoventral axes (reviewed in Johnson and Tabin, 1997; Tabin et al., 1999; Ng et al., 1999; Gilbert, 2000). As a consequence, limbs act as "morphogenetic fields" with highly regulatory properties. Normal limbs can form from parts of the normal limb primordium or from fused parts of different limb primordia (Harrison, 1918; Hinchliffe and Johnson, 1980; see also de Robertis et al., 1991; Gilbert et al., 1996).

Moreover, many classic experiments (e.g., Harrison, 1918; Hamburger, 1939) established that explanted limb buds can develop relatively normal, even when transplanted to ectopic sites (context insensitivity), while extirpation of limb buds does not greatly perturb trunk development (dispensability). A further indication of the context independence of limb development is that forelimbs and hindlimbs, despite their initiation at different axial levels with different microenvironments (for example, distinct *Hox* gene expression patterns) are known to employ very similar pattern formation mechanisms (reviewed in Duboule, 1992; Johnson and Tabin, 1997; Shubin et al., 1997; Tabin et al., 1999; Ng et al., 1999; Gilbert, 2000).

Modules of Lateral Line Development as Units of Evolution
Similar to limbs, the lateral line system has often been lost, particularly in direct-developing species that lack an aquatic larval stage and therefore have no need for sensory systems that function specifically in an aquatic environment. The lateral line system is a specialized sensory system comprising electrosensory (ampullary organs)

and mechanosensory (neuromasts) receptor organs and the sensory nerves that innervate them (Coombs et al., 1989). Caecilians and urodeles retain both sensory components, but in anurans the electroreceptive part of the system has been lost (Fritzsch, 1989). All cells of the receptor organs and the sensory ganglion cells of the lateral line nerves develop from a series of ectodermal placodes (thickenings), the lateral line placodes (Stone, 1922, 1933; Knouff, 1935; Northcutt, 1992, 1996; Northcutt et al., 1994, 1995; Northcutt and Brändle, 1995; Schlosser and Northcutt, 2000). A contribution of neural crest cells to lateral line neuromasts has also been reported (Collazo et al., 1994), but appears not to be necessary for normal neuromast development (Schlosser et al., 1999).

The lateral line system has varied greatly in amphibian evolution, and various parts of the lateral line system—such as a class of receptor organs, a subset of lateral line placodes, or individual lateral lines have—been lost in different lineages (reviewed in Fritzsch 1989; Northcutt 1989, 1992, 1997). Moreover, the entire lateral line system (comprising all receptor organs and nerves) has been repeatedly lost in many direct-developing amphibians, where it lacks functional importance, similar to the condition observed in amniotes (figure 7.2). This seems to be the case for some strictly terrestrial and viviparous caecilians (Fritzsch and Wake, 1986; Roth et al., 1993), plethodontid salamanders of the tribes Plethodontini and Bolitoglossini (D. B. Wake et al., 1987; Fritzsch, 1988; Roth et al., 1993) and frogs of the genera *Eleutherodactylus* (Lynn, 1942; Schlosser and Roth, 1997b; Schlosser et al., 1999) and probably *Leiopelma* (Stephenson, 1951).

There are possibly many more cases of complete loss in frogs, because for most direct-developing frogs the presence or absence of a lateral line system has not yet been reported. The repeated loss of the lateral line system seems to have occurred without concomitant loss or perturbation of adjacent cranial structures (such as other ectodermal placodes). This suggests that in this case the unit of evolution comprises all derivatives of lateral line placodes (plus possibly the target cells of the lateral line nerves in the central nervous system).

Further support for this suggestion comes from the observation that the metamorphic fate of the lateral line system has repeatedly changed during evolution. As a general rule, evolutionary changes in metamorphic fate affect receptor organs and nerves in a coordinated fashion, the only known exception being *S. salamandra*, which apparently retains some lateral line afferents after metamorphosis while probably losing all receptor organs. Several scenarios for the evolution of metamorphic fate are possible, the most parsimonious being that amphibians ancestrally retained lateral line receptor organs as well as nerves through metamorphosis, as most salamanders still do today. Several groups of salamanders, however, secondarily lose the entire lateral line system at metamorphosis, as do most frogs and caecilians. In a reversal to the ancestral condition, however, several lineages of caecilians and frogs (mostly frogs with aquatic adults) have reacquired persistence of the lateral line system through metamorphosis (see, for above details, Escher, 1925; Wahnschaffe et al., 1987; Fritzsch, 1988; 1990; Fritzsch and Wake, 1986; Fritzsch et al., 1987, 1988; Roth et al., 1993).

The lateral line system also acts as a module in development. All peripheral components (sensory receptors and nerves) of the lateral line system develop from the same set of lateral line placodes and are thus developmentally coupled. They are also functionally coupled because they all contribute to a common function, the sensory perception by the lateral line. Moreover, during the development of the lateral line, reciprocal interactions between different placodal cells are probably necessary for generating the different cell types of the receptor organs (as has been demonstrated for ear development; e.g., Haddon et al., 1998; Riley et al., 1999) and nerves, so that a network of both developmental and functional couplings is already operating during the embryonic development of the lateral line system.

Besides being tightly integrated, the lateral line system also develops and functions in a relatively autonomous fashion. Subsequent to their induction during gastrulation and neurulation, lateral line placodes develop relatively context independent. Even after transplantation to ectopic sites, they differentiate into lateral line ganglion cells (Stone, 1929a) and all the different cell types of lateral line receptor organs (Northcutt et al., 1995; Schlosser and Northcutt, 2001). The formation of lines (Harrison, 1904; Northcutt et al., 1995), the spacing of the receptor organs (S. C. Smith et al., 1990) and the onset of migration of the lateral line primordium (Stone, 1938) are all autonomous properties of lateral line placodes. Only the direction of migration and the polarity of receptor organs depend on external cues. Moreover, embryos develop relatively normally after extirpation of lateral line placodes (Stone, 1922; Northcutt et al., 1995; Schlosser and Northcutt, 2001), indicating that they are dispensable for the development of many other structures (including other placodes), with the known exception of pigment pattern formation (Parichy, 1996a, 1996b).

Due to the modular nature of lateral line development, the coordinated and specific loss of the entire lateral line system in evolution should be relatively easy to realize by relatively few and simple perturbations of lateral line placode induction in early embryonic development. This scenario is indeed supported by the situation observed in *Eleutherodactylus*, although for other taxa the developmental basis for lateral line loss is still unknown. In *E. coqui* (similar to amniotes) lateral line placodes specifically fail to form, while all other ectodermal placodes develop normally.

This failure is due to the loss of ectodermal competence to respond to inducing signals which are still present (Schlosser et al., 1999). A similar loss of competence underlies the loss of the cement gland in *E. coqui* (Fang and Elinson, 1996, 1999).

Cell Types as Modules and Units of Evolution

The examples of modules and units of evolution discussed so far could lead to the erroneous impression that only complex organ primordia or even entire life history stages are modular and likely to change as a unit in evolution. However, modular units can also exist at much smaller scales, including, for instance, gene regulatory networks and cell types, and these can also act as units of evolution (Schlosser, 2004). I want to briefly present three examples here. In each of those, a certain cell type acts as a unit of evolution, judged by the observation that the entire network of signaling molecules and transcription factors responsible for its differentiation has repeatedly been shifted heterotopically or heterochronically during evolution relative to other cell types. The ease of such evolutionary rearrangements of cell types may again be attributed to the fact that cell differentiation events are modules relatively insensitive to and dispensable for each other.

It is well known that most cells are "determined" (or "committed") to acquire a certain germ layer specific fate restriction (e.g., mesoderm) or to adopt a specific fate (e.g., muscle cell or neuron), well before they actually differentiate (Slack, 1991). This implies that the complex cascades or networks of events underlying differentiation of different cell types are able to proceed more or less independent from each other (and even when cells are transplanted to ectopic sites) after cell types have been determined—for instance, by one or multiple steps of stable selector gene activation (for reviews see Weintraub, 1993; Anderson, 1997; Guillemot, 1999).

The first example concerns mesoderm formation during amphibian gastrulation. In urodeles, mesoderm is derived predominantly from the superficial layer of the blastula, whereas in *Xenopus* most but not all mesoderm comes from the deep layer. Out-group comparison has shown that the presence of superficial mesoderm is the ancestral condition for amphibians and that the alteration of the fate map (heterotopy) in *Xenopus* represents a derived condition even within anurans, since most anurans have a superficial mesodermal contribution (Vogt, 1929; Pasteels, 1942; Keller, 1975, 1976; Hanken, 1986; Purcell and Keller, 1993; Bolker, 1994; Minsuk and Keller, 1996, 1997). Similar changes in the embryonic source of mesoderm have probably also occurred among teleosts (e.g., Ballard, 1981, 1982; Langeland and Kimmel, 1997).

The second example concerns the mode of cranial neural crest formation, which is similarly variable among amphibians and among vertebrates in general, possibly due to differences in the timing of crest migration relative to neural tube closure and/or differences in dorsoventral patterning mechanisms. In most urodeles the cranial neural crest emerges at the time of neural tube closure from the fusing neural folds (Platt, 1894; Landacre, 1921; Jacobson and Meier, 1984; Northcutt and Brändle, 1995). In anurans, however, prospective neural crest cells can be recognized much earlier as distinct cell masses lateral to the neural plate, and while there is some variability among anuran species, they usually start to migrate before neural tube closure (Stone, 1929b; Knouff, 1935; Schroeder, 1970; Sadaghiani and Thiébaud, 1987; Moury and Hanken, 1995; Olsson and Hanken, 1996; Collazo, 2000). The urodele pattern appears to be ancestral for amphibians because it is also found in most other vertebrates (von Kupffer, 1895, 1900; Goette, 1914; Tosney, 1982), including the Australian lungfish (Falck et al., 2000). However, modifications resembling the anuran pattern have repeatedly occurred in vertebrate evolution, for instance in teleosts and mammals (Landacre, 1910; Verwoerd and van Ostroom, 1979; Tan and Morriss-Kay, 1985; Schmitz et al., 1993; Schilling and Kimmel, 1994; Miyake et al., 1997).

The third example concerns striking differences in the mode of primordial germ cell formation between urodeles and anurans (reviewed in Nieuwkoop and Sutasurya, 1979; Hanken, 1986; Wakahara, 1996b). In anurans, primordial germ cells are derived from the endoderm during early embryonic development and their fate appears to be determined by maternally derived "germinal cytoplasm." In contrast, no such germinal cytoplasm is present in urodeles, where primordial germ cells form at much later embryonic stages from the lateral plate mesoderm and their determination depends on inductive (or permissive) influences from the endoderm. Again, both the timing and the location of origin of these cells appear to have been dramatically shifted during amphibian evolution, and they are notoriously flexible in metazoan evolution in general (reviewed in Nieuwkoop and Sutasurya, 1979; Dixon, 1994).

Conclusions

It has been hypothesized that those complexes of characters that form an integrated and context-insensitive module, while being largely dispensable from their surroundings, may act as units of evolution that tend to repeatedly coevolve, at the same time being easily dissociated from other characters. The foregoing paragraphs used amphibians as a paradigm to illustrate how such claims can be evaluated by combining experimental evidence to identify modules with a comparative phylogenetic approach to detect repeated events of dissociated coevolution. Several examples of amphibian character complexes were presented that act both as modules and as units of evolution supporting the hypothesis (although the lack of sufficient data renders these identifications preliminary in many cases).

These comprise a quite heterogeneous list, ranging from entire life history phases (thyroid-hormone-dependent metamorphosis) and organ primordia (lateral line placodes, limb buds) to cell types (mesoderm, neural crest, and primordial germ cells). Space limitations prevent the discussion of further examples, such as modules in the development of the amphibian nervous system or cranium that also act as units in amphibian evolution (see Schlosser, 2001). It also should be stressed again that a great variety of regulatory gene networks (e.g., involving *Hox*- and *Pax*-transcription factors) and signaling cascades (e.g., the Wnt-, hedgehog-, TGF-β-, or Delta-Notch pathways) could be added to this list (Schlosser, 2002, 2004).

In most of the examples discussed here, modules are individuated and integrated due to a combination of developmental and functional coupling, predominantly among their constituent characters. All components of the lateral line system (different receptor organs, nerves), for instance, are developmentally coupled because they all depend on the previous induction of lateral line placodes, and they are functionally coupled because they all contribute to sensory perception by the lateral line. In cases such as this, modules are likely to act as units of evolution for two reasons: lack of independent variability due to the developmental coupling of characters ("developmental constraints" *sensu* Schwenk, 1994; Wagner and Schwenk, 2000), and epistatic fitness interactions due to functional coupling of characters ("functional constraints" or "internal selection" *sensu* Schwenk, 1994; Wagner and Schwenk, 2000). This does not preclude, however, that in a few cases the one or the other type of couplings may predominate (i.e., modules may be defined mainly by developmental or by functional couplings and the respective units of evolution may be due mainly to developmental or functional constraints).

While case studies such as the ones presented here provide ample evidence that modules can act as units of evolution, three notes of caution need to be added. First, the congruence of a module with a unit of evolution can be demonstrated only for each character complex individually. Consequently, providing a collection of examples, where modules act as units of evolution only proves that modules *can* indeed act as units of evolution, but does not let us infer anything about the generality of this rule. Therefore, the more general claim that modules *tend* to act as units of evolution has to remain at present a working hypothesis. Only with the accumulation of evidence from many different case studies (for review see Raff, 1996; Gerhart and Kirschner, 1997; Schlosser, 2004) will we be able to assess how well this more general claim is supported.

Second, even though modules may often coincide with units of evolution, this is not necessarily true. Not all modules need to act as units of evolution and not all units of evolution need to be modules for several reasons. It has already been argued above that constraints on evolutionary change that transcend module boundaries are likely when modules are strongly and multiply coupled to characters not belonging to the module and are indispensable for their development or function. In addition, a character complex that is more inclusive than a single module may form a unit of evolution when different modules overlap by sharing some of their elements (multifunctionality of elements, pleiotropic roles of genes). At least some heritable variations of shared elements (e.g, in genes with pleiotropic roles) may have pleiotropic effects on different modules, impeding their evolutionary dissociation from each other. The complex of skeletal elements and muscles involved in feeding and breathing in many fishes and amphibians, for example, forms a unit of evolution (reviewed in Roth and Wake, 1989) because several muscles and skeletal elements play a role in both feeding and breathing, although feeding and breathing may be regarded as different (but overlapping) modules of development and function.

Third and finally, it should be emphasized again that patterns of repeated dissociated coevolution of several characters in phylogeny do not necessarily reflect the fact that they form a unit of evolution. Such patterns may, rather, be due simply to the repeated co-occurrence of environmental conditions, which leads to the inadvertent coselection of several characters, although each has independent fitness contributions (see, e.g., D. B. Wake, 1991; Maynard Smith et al., 1985). Therefore, in order to make a strong argument for a character complex as a unit of evolution, it also has to be shown that its coevolution does not correlate with particular combinations of environmental conditions (i.e., it should exhibit repeated dissociated coevolution from particular environmental conditions).

In conclusion, the increasing awareness of the modular organization of organisms within the last few years has opened a new perspective on many classical evolutionary problems. More and more case studies demonstrate that modules of development or function indeed form the "building blocks" (Wagner, 1995; Wagner and Altenberg, 1996) or units of evolution, thereby providing new insights into such phenomena as developmental and functional constraints, evolutionary trends, and homoplasies. However, we still need to broaden our empirical basis and combine detailed experimental studies in a few model organisms with a broad survey of development and function in related taxa (Raff, 1992; Hanken, 1993) in order to rigorously establish the importance of modules as units of evolution.

Acknowledgments

I am grateful to Gerhard Roth, R. Glenn Northcutt, and Chris Kintner, in whose labs most of my own studies were done, for their ongoing support of my research. I also wish to thank Alexander Haas for discussions on amphibian phylogeny; Richard Elinson, David B. Wake, and Jim Hanken for critical comments of an earlier manuscript; and Günter Wagner for helpful comments on this manuscript. My research was supported by grants of the German Science Foundation and a fellowship of the Hanse-Wissenschaftskolleg.

Notes

1. More precisely, a complex of characters will form a "unit of evolution" (Schlosser, 2002) when variants of each character cannot be generated independently from variants of another character or when the fitness of variants of each character of the complex is sensitive to the presence or absence of particular variants of the other characters of the complex, but independent from variants of characters not belonging to the complex (i.e., when there is fitness epistasis exclusively among the characters of the complex). This condition will typically be met for only a subset of all variants available to a system at a given time (e.g., a subset of all one-mutant neighbors of its constituent genes), so that for any given set of variants a character complex will act as a unit of evolution only with a certain probability (coevolution probability). This allows for the possibility of hierarchically nested units of evolution, each with a different coevolution probability. The higher the coevolution probability, the higher the expected frequency of dissociated coevolution in actual phylogeny. Conversely, therefore, the actual frequency of dissociated coevolution of a character complex (relative to its disruption frequency) in a detailed phylogeny of large taxa may serve as a guide for estimating its coevolution probability (i.e., its degree of persistence as a unit of evolution).

2. In order to act as a nondecomposable unit of selection, each character complex has to exist in several variants, these have to epistatically codetermine fitness values, and fitness has to be heritable (Lewontin, 1970). While these conditions can in principle be fulfilled by character complexes at different hierarchical levels (Lewontin, 1970, 1974; Wimsatt, 1981; Sober, 1981, 1984, 1987; Sober and Lewontin, 1982; Gould, 1982; Wilson, 1983; Eldredge, 1985; Lloyd, 1988; Brandon 1990, 1999; Sober and Wilson, 1998), in the typical case of low degrees of linkage between characters, the condition of heritability is less likely to be fulfilled the more independently variable characters there are in the complex (e.g., Lewontin, 1970; Maynard Smith, 1987). Therefore, large complexes of independently variable characters will only rarely act as actual units of selection. Nonetheless, they may act as a unit of evolution in a succession of selection processes, because the relative fitness of a character variant in a given selection process may depend on the outcome of a previous selection process among variants of another character of the complex, even though only a single character may be actually polymorphic and act as unit of selection in each selection process (Schlosser, 2002).

3. Note that the distinction between developmental (causal) and functional coupling does *not* refer to early versus late phases of ontogeny, respectively, but only to the logical pattern of interdependence between characters. Processes or structures in developing embryos may well be functionally coupled (when both are jointly necessary for the generation of a process or structure with an important func-

tion), whereas processes or structures in adults may be developmentally coupled (when their development depends on the same precursor or inducer). Consequently, developmental processes may impose functional as well as developmental constraints, and the same is true for structures and processes of the adult.

4. More precisely, neoteny is regarded (Gould, 1977; Alberch et al., 1979) as one possible cause of paedomorphosis. Paedomorphic species retain juvenile traits in the adult. This may be due to the retardation of somatic development, so that sexual maturity is achieved in animals with juvenile traits but of large size (neoteny) or due to the fact that sexual maturity is already reached in juveniles of small size and somatic development subsequently stops (progenesis). In urodeles, paedomorphosis may result either from neoteny or from progenesis (e.g., Alberch and Alberch, 1981; Hanken, 1989b). Only the former cases are considered here. It should be noted, however, that the distinction is not always clear, because both neoteny and progenesis may contribute to paedomorphosis in the same species (Denoël and Joly, 2000).

References

Adamson L, Harrison RG, Bayley I (1960) The development of the whistling frog *Eleutherodactylus martinicensis* of Barbados. Proc Zool Soc London 133: 453–469.

Alberch P (1980) Ontogenesis and morphological diversification. Am Zool 20: 653–667.

Alberch P (1982) The generative and regulatory roles of development in evolution. In: Environmental Adaptation and Evolution (Mossakowski D, Roth G, eds), 19–36. Stuttgart: Fischer.

Alberch P (1983) Morphological variation in the neotropical salamander genus *Bolitoglossa*. Evolution 37: 906–919.

Alberch P (1987) Evolution of a developmental process: Irreversibility and redundancy in amphibian metamorphosis. In: Development as as evolutionary process (Raff R, Raff E, eds), 23–46. New York: Liss.

Alberch P (1989) Development and the evolution of amphibian metamorphosis. Fortschr Zool 35: 163–173.

Alberch P, Alberch J (1981) Heterochronic mechanisms of morphological diversification and evolutionary change in the neotropical salamander, *Bolitoglossa occidentalis* (Amphibia: Plethodontidae). J Morphol 167: 249–264.

Alberch P, Gale EA (1985) A developmental analysis of an evolutionary trend: Digital reduction in amphibians. Evolution 39: 8–23.

Alberch P, Gale EA (1986) Pathways of cytodifferentiation during the metamorphosis of the epibranchial cartilage in the salamander *Eurycea bislineata*. Dev Biol 117: 233–244.

Alberch P, Gould SJ, Oster GF, Wake DB (1979) Size and shape in ontogeny and phylogeny. Paleobiology 5: 296–317.

Alley KE (1989) Myofiber turnover is used to retrofit frog jaw muscles during metamorphosis. Am J Anat 184: 1–12.

Alley KE (1990) Retrofitting larval neuromuscular circuits in the metamorphosing frog. J Neurobiol 13: 1092–1107.

Alley KE, Barnes MD (1983) Birth dates of trigeminal motoneurons and metamorphic reorganization of the jaw myoneural system in frogs. J Comp Neurol 218: 395–405.

Alley KE, Omerza FF (1998) Reutilization of trigeminal motoneurons during amphibian metamorphosis. Brain Res 813: 187–190.

Anderson DJ (1997) Cellular and molecular biology of neural crest cell lineage determination. Trends Genet 13: 276–280.

Arthur W (1997) The Origin of Animal Body Plans. Cambridge: Cambridge University Press.

Ballard WW (1981) Morphogenetic movements and fate maps of vertebrates. Am Zool 21: 391–399.

Ballard WW (1982) Morphogenetic movements and fate maps of the cypriniform teleost, *Catostomus commersoni* (Lacepede). J Exp Zool 219: 301–321.

Barnes MD, Alley KE (1983) Maturation and recycling of trigeminal motoneurons in anuran larvae. J Comp Neurol 218: 406–414.

Berry DL, Rose CS, Remo BF, Brown DD (1998) The expression pattern of thyroid hormone response genes in remodeling tadpole tissues defines distinct growth and resorption gene expression programs. Dev Biol 203: 24–35.

Berry DL, Schwartzman RA, Brown DD (1998) The expression pattern of thyroid hormone response genes in the tadpole tail identifies multiple resorption programs. Dev Biol 203: 12–23.

Bolker JA (1994) Comparison of gastrulation in frogs and fish. Am Zool 34: 313–322.

Bolt JR (1991) Lissamphibian origins. In: Origins of the Higher Groups of Tetrapods (Schultze H-P, Trueb L, eds), 194–222. Ithaca, NY: Comstock.

Bordzilovskaya NP, Dettlaff TA, Duhon ST, Malacinski GM (1989) Developmental stage series of axolotl embryos. In: Developmental Biology of the Axolotl (Armstrong JB, Malacinski GM, eds), 201–219. New York: Oxford University Press.

Brandon RN (1990) Adaptation and Environment. Princeton, NJ: Princeton University Press.

Brandon RN (1999) The units of selection revisited: The modules of selection. Biol Philos 14: 167–180.

Brown DD (1997) The role of thyroid hormone in zebrafish and axolotl development. Proc Natl Acad Sci USA 94: 13011–13016.

Callery EM, Elinson RP (1996) Developmental regulation of the urea-cycle enzyme arginase in the direct developing frog *Eleutherodactylus coqui*. J Exp Zool 275: 61–66.

Callery EM, Elinson RP (2000a) Thyroid hormone-dependent metamorphosis in a direct-developing frog. Proc Natl Acad Sci USA 97: 2615–2620.

Callery EM, Elinson RP (2000b) Opercular development and ontogenetic reorganization in a direct-developing frog. Dev Genes Evol 210: 377–381.

Callery EM, Fang H, Elinson RP (2001) Frogs without polliwogs: Evolution of anuran direct development. BioEssays 23: 233–241.

Cannatella DC, Hillis DM (1993) Amphibian relationships: Phylogenetic analysis of morphology and molecules. Herpetol. Monog 7: 1–7.

Carl TF, Richardson MK, Hanken J (1997) Limbs: An embryonic innovation in direct-developing anurans. J Morphol 232: 240.

Chibon P (1960) Développement au laboratoire d'*Eleutherodactylus martinicensis* Tschudi, batracien anoure a développement direct. Bull Soc Zool France 85: 412–418.

Coates MI, Cohn MJ (1998) Fins, limbs, and tails: Outgrowths and axial patterning in vertebrate evolution. BioEssays 20: 371–381.

Coates MI, Cohn MJ (1999) Vertebrate axial and appendicular patterning: The early development of paired appendages. Am Zool 39: 676–685.

Cohn MJ, Tickle C (1999) Developmental basis of limblessness and axial patterning in snakes. Nature 399: 474–479.

Collazo A (2000) Developmental variation, homology, and the pharyngula stage. Syst Biol 49: 3–18.

Collazo A, Fraser SE, Mabee PM (1994) A dual embryonic origin for vertebrate mechanoreceptors. Science 264: 426–430.

Collazo A, Marks SB (1994) Development of *Gyrinophilus porphyriticus*—Identification of the ancestral developmental pattern in the salamander family Plethodontidae. J Exp Zool 268: 239–258.

Coombs S, Görner P, Münz H (eds) (1989) The Mechanosensory Lateral Line: Neurobiology and Evolution. New York: Springer.

De Beer G (1958) Embryos and Ancestors. Oxford: Oxford University Press.

De Jongh HJ (1968) Functional morphology of the jaw apparatus of larval and metamorphosing *Rana temporaria* L. Neth J Zool 18: 1–103.

Denoël M, Joly P (2000) Neoteny and progenesis as two heterochronic processes involved in paedomorphosis in *Triturus alpestris* (Amphibia:Caudata). Proc Roy Soc London B267: 1481–1485.

Dent JN (1968) Survey of amphibian metamorphosis. In: Metamorphosis: A Problem in Developmental Biology (Etkin W, Gilbert LI, eds), 271–311. New York: Appleton-Century-Crofts.

De Robertis EM, Morita EA, Cho KWY (1991) Gradient fields and homeobox genes. Development 112: 669–678.

De Villiers CGS (1929) The development of a species of *Arthroleptella* from Jonkershoek, Stellenbosch. S Afr J Sci 26: 481–510.

Dixon KE (1994) Evolutionary aspects of primordial germ cell formation. Germline Dev Ciba Found Symp 182: 92–110.

Dodd MHI, Dodd JM (1976) The biology of metamorphosis. In: Physiology of the Amphibia, III (Lofts B, ed), 467–599. New York: Academic Press.

Duboule D (1992) The vertebrate limb—A model system to study the Hox/HOM gene network during development and evolution. BioEssays 14: 375–384.

Ducibella T (1974) The occurrence of biochemical metamorphic events without anatomical metamorphosis in the axolotl. Dev Biol 38: 175–186.

Duellman WE (1992) Reproductive strategies of frogs. Sci Am 267: 80–87.

Duellman WE, Maxson LR, Jesiolowski CA (1988) Evolution of marsupial frogs (Hylidae:Hemiphractinae): Immunological evidence. Copeia 1988: 527–543.

Duellman WE, Trueb L (1986) Biology of Amphibians. New York: McGraw-Hill.

Dünker N, Wake MH, Olson WM (2000) Embryonic and larval development in the caecilian *Ichthyophis kohtaoensis* (Amphibia, Gymnophiona): A staging table. J Morphol 243: 3–34.

Eisen MB, Spellman PT, Brown PO, Botstein D (1998) Cluster analysis and display of genome-wide expression patterns. Proc Natl Acad Sci USA 95: 14863–14868.

Eldredge N (1985) Unfinished Synthesis. New York: Oxford University Press.

Eldredge N, Cracraft J (1980) Phylogenetic Patterns and the Evolutionary Process. New York: Columbia University Press.

Elinson RP (1987) Change in developmental patterns: Embryos of amphibians with large eggs. In: Development as an Evolutionary Process (Raff RA, Raff EC, eds), 1–21. New York: Liss.

Elinson RP (1990) Direct development in frogs: Wiping the recapitulationist slate clean. Semi Devel Biol 1: 263–270.

Elinson RP (1994) Leg development in a frog without a tadpole (*Eleutherodactylus coqui*). J Exp Zool 270: 202–210.

Escher K (1925) Das Verhalten der Seitenorgane der Wirbeltiere und ihrer Nerven beim Übergang zum Landleben. Acta Zool 6: 307–414.

Falck P, Joss J, Olsson L (2000) Cranial neural crest cell migration in the australian lungfish, *Neoceratodus forsteri*. Evol Dev 2: 179–185.

Fang H, Elinson RP (1996) Patterns of distal-less gene expression and inductive interactions in the head of the direct developing frog *Eleutherodactylus coqui*. Dev Biol 179: 160–172.

Fang H, Elinson RP (1999) Evolutionary alteration in anterior patterning: otx2 expression in the direct developing frog *Eleutherodactylus coqui*. Dev Biol 205: 233–239.

Felsenstein J (1985) Phylogenies and the comparative method. Amer Nat 125: 1–15.

Fink WL (1982) The conceptual relationship between ontogeny and phylogeny. Paleobiology 8: 254–264.

Ford LS, Cannatella DC (1993) The major clades of frogs. Herpetol Monog 7: 94–117.

Fox H (1981) Cytological and morphological changes during amphibian metamorphosis. In: Metamorphosis: A Problem in Developmental Biology (Gilbert LI, Frieden E, eds), 327–362. New York: Plenum Press.

Fritzsch B (1988) The lateral-line and inner-ear afferents in larval and adult urodeles. Brain Behav Evol 31: 325–348.

Fritzsch B (1989) Diversity and regression in the amphibian lateral line and electrosensory system. In: The Mechanosensory Lateral Line (Coombs S, Görner P, Münz H, eds), 99–114. New York: Springer.

Fritzsch B (1990) The evolution of metamorphosis in amphibians. J Neurobiol 21: 1011–1021.

Fritzsch B, Drewes RC, Ruibal R (1987) The retention of the lateral-line nucleus in adult anurans. Copeia 1987: 127–135.

Fritzsch B, Wahnschaffe U, Bartsch U (1988) Metamorphic changes in the octavolateralis system of amphibians. In: The Evolution of the Amphibian Auditory System (Fritzsch B, Ryan M, Wilczynski W, Hetherington TE, Walkowiak W, eds), 359–376. New York: Wiley.

Fritzsch B, Wake M (1986) The distribution of ampullary organs in Gymnophiona. J Herpetol 20: 90–93.

García-Bellido A (1996) Symmetries throughout organic evolution. Proc Natl Acad Sci USA 93: 14229–14232.

Gellon G, McGinnis W (1998) Shaping animal body plans in development and evolution by modulation of Hox expression patterns. BioEssays 20: 116–125.

Gerhart J (1999) 1998 Warkany lecture: Signaling pathways in development. Teratology 60: 226–239.

Gerhart J, Kirschner M (1997) Cells, Embryos, and Evolution. Malden, MA: Blackwell Science.

Gilbert SF (1998) Conceptual breakthroughs in developmental biology. J Biosci 23: 169–176.

Gilbert SF (2000) Developmental Biology. Sunderland, MA: Sinauer.

Gilbert SF, Opitz JM, Raff RA (1996) Resynthesizing evolutionary and developmental biology. Dev Biol 173: 357–372.

Gitlin D (1944) The development of *Eleutherodactylus portoricensis*. Copeia 1944: 91–98.

Goette A (1914) Die Entwicklung der Kopfnerven bei Fischen und Amphibien. Arch Mikro und Anat 85: 1–165.

Goin CJ (1947) Studies on the life history of *Eleutherodactylus ricordii planirostris* (Cope) in Florida. Univ Florida Stud, Biol Sci ser 4: 1–66.

Gould SJ (1977) Ontogeny and phylogeny. Cambridge, MA: Belknap Press of Harvard University Press.

Gould SJ (1982) Darwinism and the expansion of the evolutionary theory. Science 216: 380–387.

Gould SJ, Lewontin RC (1979) The spandrels of San Marco and the Panglossian paradigm: A critique of the adaptationist programme. Proc Roy Soc London B205: 581–598.

Guillemot F (1999) Vertebrate bHLH genes and the determination of neuronal fates. Exper Cell Res 253: 357–364.

Haddon C, Jiang YJ, Smithers L, Lewis J (1998) Delta-Notch signalling and the patterning of sensory cell differentiation in the zebrafish ear: Evidence from the mind bomb mutant. Development 125: 4637–4644.

Hadfield KA, Swalla BJ, Jeffery WR (1995) Multiple origins of anural development in ascidians inferred from rDNA sequences. J Molec Evol 40: 413–427.

Hall BK (1992) Evolutionary Developmental Biology. London: Chapman and Hall.

Hamburger V (1939) The development and innervation of transplanted limb primordia of chick embryos. J Exp Zool 80: 347–389.

Hamburger V, Hamilton HL (1951) A series of normal stages in the development of the chick embryo. J Morphol 88: 49–92.

Hanken J (1986) Developmental evidence for amphibian origins. In: Evolutionary Biology (Hecht MK, Wallace B, Prance GT, eds), 389–417. New York: Plenum Press.

Hanken J (1989a) Developmental characters in phylogenetic inference: A test case using amphibians. Fortschr Zool 35: 174–179.

Hanken J (1989b) Development and evolution in amphibians. Amer Sci 77: 336–343.

Hanken J (1992) Life history and morphological evolution. J Evol Biol 5: 549–557.

Hanken J (1993) Model systems versus outgroups—Alternative approaches to the study of head development and evolution. Amer Zool 33: 448–456.

Hanken J, Hall BK (1988) Skull development during anuran metamorphosis: II. Role of thyroid hormone in osteogenesis. Anat Embryol 178: 219–227.

Hanken J, Jennings DH, Olsson L (1997) Mechanistic basis of life-history evolution in anuran amphibians: Direct development. Amer Zool 37: 160–171.

Hanken J, Klymkowsky MW, Alley KE, Jennings DH (1997) Jaw muscle development as evidence for embryonic repatterning in direct-developing frogs. Proc Roy Soc London B264: 1349–1354.

Hanken J, Klymkowsky MW, Summers CH, Seufert DW, Ingebrigtsen N (1992) Cranial ontogeny in the direct-developing frog, *Eleutherodactylus coqui* (Anura, Leptodactylidae), analyzed using whole-mount immunohistochemistry. J Morphol 211: 95–118.

Hanken J, Summers CH (1988) Skull development during anuran metamorphosis. III: Role of thyroid hormone in chondrogenesis. J Exp Zool 246: 156–170.

Hanken J, Summers CH, Hall BK (1989) Morphological integration in the cranium during anuran metamorphosis. Experientia 45: 872–875.

Harrison RG (1904) Experimentelle Untersuchungen über die Entwicklung der Sinnesorgane der Seitenlinie bei den Amphibien. Arch Mikrosk Anat Entw Gesch 63: 35–149.

Harrison RG (1918) Experiments on the development of the forelimb of *Amblystoma,* a self-differentiating equipotential system. J Exp Zool 25: 413–461.

Hartwell LH, Hopfield JJ, Leibler S, Murray AW (1999) From molecular to modular cell biology. Nature 402 supp:C47–C52.

Harvey PH, Pagel MD (1991) The Comparative Method in Evolutionary Biology. Oxford: Oxford University Press.

Hay JM, Ruvinsky I, Hedges SB, Maxson LR (1995) Phylogenetic relationships of amphibian families inferred from DNA sequences of mitochondrial 12S and 16S ribosomal RNA genes. Molec Biol Evol 12: 928–937.

Hayes TB (1997) Hormonal mechanisms as potential constraints on evolution: Examples from the anura. Amer Zool 37: 482–490.

Heanue TA, Reshef R, Davis RJ, Mardon G, Oliver G, Tomarev S, Lassar AB, Tabin CJ (1999) Synergistic regulation of vertebrate muscle development by Dach2, Eya2, and Six1, homologs of genes required for *Drosophila* eye formation. Genes Devel 13: 3231–3243.

Hetherington TE (1988) Metamorphic changes in the middle ear. In: The Evolution of the Amphibian Auditory System (Fritzsch B, Ryan M, Wilczynski W, eds), 339–357. New York: Wiley.

Heyer WR, McDiarmid RW, Weigmann DL (1975) Tadpoles, predation and pond habitats in the tropics. Biotropica 7: 100–111.

Hillis DM (1991) The phylogeny of amphibians—Current knowledge and the role of cytogenetics. In: Amphibian Cytogenetics and Evolution (Green DM, Sessions SK, eds), 7–31. San Diego: Academic Press.

Hillis DM, Ammerman LK, Dixon MT, De Sá RO (1993) Ribosomal DNA and the phylogeny of frogs. Herpetol Monog 7: 118–131.

Hinchliffe JR, Johnson DR (1980) The Development of the Vertebrate Limb. Oxford: Clarendon Press.

Hoskins SG, Grobstein P (1985) Development of the ipsilateral projection in the frog *Xenopus laevis*. II. Ingrowth of optic nerve fibers and production of ipsilaterally projecting cells. J Neurosci 5: 920–929.

Hourdry J, Lhermite A, Ferrand R (1996) Changes in the digestive tract and feeding behavior of anuran amphibians during metamorphosis. Physiolog Zool 69: 219–251.

Hughes A (1959a) Studies in embryonic and larval development in Amphibia. I. The embryology of *Eleutherodactylus ricordii*, with special reference to the spinal cord. J Embryol Exp Morphol 7: 22–38.

Hughes A (1959b) Studies in embryonic and larval development in amphibia. II. The spinal motor-root. J Embryol Exp Morphol 7: 128–145.

Hughes A (1962) An experimental study on the relationships between limb and spinal cord in the embryo of *Eleutherodactylus martinicensis*. J Embryol Exp Morphol 10: 575–601.

Hughes A (1965a) The development of behavior in *Eleutherodactylus martinicensis* (Amphibia, Anura). Proc Zool Soc London 144: 153–161.

Hughes A (1965b) A quantitative study of the development of the nerves in the hind-limb of *Eleutherodactylus martinicensis*. J Embryol Exp Morphol 13: 9–34.

Hughes A (1966) The thyroid and the development of the nervous system in *Eleutherodactylus martinicensis*: An experimental study. J Embryol Exp Morphol 16: 401–430.

Ishizuya-Oka A (1996) Apoptosis of larval cells during amphibian metamorphosis. Micro Res Tech 34: 228–235.

Jacobson AG, Meier S (1984) Morphogenesis of the head of a newt: Mesodermal segments, neuromeres, and distribution of neural crest. Dev Biol 106: 181–193.

Jeffery WR, Swalla BJ (1990) Anural development in ascidians: Evolutionary modification and elimination of the tadpole larva. Sem Devel Biol 1: 253–261.

Jennings DH (1997) Evolution of Endocrine Control of Development in Direct-developing Amphibians. PhD thesis, University of Colorado.

Jennings DH, Hanken J (1998) Mechanistic basis of life history evolution in anuran amphibians: Thyroid gland development in the direct-developing frog, *Eleutherodactylus coqui*. Gen Comp Endocrin 111: 225–232.

Johnson RL, Tabin CJ (1997) Molecular models for vertebrate limb development. Cell 90: 979–990.

Kaufman MH (1992) The Atlas of Mouse Development. London: Academic Press.

Keller RE (1975) Vital dye mapping of the gastrula and neurula of *Xenopus laevis*. I. Prospective areas and morphogenetic movements of the superficial layer. Dev Biol 42: 222–241.

Keller RE (1976) Vital dye mapping of the gastrula and neurula of *Xenopus laevis*. II. Prospective areas and morphogenetic movements of the deep layer. Dev Biol 51: 118–137.

Kemp A (1982) The embryological development of the Queensland lungfish, *Neoceratodus forsteri* (Krefft). Mem Queensland Mus 20: 553–597.

Kemp A (1999) Ontogeny of the skull of the Australian lungfish *Neoceratodus forsteri* (Osteichthyes: Dipnoi). J Zool (London) 248: 97–137.

Kluge AG, Strauss RE (1985) Ontogeny and systematics. Ann Rev Ecol Syst 16: 247–268.

Knouff RA (1935) The developmental pattern of ectodermal placodes in *Rana pipiens*. J Comp Neurol 62: 17–71.

Kollros JJ (1981) Transitions in the nervous system during amphibian metamorphosis. In: Metamorphosis: A Problem in Developmental Biology (Gilbert LI, Frieden E, eds), 445–459. New York: Plenum Press.

Lamborghini JE (1987) Disappearance of Rohon–Beard neurons from the spinal cord of larval *Xenopus laevis*. J Comp Neurol 264: 47–55.

Landacre FL (1910) The origin of the cranial ganglia in *Ameiurus*. J Comp Neurol 20: 309–411.

Landacre FL (1921) The fate of the neural crest in the head of the urodeles. J Comp Neurol 33: 1–43.

Lande R (1978) Evolutionary mechanisms of limb loss in tetrapods. Evolution 32: 73–92.

Langeland JA, Kimmel CB (1997) Fishes. In: Embryology (Gilbert SF, Raunio AM, eds), 383–407. Sunderland, MA: Sinauer.

Larson A, Dimmick WW (1993) Phylogenetic relationships of the salamander families: An analysis of congruence among morphological and molecular characters. Herpetol Monog 7: 77–93.

Lewontin RC (1970) The units of selection. Ann Rev Ecol Syst 1: 1–18.

Lewontin RC (1974) The Genetic Basis of Evolutionary Change. New York: Columbia University Press.

Lloyd EA (1988) The Structure and Confirmation of Evolutionary Theory. Princeton, NJ: Princeton University Press.

Lumsden A (1987) The neural crest contribution to tooth development in the mammalian embryo. In: Developmental and Evolutionary Aspects of the Neural Crest (Maderson PFA, ed), 261–300. New York: Wiley.

Lutz B (1947) Trends towards non-aquatic and direct development in frogs. Copeia 1947: 242–252.

Lutz B (1948) Ontogenetic evolution in frogs. Evolution 2: 29–39.

Lynn WG (1936) A study of the thyroid in embryos of *Eleutherodactylus nubicola*. Anat Rec 64: 525–539.

Lynn WG (1942) The embryology of *Eleutherodactylus nubicola*, an anuran which has no tadpole stage. Contrib Embryol Carnegie Inst Washington 190: 27–62.

Lynn WG (1948) The effects of thiourea and phenylthiourea upon the development of *Eleutherodactylus ricordii*. Biol Bull 94: 1–15.

Lynn WG (1961) Types of amphibian metamorphosis. Amer Zool 1: 151–161.

Lynn WG, Lutz B (1946) The development of *Eleutherodactylus nasutus* Lutz. Bolet Mus Nac 79: 1–30.

Lynn WG, Peadon M (1955) The role of the thyroid gland in direct development in the anuran, *Eleutherodactylus martinicensis*. Growth 19: 263–286.

Mabee PM (1993) Phylogenetic interpretation of ontogenetic change—Sorting out the actual and artefactual in an empirical case study of centrarchid fishes. Zool J Linn Soc 107: 175–291.

Maddison WP (1990) A method for testing the correlated evolution of two binary characters: Are gains or losses concentrated on certain branches of a phylogenetic tree? Evolution 44: 539–557.

Matsuda R (1987) Animal Evolution in Changing Environments with Special Reference to Abnormal Metamorphosis. New York: Wiley.

Maynard Smith J (1987) How to model evolution. In: The Latest on the Best (Dupré J, ed), 119–131. Cambridge, MA: MIT Press.

Maynard Smith J, Burian R, Kauffman S, Alberch P, Campbell J, Goodwin B, Lande R, Raup D, Wolpert L (1985) Developmental constraints and evolution. Quart Rev Biol 60: 265–287.

McKinney ML, McNamara KJ (1991) Heterochrony: The Evolution of Ontogeny. New York: Plenum Press.

Minsuk SB, Keller RE (1996) Dorsal mesoderm has a dual origin and forms by a novel mechanism in *Hymenochirus*, a relative of *Xenopus*. Dev Biol 174: 92–103.

Minsuk SB, Keller RE (1997) Surface mesoderm in *Xenopus*: A revision of the stage 10 fate map. Dev Genes Evol 207: 389–401.

Miyake T, von Herbing IH, Hall BK (1997) Neural ectoderm, neural crest, and placodes: Contribution of the otic placode to the ectodermal lining of the embryonic opercular cavity in Atlantic cod (Teleostei). J Morphol 231: 231–252.

Moser H (1950) Ein Beitrag zur Analyse der Thyroxinwirkung im Kaulquappenversuch und zur Frage nach dem Zustandekommen der Frühbereitschaft des Metamorphose-Reaktionssystems. Rev Suisse Zool 57: 1–144.

Moury JD, Hanken J (1995) Early cranial neural crest migration in the direct-developing frog, *Eleuthero-dactylus coqui*. Acta Anat 153: 243–253.

Müller GB (1991) Experimental strategies in evolutionary embryology. Amer Zool 31: 605–615.

Needham J (1933) On the dissociability of the fundamental processes in ontogenesis. Biol Rev 8: 180–223.

Ng JK, Tamura K, Buscher D, Izpisuabelmonte JC (1999) Molecular and cellular basis of pattern formation during vertebrate limb development. Curr Top Dev Biol 41: 37–66.

Niehrs C, Pollet N (1999) Synexpression groups in eukaryotes. Nature 402: 483–487.

Nieuwkoop PD, Sutasurya LA (1979) Primordial Germ Cells in the Chordates. Cambridge: Cambridge University Press.

Nishikawa A, Hayashi H (1994) Isoform transition of contractile proteins related to muscle remodeling with an axial gradient during metamorphosis in *Xenopus laevis*. Dev Biol 165: 86–94.

Nishikawa A, Hayashi H (1995) Spatial, temporal and hormonal regulation of programmed muscle cell death during metamorphosis of the frog *Xenopus laevis*. Differentiation 59: 207–214.

Noble GK (1927) The value of life history data in the study of the evolution of the amphibia. Ann NY Acad Sci 30: 31–128.

Noll M (1993) Evolution and role of Pax genes. Curr Opin Genet Devel 3: 595–605.

Northcutt RG (1989) The phylogenetic distribution and innervation of craniate mechanoreceptive lateral lines. In: The Mechanosensory Lateral Line (Coombs S, Görner P, Münz H, eds), 17–78. New York: Springer.

Northcutt RG (1990) Ontogeny and phylogeny: A re-evaluation of conceptual relationships and some applications. Brain Behav Evol 36: 116–140.

Northcutt RG (1992) The phylogeny of octavolateralis ontogenies: A reaffirmation of Garstang's hypothesis. In: The Evolutionary Biology of Hearing (Webster DB, Fay RR, Popper AN, eds), 21–47. New York: Springer.

Northcutt RG (1996) The origin of craniates: Neural crest, neurogenic placodes, and homeobox genes. Israel J Zool 42: 273–313.

Northcutt RG (1997) Evolution of gnathostome lateral line ontogenies. Brain Behav Evol 50: 25–37.

Northcutt RG, Brändle K (1995) Development of branchiomeric and lateral line nerves in the axolotl. J Comp Neurol 355: 427–454.

Northcutt RG, Brändle K, Fritzsch B (1995) Electroreceptors and mechanosensory lateral line organs arise from single placodes in axolotls. Dev Biol 168: 358–373.

Northcutt RG, Catania KC, Criley BB (1994) Development of lateral line organs in the axolotl. J Comp Neurol 340: 480–514.

Olsson L, Hanken J (1996) Cranial neural-crest migration and chondrogenic fate in the oriental fire-bellied toad *Bombina orientalis*—Defining the ancestral pattern of head development in anuran amphibians. J Morphol 229: 105–120.

Omerza FF, Alley KE (1992) Redeployment of trigeminal motor axons during metamorphosis. J Comp Neurol 325: 124–134.

Orton GL (1949) Larval development of *Nectophrynoides tornieri* (Roux) with comments on direct development in frogs. Ann Carnegie Mus 32: 257–277.

Orton GL (1951) Direct development in frogs. Turtox News 292: 2–6.

Orton GL (1953) The systematics of vertbrate larvae. Syst Zool 2: 63–75.

Pagel M (1999) Inferring the historical patterns of biological evolution. Nature 401: 877–884.

Parichy DM (1996a) Pigment patterns of larval salamanders (Ambystomatidae, Salamandridae)—The role of the lateral line sensory system and the evolution of pattern-forming mechanisms. Dev Biol 175: 265–282.

Parichy DM (1996b) When neural crest and placodes collide—Interactions between melanophores and the lateral lines that generate stripes in the salamander *Ambystoma tigrinum tigrinum* (Ambystomatidae). Dev Biol 175: 283–300.

Pasteels J (1942) New observations concerning the maps of presumptive areas of the young amphibian gastrula (*Amblystoma* and *Discoglossus*). J Exp Zool 89: 255–281.

Patil NS, Kanamadi RD (1997) Direct development in the rhacophorid frog, *Philautus variabilis* (Gunther). Curr Sci 73: 697–701.

Platt JB (1894) Ontogenetische Differenzierung des Ektoderms in *Necturus*. Arch Mikrosk Anat 43: 911–966.

Purcell SM, Keller R (1993) A different type of amphibian mesoderm morphogenesis in *Ceratophrys ornata*. Development 117: 307–317.

Raff RA (1987) Constraint, flexibility, and phylogenetic history in the evolution of direct development in sea urchins. Dev Biol 119: 6–19.

Raff RA (1992) Evolution of developmental decisions and morphogenesis: The view from two camps. Development (supp.): 15–22.

Raff RA (1996) The Shape of Life. Chicago: University of Chicago Press.

Raff RA, Parr BA, Parks AL, Wray GA (1990) Heterochrony and other mechanisms of radical evolutionary change in early development. In: Evolutionary Innovations (Nitecki MH, ed), 71–98. Chicago: University of Chicago Press.

Raff RA, Wray GA (1989) Heterochrony: Developmental mechanisms and evolutionary results. J Evol Biol 2: 409–434.

Raynaud A (1985) Development of limbs and embryonic limb reduction. In: Biology of the Reptilia (Gans C, ed), 59–148. New York: Wiley.

Relaix F, Buckingham M (1999) From insect eye to vertebrate muscle: Redeployment of a regulatory network. Genes Devel 13: 3171–3178.

Richardson MK (1995) Heterochrony and the phylotypic period. Dev Biol 172: 412–421.

Richardson MK, Carl TF, Hanken J, Elinson RP, Cope C, Bagley P (1998) Limb development and evolution: A frog embryo with no apical ectodermal ridge (AER). J Anat 192: 379–390.

Riley BB, Chiang MY, Farmer L, Heck R (1999) The deltaA gene of zebrafish mediates lateral inhibition of hair cells in the inner ear and is regulated by pax2.1. Development 126: 5669–5678.

Rose CS (1995a) Skeletal morphogenesis in the urodele skull. 1. Postembryonic development in the Hemidactyliini (Amphibia: Plethodontidae). J Morphol 223: 125–148.

Rose CS (1995b) Skeletal morphogenesis in the urodele skull. 2. Effect of developmental stage in thyroid hormone-induced remodeling. J Morphol 223: 149–166.

Rose CS (1995c) Skeletal morphogenesis in the urodele skull. 3. Effect of hormone dosage in TH-induced remodeling. J Morphol 223: 243–261.

Rose CS (1996) An endocrine-based model for developmental and morphogenetic diversification in metamorphic and paedomorphic urodeles. J Zool 239: 253–284.

Rosenkilde P, Ussing AP (1996) What mechanisms control neoteny and regulate induced metamorphosis in urodeles? Int J Devel Biol 40: 665–673.

Roth G, Nishikawa KC, Naujoks-Manteuffel C, Schmidt A, Wake DB (1993) Paedomorphosis and simplification in the nervous system of salamanders. Brain Behav Evol 42: 137–170.

Roth G, Wake DB (1985) Trends in the functional morphology and sensorimotor control of feeding behavior in salamanders: An example of the role of internal dynamics in evolution. Acta Biotheor 34: 175–192.

Roth G, Wake DB (1989) Conservatism and innovation in the evolution of feeding in vertebrates. In: Complex Organismal Functions: Integration and Evolution in Vertebrates (Wake DB, Roth G eds), 7–21. Chichester, UK: Wiley.

Sadaghiani B, Thiébaud CH (1987) Neural crest development in the *Xenopus laevis* embryo, studied by interspecific transplantation and scanning electron microscopy. Dev Biol 124: 91–110.

Safi R, Begue A, Hanni C, Stehelin F, Tata JR, Laudet V (1997) Thyroid hormone receptor genes of neotenic amphibians. J Molec Evol 44: 595–604.

Salthe SN, Duellman WE (1973) Quantitative constraints associated with reproductive mode in anurans. In: Biology of the Anurans (Vial JL, ed), 229–249. Columbia: University of Missouri Press.

Salthe SN, Mecham JS (1974) Reproductive and courtship patterns. In: Physiology of the Amphibia II (Lofts B, ed), 309–521. New York: Academic Press.

Schilling TF, Kimmel CB (1994) Segment and cell type lineage restrictions during pharyngeal arch development in the zebrafish embryo. Development 120: 483–494.

Schlosser G (1998) Self-reproduction and functionality. A systems-theoretical approach to teleological explanation. Synthèse 116: 303–354.

Schlosser G (2001) Using heterochrony plots to detect the dissociated coevolution of characters. J Exp Zool (Mol Dev Evol) 291: 282–304.

Schlosser G (2002) Modularity and the units of evolution. Theory in Biosciences 121: 1–80.

Schlosser G (2004) The role of modules in development and evolution. In: Modularity in Development and Evolution (Schlosser G, Wagner G, eds), 519–582. Chicago: University of Chicago Press.

Schlosser G, Kintner C (1999) Monophasic development of the spinal cord in direct developing frogs. Amer Zool 39: 14A.

Schlosser G, Kintner C, Northcutt RG (1999) Loss of ectodermal competence for lateral line placode formation in the direct developing frog *Eleutherodactylus coqui*. Dev Biol 213: 354–369.

Schlosser G, Northcutt RG (2000) Development of neurogenic placodes in *Xenopus laevis*. J Comp Neurol 418: 121–146.

Schlosser G, Northcutt RG (2001) Lateral line placodes are induced during neurulation in the axolotl. Dev Biol 234: 55–71.

Schlosser G, Roth G (1997a) Evolution of nerve development in frogs. I. The development of the peripheral nervous system in *Discoglossus pictus* (Discoglossidae). Brain Behav Evol 50: 61–93.

Schlosser G, Roth G (1997b) Evolution of nerve development in frogs. II. Modified development of the peripheral nervous system in the direct-developing frog *Eleutherodactylus coqui* (Leptodactylidae). Brain Behav Evol 50: 94–128.

Schlosser G, Roth G (1997c) Development of the retina is altered in the directly developing frog *Eleutherodactylus coqui* (Leptodactylidae). Neurosci Lett 224: 153–156.

Schlosser G, Thieffry D (2000) Modularity in development and evolution. BioEssays 22: 1043–1045.

Schmidt A, Böger T (1993) Thyroxine induces a decoupling of ontogenetic processes in urodeles: Consequences for the visual system. In: Proc. 21st Göttingen Neurobiology Conference (Elsner N, Heisenberg M, eds), 762. Stuttgart: Thieme.

Schmidt A, Roth G (1996) Differentiation processes in the amphibian brain with special emphasis on heterochronies. Int Rev Cytol 169: 83–150.

Schmitz B, Papan C, Campos-Ortega JA (1993) Neurulation in the anterior trunk region of the zebrafish *Brachydanio rerio*. Roux Arch Devel Biol 202: 250–259.

Schroeder TE (1970) Neurulation in *Xenopus laevis:* An analysis and model based upon light and electron microscopy. J Embryol Exp Morphol 23: 427–462.

Schwenk K (1994) A utilitarian approach to evolutionary constraint. Zoology 98: 251–262.

Shaffer HB, Voss SR (1996) Phylogenetic and mechanistic analysis of a developmentally integrated character complex: Alternate life history modes in ambystomatid salamanders. Amer Zool 36: 24–35.

Shi YB (1996) Thyroid hormone-regulated early and late genes during amphibian metamorphosis. In: Metamorphosis: Postembryonic Reprogramming of Gene Expression in Amphibian and Insect Cells (Gilbert LI, Tata JR, Atkinson BG, eds), 505–538. San Diego: Academic Press.

Shi YB, Su Y, Li Q, Damjanovski S (1998) Auto-regulation of thyroid hormone receptor genes during metamorphosis: Roles in apoptosis and cell proliferation. Int J Devel Biol 42: 107–116.

Shi YB, Wong J, Puzianowska-Kuznicka M, Stolow MA (1996) Tadpole competence and tissue-specific temporal regulation of amphibian metamorphosis: Roles of thyroid hormone and its receptors. BioEssays 18: 391–399.

Shubin N (1998) Evolutionary cut and paste. Nature 394: 12–13.

Shubin N, Tabin C, Carroll S (1997) Fossils, genes and the evolution of animal limbs. Nature 388: 639–648.

Shubin N, Wake D (1996) Phylogeny, variation, and morphological integration. Amer Zool 36: 51–60.

Slack JMW (1991) From Egg to Embryo. Cambridge: Cambridge University Press.

Slack JMW, Holland PWH, Graham CF (1993) The zootype and the phylotypic stage. Nature 361: 490–492.

Smith KK (1996) Integration of craniofacial structures during development in mammals. Amer Zool 36: 70–79.

Smith KK (1997) Comparative patterns of craniofacial development in eutherian and metatherian mammals. Evolution 51: 1663–1678.

Smith SC, Lannoo MJ, Armstrong JB (1990) Development of the mechanoreceptive lateral-line system in the axolotl: Placode specification, guidance of migration, and the origin of neuromast polarity. Anat Embryol 182: 171–180.

Sober E (1981) Holism, individualism and the units of selection. PSA 1981: 93–121.

Sober E (1984) The Nature of Selection. Chicago: University of Chicago Press.

Sober E (1987) Comments on Maynard Smith's "How to model evolution." In: The Latest on the Best (Dupré J, ed), 133–149. Cambridge, MA: MIT Press.

Sober E, Lewontin RC (1982) Artifact, cause and genic selection. Phil Sci 49: 157–180.

Sober E, Wilson DS (1998) Unto Others. Cambridge, MA: Harvard University Press.

Stephenson WG (1951) Observations on the development of the amphicoelus frogs, *Leiopelma* and *Ascaphus*. Zool J Linn Soc 42: 18–28.

Stone LS (1922) Experiments on the development of the cranial ganglia and the lateral line sense organs in *Amblystoma punctatum*. J Exp Zool 35: 421–496.

Stone LS (1929a) Experiments on the transplantation of placodes of the cranial ganglia in the amphibian embryo. IV. Heterotopic transplantation of the postauditory placodal material upon the head and body of *Amblystoma punctatum*. J Comp Neurol 48: 311–330.

Stone LS (1929b) Experiments showing the role of migrating neural crest (mesectoderm) in the formation of head skeleton and loose connective tissue in *Rana palustris*. Roux Arch Entw Mech 118: 40–77.

Stone LS (1933) The development of lateral-line sense organs in amphibians observed in living and vital-stained preparations. J Comp Neurol 57: 507–540.

Stone LS (1938) Further experimental studies of the development of lateral line sense organs in amphibians observed in living preparations. J Comp Neurol 68: 83–115.

Su Y, Damjanovski S, Shi Y, Shi YB (1999) Molecular and cellular basis of tissue remodeling during amphibian metamorphosis. Histol Histopathol 14: 175–183.

Szarski H (1957) The origin of the larva and metamorphosis in amphibia. Am Nat 91: 283–301.

Tabin CJ, Carroll SB, Panganiban G (1999) Out on a limb: Parallels in vertebrate and invertebrate limb patterning and the origin of appendages. Amer Zool 39: 650–663.

Tan SS, Morriss-Kay G (1985) The development and distribution of the cranial neural crest in the rat embryo. Cell Tiss Res 240: 403–416.

Tata JR (1996) Metamorphosis: An exquisite model for hormonal regulation of post-embryonic development. Biochem Soc Symp 62: 123–136.

Tata JR (1998) Amphibian metamorphosis as a model for studying the developmental actions of thyroid hormone. Ann Endocrin 59: 433–442.

Tata JR, Baker BS, Machuca I, Rabelo EML, Yamauchi K (1993) Autoinduction of nuclear receptor genes and its significance. J Ster Biochem Molec Biol 46: 105–119.

Titus TA, Larson A (1996) Molecular phylogenetics of desmognathine salamanders (Caudata:Plethodontidae): A reevaluation of evolution in ecology, life history, and morphology. Syst Biol 45: 451–472.

Tosney KW (1982) The segregation and early migration of cranial neural crest cells in the avian embryo. Dev Biol 89: 13–24.

Townsend DS, Stewart MM (1985) Direct development in *Eleutherodactylus coqui* (Anura: Leptodactylidae): A staging table. Copeia 1985: 423–436.

Trueb L, Cloutier R (1991) A phylogenetic investigation of the inter- and intrarelationships of the Lissamphibia (Amphibia: Temnospondyli). In: Origins of the Higher Groups of Tetrapods (Schultze H-P, Trueb L, eds), 223–313. Ithaca, NY: Comstock.

Verwoerd CDA, van Oostrom CG (1979) Cephalic neural crest and placodes. Adv Anat Embryol Cell Biol 58: 1–75.

Vogt W (1929) Gestaltungsanalyse am Amphibienkeim mit örtlicher Vitalfärbung. II. Gastrulation und Mesodermbildung bei Urodelen und Anuren. Roux Arch Devel Biol 120: 384–706.

Von Dassow G, Munro E (1999) Modularity in animal development and evolution: Elements of a conceptual framework for EvoDevo. J Exp Zool (Mol Dev Evol) 285: 307–325.

Von Kupffer C (1895) Studien zur vergleichenden Entwicklungsgeschichte des Kopfes der Kranioten. III. Die Entwicklung der Kopfnerven von *Ammocoetes planeri*. Munich: J.F. Lehmann.

Von Kupffer C (1900) Studien zur vergleichenden Entwicklungsgeschichte des Kopfes der Kranioten. IV. Zur Kopfentwicklung von *Bdellostoma*. Munich: J.F. Lehmann.

Voss SR (1995) Genetic basis of paedomorphosis in the axolotl, *Ambystoma mexicanum*: A test of the single-gene hypothesis. J Hered 86: 441–447.

Voss SR, Shaffer HB (1997) Adaptive evolution via a major gene effect: Paedomorphosis in the Mexican axolotl. Proc Natl Acad Sci USA 94: 14185–14189.

Voss SR, Shaffer HB (2000) Evolutionary genetics of metamorphic failure using wild-caught vs. laboratory axolotls (*Ambystoma mexicanum*). Molec Ecol 9: 1401–1407.

Wagner GP (1995) The biological role of homologues: A building block hypothesis. N Jb Geol Paläont Abh 195: 279–288.

Wagner GP (1996) Homologues, natural kinds, and the evolution of modularity. Amer Zool 36: 36–43.

Wagner GP, Altenberg L (1996) Perspective: Complex adaptations and the evolution of evolvability. Evolution 50(3): 967–976.

Wagner GP, Schwenk K (2000) Evolutionarily stable configurations: Functional integration and the evolution of phenotypic stability. Evol Biol 31: 155–217.

Wahnschaffe U, Bartsch U, Fritzsch B (1987) Metamorphic changes within the lateral-line system of Anura. Anat Embryol 175: 431–442.

Wakahara M (1996a) Heterochrony and neotenic salamanders: Possible clues for understanding the animal development and evolution. Zool Sci 13: 765–776.

Wakahara M (1996b) Primordial germ cell development: Is the urodele pattern closer to mammals than to anurans? Int J Devel Biol 40: 653–659.

Wakahara M, Yamaguchi M (1996) Heterochronic expression of several adult phenotypes in normally metamorphosing and metamorphosis-arrested larvae of a salamander *Hynobius retardatus*. Zool Sci 13: 483–488.

Wake DB (1991) Homoplasy: The result of natural selection, or evidence of design limitations? Amer Nat 138: 543–567.

Wake DB, Hanken J (1996) Direct development in the lungless salamanders—What are the consequences for developmental biology, evolution and phylogenesis? Int J Devel Biol 40: 859–869.

Wake DB, Larson A (1987) Multidimensional analysis of an evolving lineage. Science 238: 42–48.

Wake DB, Roth G (1989) The linkage between ontogeny and phylogeny in the evolution of complex systems. In: Complex Organismal Functions: Integration and Evolution in Vertebrates (Wake DB, Roth G, eds), 361–377. Chichester, UK: Wiley.

Wake DB, Roth G, Nishikawa KC (1987) The fate of the lateral line system in plethodontid salamanders. Amer Zool 27: 166A.

Wake MH (1980) The reproductive biology of *Nectophrynoides malcolmi* (Amphibia: Bufonidae), with comments on the evolution of reproductive modes in the genus *Nectophrynoides*. Copeia 1980: 193–209.

Wake MH (1989) Phylogenesis of direct development and viviparity in vertebrates. In: Complex Organismal Functions: Integration and Evolution in Vertebrates (Wake DB, Roth G, eds), 235–250. Chichester, UK: Wiley.

Warren E (1922) Observations on the development of the non-aquatic tadpole of *Anhydrophryne tarrrayi* Hewitt. S Afr J Sci 19: 254–262.

Wassersug RJ (1974) Evolution of anuran life cycles. Science 185: 377–378.

Wassersug RJ, Hoff K (1982) Developmental changes in the orientation of the anuran jaw suspension. Evol Biol 15: 223–246.

Weintraub H (1993) The MyoD family and myogenesis: Redundancy, networks and thresholds. Cell 75: 1241–1244.

Wen X, Fuhrman S, Michaels GS, Carr DB, Smith S, Barker JL, Somogyi R (1998) Large-scale temporal expression mapping of central nervous system development. Proc Natl Acad Sci USA 95: 334–339.

White BA, Nicoll CS (1981) Hormonal control of amphibian metamorphosis. In: Metamorphosis: A Problem in Developmental Biology (Gilbert LI, Frieden E, eds), 363–396. New York: Plenum Press.

Wilbur HM, Collins JP (1973) Ecological aspects of amphibian metamorphosis. Science 182: 1305–1314.

Wiley EO (1981) Phylogenetics: The Theory and Practice of Phylogenetic Systematics. New York: Wiley.

Wilson DS (1983) The group selection controversy: History and current status. Ann Rev Ecol Syst 14: 159–187.

Wimsatt WC (1981) Units of selection and the structure of the multilevel genome. PSA 2: 122–183.

Wray GA, Bely AE (1994) The evolution of echinoderm development is driven by several distinct factors. Development (supp.) 97–106.

Wray GA, Raff RA (1991) Rapid evolution of gastrulation mechanisms in a sea urchin with lecithotrophic larvae. Evolution 45: 1741–1750.

Yaoita Y, Brown DD (1990) A correlation of thyroid hormone receptor gene expression with amphibian metamorphosis. Genes Devel 4: 1917–1924.

Yoshizato K (1992) Death and transformation of larval cells during metamorphosis of Anura. Devel Growth Different 34: 607–612.

III EVO-PATTERNS: WORKING TOWARD A GRAMMAR OF FORMS

Patterns permeate nature at all levels of organization. From molecules in a cell to organs in a body, from animals in a colony to ecosystems in the biosphere, patterns exist everywhere. But patterns are also the realm of art and human enterprise. Thus, we recognize a sense of universality embedded in patterns, which have permeated human culture through an inner necessity to comprehend natural phenomena. The fabulous limb of a dinosaur, the mighty limb of an elephant, and the gentle hand of Mona Lisa are all product of million of years of evolution, are all the same pattern. Patterns provide the necessary clues for understanding the processes that shape physical entities, allowing their study, and, in the case of biology, they provide the link between development and evolution. This section underscores a middle ground between fundamental biological processes as documented in the second section and human experience as portrayed in the next one. Modules here are physical entities, building blocks of organisms that can be isolated and scrutinized scientifically.

A striking feature of nature is the existence of common themes that recur over and over in fundamentally different systems. Patterns appear to be constrained by several physical and geometrical properties of matter, making the form and size of organisms predictable. Or that is what we would like, for although several "lawlike rules" have been predicted to account for the evolution of form, such as "Cope's rule" and several allometric principles of size and shape, we are far from having a complete understanding of the "logic of organic form." Ideally, we would like to have a set of rules that would make it trivial to know which sort of species can arise from an ancestor, so that the phylogenetic relationships between clades would be a matter of applying these rules.

But not everything is lost. The universality of patterns and their restrictive repertoire indicates that there might be such a set of rules. Perhaps the concept of modularity would open the door to the elaboration of a morphological grammar to understand the meaning of organic form.

Modularity is defined through a process that starts by recognizing patterns, shapes, or events that repeat at some scale of observation. The way we partition an object in order to study it determines our perception of its modularity. This section of the book is devoted to making sense of modularity as a recognizable, observable feature in nature. We do this at very different levels of organization.

Daniel McShea and Carl Anderson (chapter 8) consider "parts" (cf. the "subsystems" of nearly decomposable systems discussed by Simon in the foreword to this volume) at all levels of the biological hierarchy and present a very elaborated case to account for patterns in the evolution of multicellularity. A part, on their account, is a module in the "operation" of an organism (e.g., in its physiology or behavior) rather than its development. Two hypotheses about loss and gain of parts and their relation to functionality are the core of their contribution. They conclude by noting a phenomenon they call "remodularization," as organisms evolve parts that are transferred across levels of organization.

Chapter 9, by Diego Rasskin-Gutman, makes a case for modularity as a bridge between form and function and speculates about the need to quantify modularity, recognizing the existence of "degrees of modularity." Rasskin-Gutman seeks to counterbalance the tendency to view modularity almost exclusively as an expression of adaptation and the neo-Darwinian emphasis on population dynamics this implies. Taking modularity as the mark of *organization* in living systems, he argues that it provides the long-sought nexus between form and function. The result of the iterative process of making, repeating, and changing parts is the segmentation of the body and a separation of tightly integrated repetitive patterns. The generative competence of embryonic developmental processes defines a theoretical space of possible morphologies. Rasskin-Gutman suggests that the evolutionary significance of modularity can be best appreciated in the framework of morphospace. If life takes modularity as a set of construction rules, then evolution must proceed by moving around regions of modular design, leaving gaps in nonmodular regions. He proposes the notion of modular space as an intermediary between the all-encompassing theoretical morphospace and empirical morphospace, which elicits a portion of the actual realization of the former in nature.

Gunther Eble (chapter 10) views morphological modules as hypotheses of individuation that may find validation in separate mechanistic or theoretical contexts, but that can also be justified on their own, in terms of the distinct evolutionary and developmental dynamics that morphology entails. Morphological modules are, minimally, cohesive units of organismal integration; this definition is consistent with, but not equivalent to, general definitions of developmental module. Referring to his own analysis of data on sea urchins, Eble reasons that modularity is an evolutionary proxy that can be empirically discovered through the usage of landmarks. Groups of landmarks that vary in a correlated manner are putative modules that can be conveniently analyzed across taxa. Eble emphasizes the importance of modularity for understanding macroevolution.

Roger Thomas (chapter 11) uses the skeleton space as an approximation to modular design and argues that complexity has evolved following a logistic curve, with most designs already "discovered" by the time of the Cambrian explosion. The hard-part skeletons of living, crown-group metazoans evolved by integration and specialization of simple, modular elements that emerged first in disparate groups of small organisms, at the end of the Proterozoic and in earliest Cambrian time. He identifies five stages of development and permutation of skeletal elements in the evolution of metazoan skeletons. Metabolic processes that evolved at lower structural levels, within cells, were essential precursors of development in which they would be co-opted to secrete hard parts. Skeletal elements coevolved with the soft tissues by which they were formed. Thomas interprets the exploitation by metazoans of the design options available to form skeletons as one of a series of increasingly rapid structural bifurcations in the history of life, ranging from the gradual diversification of cell types among protists through most of the Proterozoic to the explosive proliferation of cultural artifacts unleashed by human behavior.

Slavik Jablan and Buscalioni et al. close this section by putting patterns and modularity in a human perspective. In mathematics and physics, painting and dancing, modularity is not only a physical reality, but also a perceptual and creative one. Jablan (chapter 12) introduces the mathematical regularities behind tiling and tessellations. He uses the notion of "prototiles" as modules that through symmetry rules make up modular, repetitive patterns, and compares this theory against ancient art and modern op art. An interesting notion he introduces is that of "degree of impossibility," to account for the perception of impossibility in impossible forms, such as the artwork of Escher.

Angela Buscalioni, Alicia de la Iglesia, Rafael Delgado-Buscalioni, and Anne Dejoan (chapter 13) go on to elaborate a whole theory of modularity as a tripartite phenomenon involving the entire modular structure, the individual modules, and the model that appears through the interactions between modules. They propose an enthralling journey across the boundary between science and art, spanning fluid dynamics, turbulences, animal anatomy, painting, architecture, music, and more. By looking at modular patterns in science and art, the last two chapters prepare the ground for the final section of this book, on mind and culture.

8 The Remodularization of the Organism

Daniel W. McShea and Carl Anderson

The evolutionary transitions from free-living, single-celled existence to full multi-cellularity involved both gains and losses. In each such transition, a multicellular organism was gained at a higher level. And at the same time, at a lower level, an organism was lost, in a sense, with the transformation of a free-living protist into a mere part in a larger whole. A similar point could be made for hierarchical transitions at other levels, especially the emergence of the first eukaryotic cell from symbiotic associations of prokaryotic cells (one level down from the cell–multicell transition) and the origin of individuated colonies from associations of multi-cellular organisms (one level up).

How—in structural terms—were these gains and losses achieved? Here, we offer a partial answer in the form of two hypotheses. The first is that the emergence of a higher-level entity with functional capabilities is ordinarily accompanied by the loss of part types within the lower-level organisms that constitute it. Thus, the sugges-tion is that cells in multicellular organisms will have fewer part types than free-living protists. The second hypothesis is that the lower-level organisms are transformed into differentiated parts within the higher-level entity. Along with this—as size increases—parts emerge at an intermediate scale, between the lower-level organ-isms and the higher-level entity. Thus, for example, in the evolution of multi-cellularity, cells are transformed from organisms (i.e., protists) into differentiated parts. Then, as the size of the multicellular entity increased, cells combined to form larger parts, intermediate in scale between a cell and the multicellular organism as a whole (e.g., organs). As we will show, these changes amount to a vertical shift in hierarchical structure, a transfer of parts from a lower level to a higher one, or a remodularization of the organism. We also offer a simple rationale for why this sort of remodularization might occur and discuss some of its possible implications for evolutionary trends in complexity.

It is worth stressing at the outset that in dividing these transformations into two categories (losses of parts at a lower level and gains at a higher level), and stating the overall pattern as two distinct hypotheses, we are making a conceptual separa-tion only. In fact, with the advent of functionality in the higher-level entity, losses and gains (and many other changes) presumably occur in concert. Indeed, as will be seen, the loss of parts within lower-level organisms is understood to be a conse-quence of, or even an aspect of, their transformation into parts of the higher-level whole. In other words, the lower-level organisms *lose* parts as they *become* parts.

Hierarchical transitions can be studied from a number of different angles. In one common approach, the goal is to understand how selection at the level of the lower-level organisms can be overcome by selection at the level of the whole—or, in other words, how the reproductive capacities of the lower-level organisms can be brought under control, or tamed (Leigh, 1983, 1991; Maynard Smith and Szathmáry, 1995; Michod, 1999). Alternatively, one could study the unique structural and organizational changes that accompany particular transitions (e.g., Beklemishev, 1969; Boardman and Cheetham, 1973; Buss, 1987; Maynard Smith and Szathmáry, 1995). Finally, one could investigate the common or generic structural and organizational features of all hierarchical transitions in organisms (e.g., Wimsatt, 1974, 1994; Salthe, 1985, 1993; Bonner, 1988; Anderson and McShea, 2001a); this last approach is the one we take here. This chapter extends earlier work on parts and hierarchy (McShea, 2001, 2002), in particular, earlier treatments of intermediate-level parts (McShea, 2001; see also Anderson and McShea, 2001a,b). The hypothesis that parts will be lost in lower-level organisms as functional higher-level entities emerge was proposed and tested in McShea (2002).

In the discussion that follows, we make two assumptions that here need to be made explicit. The first is that hierarchy is a matter of degree. In other words, the extent to which a higher-level entity constitutes a unified whole, or its degree of "individuation" (Salthe, 1985; also Beklemishev, 1969; Hull, 1980; Mishler and Brandon, 1987; Wilson, 1999; Dewel, 2000; McShea, 2001) is a continuous variable. The thinking is that the emergence of a higher-level entity entails a cluster of structural changes, such as the increase in size of the higher-level entity and the increase in connectedness among lower-level organisms (Anderson and McShea, 2001a), as well as the changes suggested by the hypotheses above, and that all of these are continuous variables. No empirical study has been done, but it seems likely that a mammal, a magnolia, and a mushroom are more individuated at the multicellular level than a sponge, a seaweed, and a slime mold, respectively (see discussion in McShea, 2001). The second assumption has to do with the emergence of higher-level function, or the emergence in a higher-level entity of the ability to feed, move, reproduce, and so on, as a unified whole, presumably as a result of selection acting at the higher level. The assumption is that individuation and function are related to each other, and further that the relationship is causal, that individuation is the result of selection for functional capability.

It is hard to know what to call entities that are in transition, entities that are thought to have either gained or lost functional capabilities, and thus have become either more or less organism-like. We have chosen consistency over accuracy. That is, we consistently refer to lower-level entities as organisms, even after they have lost a number of functional capabilities (as predicted by hypothesis one; see below). And we consistently refer to higher-level entities merely as entities, even after they have gained considerable functional capabilities (as assumed by both hypotheses; see below). Notice that in doing so, we avoid having to take a position on whether or not a functional colony of multicellular organisms is properly called an organism (or, rather, a superorganism; e.g., Wheeler, 1911; 1928; Seeley, 1989; Wilson and Sober, 1989), an issue that is decidedly beside the point in our treatment.

Hypothesis One: Parts Lost

The first hypothesis is that as a higher-level entity arises from an association of lower-level organisms, and as the higher-level entity acquires the ability to perform functions, the number of part types *within* the lower-level organisms (i.e., at a level just below the lower-level organisms) decreases, on average (figure 8.1). (A somewhat more technical discussion of the terms "part" and "level" will be offered shortly; for present purposes, the colloquial meanings will suffice.) More concretely, the prediction is that as an association of eukaryotic cells was transformed in evolution into a functional multicellular organism (as occurred in at least three cases: the plants, animals, and fungi), the number of part types within the cells would have been reduced, on average. An example of a part type that might be lost is an organelle, such as a mitochondrion, although cells have many other part types besides those conventionally called organelles (see below).

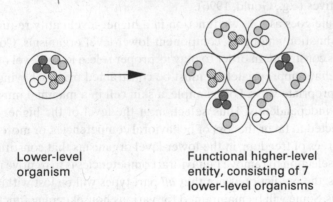

Lower-level
organism

Functional higher-level
entity, consisting of 7
lower-level organisms

Figure 8.1
Hypothesis One, showing the transformation of a lower-level organism (left) with five part types (the four groups of similarly shaded small circles, plus the outer membrane) into a functional higher-level entity (right, the group of seven lower-level organisms). The hypothesis is that the number of part types *within* the lower-level organisms decreases, on average, in this case, from exactly five part types (left) to about 2.9 (right), which is the average for the seven lower-level organisms (including their outer membranes).

Similarly, at a lower level, the suggestion is that in the evolution of the eukaryotic cell from a symbiotic association of prokaryotic cells, the number of part types within the prokaryotic symbionts would have been reduced. And at a higher level, the hypothesis is that as functional colonies arose from associations of multicellular organisms, the number of part types within those organisms would have declined. (For a longer discussion, see McShea, 2002).

The hypothesis has at least two rationales, two possible justifications. First, organisms in a free-living condition ought to be able to perform all survival- and reproduction-related functions for themselves. However, when they abandon a free-living existence and become incorporated into a higher-level entity, and as that entity acquires the ability to perform functions—to feed, move, defend itself, reproduce, and so on—the functional demands on the component lower-level organisms are reduced. The expectation then is that selection would favor the loss of part types within those entities in the interest of economy. Thus, a skin cell in a mammal experiences few functional demands, because most functions are performed by the animal as a whole, and therefore the cell requires few internal parts to perform those functions. A free-living protist, however, must perform all functions for itself, and therefore requires more internal parts. A required assumption, of course, is that the number of part types reflects, or correlates well with, the number of functions (McShea, 2000). The argument is essentially a generalization of the well-known argument for the putative reduction in parts in many parasites relative to those in their free-living relatives (e.g., Gould, 1996).

The second rationale goes as follows: function in a higher-level entity requires a high degree of coordination among its component lower-level organisms. Coordination in turn implies constraint. In order to play its proper role, a lower-level organism must not only behave appropriately, it must be constrained from behaving in a wide variety of inappropriate ways. For example, a skin cell in a mammal must not be allowed to move independently. Thus, selection at the level of the higher-level entity would be expected to favor the loss of behavioral competencies, or more generally the loss of degrees of freedom, in the lower-level organisms that constitute it. And a direct way for selection to remove behavioral competencies is to remove parts.

Some clarifications: the prediction is not that *all* part types will be lost within the lower-level organisms. Some will be maintained for various housekeeping functions. Others will be ineliminable due to constraints. Also, notice that the prediction is only that the *average* number of part types will be reduced. In a multicellular organism, for example, some specialized cells will require many part types (e.g., retinal cells, with their complex light-receiving apparatus) and some very few (e.g., mature human hemocytes, with essentially no internal macroscopic structures). Finally, the argument is framed here in functional terms, and therefore the loss of parts is assumed to be driven by selection.

Parts

A part is understood here as a set of components that are relatively well integrated or connected with each other and also relatively well isolated from other components outside the set. A connection between components refers to any form of interaction that produces correlations in their behavior, including bonds, collisions, signals, and so on. Thus, a solid object—such as a cell or an organ—is a part, in that its atomic or molecular components are ordinarily well bonded to each other and less well bonded to external entities. But a more dispersed set of components is also a part if their activities are well correlated, perhaps on account of signals exchanged among them. More concretely, within a multicellular organism, the components of a hormone-mediated control system might be a part. Or a behavior might be a part, if the nerve and muscle cells mediating the behavior are sufficiently well connected.

Here, a part refers only to a pattern of connectedness and isolation divorced from function. In other words, in principle, parts need not be functional. However, there is some reason to think that in organisms, parts and functions are in fact closely related; in particular, that the number of part types is well correlated with the number of functions. Briefly, the argument is that in order for a system to function, it requires a certain amount of internal coordination, and therefore connectedness to achieve that coordination, and also some degree of isolation, to limit interference from other systems. Thus, for example, for an organism to move and feed at the same time, the components involved in movement should be isolated from those involved in feeding.

More generally, in organisms, selection is expected to have isolated functions in parts to some degree. Of course, some degree of overlap is permitted; multiple functions may overlap in their use of the same part, so that the relationship between parts and functions is not expected to be one to one. But the expectation is nevertheless that the number of part types should be well correlated with the number of functions (For a longer discussion, see McShea, 2000). The first hypothesis relies heavily on this correlation. That is, it assumes that a reduction in number of functional demands on lower-level entities will be manifested as a reduction in number of part types.

Parts correspond closely with what Campbell (1958) called "entities," and what Simon (1962; also, Foreword in this volume) called "subsystems" in his discussion of "nearly decomposable systems." A part is also a kind of "module," but a module in a different sense than that which has become standard in evolutionary–

developmental biology in recent years. In that field, "module" has been used to refer to a more or less independent unit in the development of an organism (e.g., Wagner and Altenberg, 1996). A part, however, is a module in what might be called the "operation" of an organism (for example, in its physiology or behavior, rather than its development). The "dynamic modules" of Mittenthal et al. (1992) include both operational and developmental entities. Finally, technically a whole organism is a part (although the usage is an odd one when the organism is not part of any larger entity). But it is a special kind of part, a part with many functions all occurring within a common boundary (among other properties), presumably on account of strong or persistent selection acting on it at its own level.

Structural Hierarchy

The term "hierarchy" as used here refers to a structural relationship among parts, what Salthe (1985, 1993) called a scalar hierarchy, Wimsatt (1994) described as compositional levels of organization, and Valentine and May (1996) called a cumulative constitutive hierarchy (see also Pettersson, 1996; McShea, 1996). Hierarchical levels are understood as levels of nestedness: lower-level parts are physically contained within and partly constitute higher-level parts. It will be useful here to distinguish major levels of nesting from intermediate levels. The major levels are occupied by prokaryotic cells, eukaryotic cells, multicellular organisms, and colonies. A set of (former) prokaryotic cells is contained within and partly constitutes a eukaryotic cell, which is contained within and partly constitutes a metazoan, which in turn is part of a colony. In other words, the major levels are essentially the present and former "levels of selection" (e.g., Brandon, 1996).

Intermediate levels of nestedness also exist, and parts can be identified at those levels as well. Tissues and organs, for example, are intermediate-level parts between the eukaryotic cell and multicellular major levels. Intermediate-level parts will figure prominently in the discussion of the second hypothesis (see below).

For purposes of testing the hypothesis, attention to levels is crucial. The first hypothesis is that as function emerges at a higher level, lower-level organisms lose part types. Thus, cells in a functional multicellular organism should have fewer internal parts than their free-living relatives. More precisely, the prediction is that part types will be lost *at the level just below* the lower-level organisms, not at all lower levels. That is, the loss of part types does not necessarily entail the loss of part-types-within-part-types, or subpart types (see McShea and Venit, 2001). For example, in a eukaryotic cell, the chromosomes lie within the nucleus and are therefore subparts, not parts, of the cell. And in principle, the nucleus could be lost while the chromosomes are retained. Likewise, a flagellum contains microtubules as subparts, but the loss of a flagellum does not necessarily entail the loss of that subpart type, of all microtubules in the cell.

It may seem that the "true" parts of an organism are its genes, and therefore in principle the first hypothesis really predicts loss of gene types. In fact, however, it does not. One reason is that our understanding of parts here is structural, not generative. But even employing the generative sense, the loss of parts at a level just below the cell does not necessarily entail the loss of genes. A flagellum, for example, (apparently) contains no genes as subparts, and therefore the loss of a flagellum entails no loss of genes. Indeed, eliminating a flagellum might require additional genes. The conclusion is that in order to test the first hypothesis, we should count parts only at the level just below that of the cell, and no lower. (Notice that the issue is hierarchical level and not absolute size; some molecular species, such as those that are free in the cytoplasm, not contained within any structure, lie at a level just below the cell and therefore count as cell parts.)

It may be obvious—but is nonetheless worth stating clearly—that the first hypothesis also does not predict the loss of parts at higher levels. To see this, compare a multicellular organism having the capacity to perform many functions, perhaps a mammal, with another having fewer capacities, perhaps a cephalochordate such as *Amphioxus*. The hypothesis predicts that the mammal's cells should have fewer part types than those of *Amphioxus*. However, it does *not* predict that the mammal should have fewer cell types than *Amphioxus*, nor that it should have fewer tissue and organ types. (Indeed, as will be seen, the second hypothesis predicts it should have *more* of all of these types of structures.) In sum, the prediction of the first hypothesis is only that part types will be reduced at a single level, that just below the lower-level organisms.

A final clarification is in order: generalizing the point above about parts and genes, the concern here is with hierarchies of objects, *not* hierarchies of processes—in particular, not with the hierarchy of events in development (Riedl, 1978; Wimsatt, 1986; Valentine and Erwin, 1987; Salthe, 1993; Raff, 1996; Arthur, 1997). A developmental sequence is hierarchical in the sense that (and to the extent that) later events are dependent on earlier events, and also in that a one-to-many relationship exists between early and later events (McShea, 1996). Genealogical relationships can also be understood as hierarchical in this sense, with early arising individuals or taxa giving rise to (or causing) later arising individuals or taxa, producing in some cases a one-to-many relationship between early and late. In the present discussion, however, hierarchy refers only to physical nestedness, not to the structure of causal relationships among steps in a process (although as an empirical matter the two could be related). In the terms used by Eldredge and Salthe (1984), the concern is with the ecological hierarchy, not the genealogical hierarchy.

A Test at the Cell Level

Part-type counts within cells are consistent with the prediction of the first hypothesis. In particular, cells in metazoans and land plants have fewer part types, on average, than free-living protists (McShea, 2002). In that study, counting parts required two major assumptions. First, the number of types of "object-parts" is a good proxy for the number of all part types (i.e., including the nonobject parts, such as behaviors and physiological cycles; McShea and Venit, 2001). In other words, the assumption was that the parts that are manifested as objects are an unbiased sample of all true parts, including those with more dispersed components. Within cells, examples of object-part types include the nucleus, mitochondria, chloroplasts, cell membranes, cell walls, the Golgi apparatus, peroxisomes, contractile vacuoles, and so on. Notice that all of these objects lie at a level just below the cell, consistent with the terms of the first hypothesis (discussed in the section above). Objects lying topologically within these—such as cristae within mitochondria and chromosomes within the nucleus—occupy a hierarchical level too low (i.e., they are subparts of a cell rather than parts), and therefore are not counted.

Second, object-part counts were based on descriptions and electron micrographs in the cytological literature. Thus, very small, molecule-sized object parts were invisible and not counted, and the assumption was that the large object parts visible in electron micrographs are an unbiased sample of all true parts. This assumption is not clearly justified. One might argue, contrary to the rationale above, that the internal environment of a multicellular organism places a considerable number of demands on cells, especially for signaling and signal reception in the development, behavior, physiology, and so on of the higher-level entity. If so, the argument continues, signaling parts are likely to be molecule-sized, and therefore to be invisible in electron micrographs. Thus, in metazoans and land plants, counting only the large, visible parts biases the data in favor of the hypothesis. The possibility is worth raising, but it is not obvious that such a bias is present. The external environment of a protist is probably very complex, and its signaling and signal-detection requirements are likely to be considerable, perhaps even greater than those for a cell in a multicellular organism. In any case, the assumption in this test was that the large, visible parts are representative of all parts.

Tests are needed at other levels. At a higher level, the prediction is that as multicellular organisms combine to form a colony, and as the colony acquires the ability to perform functions, the number of part types within the organisms decreases. In this case, the relevant parts would be those that lie at a hierarchical level just below that of the multicellular organism, or roughly at the tissue and organ level (McShea and Venit, 2001). Thus, multicellular organisms in a eusocial insect colony, for example, should have fewer part types, on average, than their solitary or less social relatives. This could be tested using object parts, and in fact, consistent with the hypothesis, workers in the more differentiated ant colonies sometimes lack a key part, ovaries (Oster and Wilson, 1978; Noll, 1999).

An alternative is to use "behavior parts," rather than object parts, as a proxy for all true parts. Then the prediction would be that ants in complex, functional colonies should have smaller behavioral repertoires (i.e., fewer behavior part types) than ants in simple colonies (see Anderson and McShea, 2001, for further discussion). There is some evidence that this is the case. For example, in obligate slave-making ants, there is evidence of behavioral degeneration: all five *Polyergus* species have lost the ability to feed themselves and rely totally on their slaves for food, as well as for brood care and general housekeeping (E. O. Wilson, 1975; Mori and Le Moli, 1988; Topoff, 1999). Testing is also desirable at a lower level. In a eukaryotic cell, the former endosymbionts—mitochondria, chloroplasts, and perhaps others—should have fewer part types than free-living prokaryotes, on average.

Hypothesis Two: Parts Regained

The second hypothesis is that as a higher-level entity arises and acquires the ability to perform functions, the lower-level organisms are transformed into differentiated parts. And as size increases, the prediction is that parts arise at an intermediate level between the organisms and the higher-level entity (figure 8.2). Thus, a prediction of the second hypothesis is that the more functional and larger multicellular organisms will have more cell types and also more organ and tissue types, that is, more intermediate-level parts. The following section describes the various types of intermediate-level parts, and offers a possible logic that predicts their origin.

A connection between the two hypotheses is worth recalling here: that the loss of parts within the lower-level organisms predicted by the first hypothesis is an aspect of the transformation of lower-level organisms into differentiated and specialized parts, predicted by the second hypothesis. In other words, the loss of internal parts is one aspect of the more general process of specializing that the lower-level organisms undergo. Thus, the decision to include this specialization under the second hypothesis rather than the first was somewhat arbitrary.

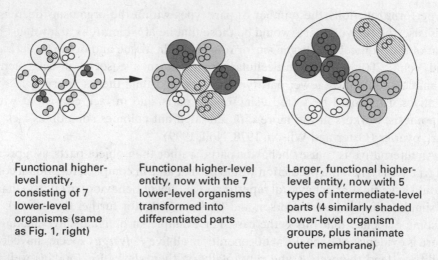

Functional higher-level entity, consisting of 7 lower-level organisms (same as Fig. 1, right)

Functional higher-level entity, now with the 7 lower-level organisms transformed into differentiated parts

Larger, functional higher-level entity, now with 5 types of intermediate-level parts (4 similarly shaded lower-level organism groups, plus inanimate outer membrane)

Figure 8.2
Hypothesis two, showing the transformation of the functional group of lower-level organisms (left, seven medium-sized circles; identical to figure 8.1, right) into differentiated parts of the higher-level entity (middle, with shading showing differentiation). Then, with an increase in size of the higher-level entity (right, shown here as only a slight increase in size, from seven to ten lower-level organisms), five intermediate-level parts emerge (right, shown as four groups of medium-sized circles, plus the new, larger, outer membrane). Notice that the medium-sized circles representing the four internal intermediate-level parts are similarly shaded, suggesting that the components of an intermediate-level part must be identical. But this need not be the case. See text.

Intermediate-Level Parts

These have been defined elsewhere (McShea, 2001) as parts within the higher-level entity consisting of either (1) a single lower-level organism that is enlarged or elaborated relative to a typical lower-level organism or (2) an association of two or more lower-level organisms. The definition was extended by Anderson and McShea (2001) to include (3) inanimate structures that are intermediate in size between lower- and higher-level entities.

Examples of the first type of intermediate-level parts include the following: the central gastrozooid in chondrophorines (colonial hydroids) consists of a single zooid supporting a large pneumatophore, or float, with a number of smaller gonozooids attached at the outer margin (Hyman, 1940; Kozloff, 1990). The central gastrozooid is a single, hypertrophied individual, and thus constitutes a part that is intermediate between the zooid level (i.e., historically, a multicellular organism) and the colony level. At a much lower hierarchical level, in a eukaryotic cell, the cell body, excluding the former endosymbionts—mitochondria, chloroplasts, and perhaps others—is an intermediate-level part. The cell body is presumably homologous with a prokaryotic cell, the ancestral archaebacterial host, and therefore can be considered a single, enlarged, and elaborated lower-level entity.

Examples of the second type, associations of two or more lower-level entities, include a variety of structures in multicellular organisms. The multicellular chlorophyte *Volvox* consists in the vegetative phase of a spherical shell of small, nonreproductive flagellated cells enclosing a smaller number of larger reproductive cells (Kirk, 1998). The shell of flagellated cells is an association of lower-level organisms and constitutes an intermediate-level part. The tissues, organs, and organ systems of the larger multicellular organisms with more functional capacities, such as bilaterian metazoans, also qualify as intermediate-level parts.

At a higher major level, the colony level, many colonial marine invertebrates have such collaborative associations. For example, in cyclostome bryzoans, the lower-level individuals, the zooids, feed by drawing in sea water and extracting food particles from it using a specialized feeding organ (lophophore); in many of the larger colonies, especially in encrusting forms, the immense volume of filtered or food-depleted water generated by the feeding zooids is expelled through excurrent chimneys called maculae (Banta et al., 1974; McKinney, 1990; Taylor, 1999). Maculae are often manifest not as objects but as regions, regularly spaced on the surface of the colony, where only nonfeeding zooids are present, or where the density of nonfeeding zooids is relatively high. Thus, a macula is the result of the coordinated activity—or more precisely, the coordinated *in*activity—of a subset of the lower-level entities (zooids), and therefore constitutes an intermediate-level part. Most of the structures that Beklemishev (1969) called "cormidia" are intermediate-level parts; for other examples in colonial marine invertebrates, see Gould (1985).

Also at the colony level, the "groups" and "teams" in insect societies are intermediate-level parts of this type. A group task (*sensu* Anderson and Franks, 2001) is one that is accomplished by many lower-level organisms acting in concert and performing roughly the same behavior; for example, many ants retrieve forage items as a group, using the combined strength of many ants pulling on an item at once to overcome frictional forces (Sudd 1963, 1965). In contrast, in a team (*sensu* Anderson and Franks, 2004), each lower-level organism has its own specialized job. For example, in colony defense, *Pheidole pallidula* ants may act as a team, with a group of minors pinning down an intruder while a major (with larger and stronger mandibles) decapitates it (Detrain and Pasteels, 1992). Team tasks are also known in a variety of vertebrate groups, including lions, chimpanzees, and humpback whales (see Anderson and Franks, 2004). These groups and teams are parts in that, while they are in progress, they consist of sets of components (the individual ants) that are well connected internally (via their interactions) and relatively well isolated externally from other components.

Two comments are in order regarding this second type of intermediate-level part. First, as should be evident from these examples, the components of an intermediate-level part of this type may or may not be differentiated; the behaviors of the ants in a team task are differentiated, while the morphologies of the cells of the flagellated shell in *Volvox* are not. (Figure 8.2, right, represents intermediate-level parts as undifferentiated.) Second, this notion of an intermediate-level part allows us to formalize a connection that has long been recognized on an intuitive level, i.e., that between tissues and organs in multicellular entities on the one hand, and groups and teams in colonies on the other. For example, in a 1985 paper on "sociogenesis" in insect societies, Wilson wrote: "The workers of advanced insect societies are not unlike cells that emigrate to new positions, transform into new types, and aggregate to form tissues and organs" (Wilson, 1985, p. 1492).

The third type of intermediate-level part consists of inanimate objects that are intermediate between the lower-level organisms and the higher-level entity. For example, the sheaths or envelopes surrounding many colonial cyanobacteria (found, for example, in filamentous forms such as *Anabaena*) are intermediate between the bacteria and the colony or filament as a whole. One hierarchical level up, the shells of certain free-living protists, such as those in radiolarians and other testate amoebae, are intermediate-level parts. At a still higher level, the skeletal elements of multicellular organisms are intermediate-level parts. At the colony level, the nests of various social vertebrate and invertebrate species count as intermediate-level parts. In insect societies, intermediate parts include the walled and sometimes roofed foraging trails produced in certain army ant species, the protective shelters built by various ant species to house aphids and other honeydew-producing insects, and elevated corridors and bridges (see Anderson and McShea, 2001b).

Inanimate intermediate-level parts in an organism are often produced by the activities of either of the other two types of intermediate-level parts in the same organism: single hypertrophied lower-level entities or associations of two or more entities. For example, skeletal elements, sheaths, and so on are normally secreted by one or more of the lower-level entities they support and protect. This is not required, however. A snail shell found and adopted by a hermit crab becomes an intermediate-level part (of the crab). And the bubble carried underwater as an air supply and gas exchanger by the so-called bubble-carrying beetle, *Notonecta* (Turner, 2000) is one of the beetle's intermediate-level parts.

Rationale

Differentiation of the lower-level organisms in a higher-level entity is expected when division of labor produces increases in efficiency (Smith, 1776; Bonner, 1988; Bell and Mooers, 1997)—or when it produces synergistic effects (Corning, 1983)—that improve the functioning of the whole. In principle, increases in numbers of lower-level organisms (i.e., increases in size of the whole) permit greater division of labor, and therefore the number of differentiated types should increase with size (Bell and Mooers, 1997). Further, Bonner (1988) has argued that large size may actually require greater division of labor, and therefore greater differentiation. For example, large size reduces surface-area-to-volume ratios for an organism, which changes the performance requirements for certain tasks, and these changes in turn necessitate the evolution of specialized devices; for example, a large multicellular organism, unlike a solitary protist, may require specialized parts for gas exchange. Thus, as higher-level entities form from lower-level organisms, acquire the ability to perform functions, and increase in size, the lower-level organisms are expected to differentiate and to become transformed into specialized parts in the larger whole.

One might imagine that specialization ought to be substantial, so that each lower-level organism performs a unique function, or at least that the number of specialists should be large. And indeed, the ergonomic model of Oster and Wilson (1978) predicts that, in an insect society, if no constraints are present, number of castes should equal number of tasks. Generalizing, the expectation is that the number of differentiated types of lower-level organisms is expected to be equal to the number of functions, even when the number of lower-level organisms is large (i.e., when the higher-level entity is large). In fact, this is not observed; instead, as size increases, differentiation among lower-level organisms lags behind the increase in their numbers (Bell and Mooers, 1997). One reason is undoubtedly that differentiation is limited by numerous developmental and physiological constraints (Oster and Wilson, 1978).

A second reason may be that extreme specialization is somewhat risky. If each lower-level organism performs a single, unique function, loss of that key individual eliminates a functional capability for the whole; safety probably requires some redundancy (Anderson and McShea, 2001). Related to this, if each lower-level organism is uniquely specialized, so that task switching is impossible, then the higher-level entity may be unable to track changes in functional demands that occur when the environment fluctuates.

The second hypothesis also predicts the emergence of intermediate-level parts. The thinking is that as size increases, single lower-level organisms may become too small to make a significant contribution to the whole (Bell and Mooers, 1997). Thus, to accomplish a function at the scale of the whole, when the whole is large, selection may favor the enlargement of single lower-level organisms, collaborations among multiple identical lower-level organisms, or large, inanimate devices, that is, intermediate-level parts. Further, with the advent of differentiation among the lower-level organisms, selection may favor the performance of functions using intermediate-level parts consisting of multiple lower-level types, for example, teams in ant colonies (see above). Indeed, increases in number of types of lower-level organisms should produce dramatic increases in combinatorial possibilities, following a power law (Changizi, 2001), and thus also in the possible number of types of intermediate-level parts.

Possible Tests
The prediction that differentiation at a lower level increases with size at the higher level has been demonstrated for the cell–multicell transition by Bell and Mooers (1997), who found a power-law relationship between size and number of cell types in multicellular organisms. Also, at a higher level, the degree of polyphenism—the degree of morphological, physiological, and/or behavioral differentiation—in insect societies increases with colony size (Anderson and McShea, 2001).

Regarding intermediate-level parts, some evidence exists that supports the second hypothesis. First, it is widely acknowledged that complexity—understood as diversity of internal organs and other intermediate-level parts—is lower, on average, in the smaller multicellular organisms, especially in the so-called interstitial organisms, or meiofauna (Swedmark, 1964; Westheide, 1987; Hanken and Wake, 1993). At a higher level, in ant colonies, certain types of intermediate-level parts, such as groups and teams, appear only at large colony size (Anderson and McShea, 2001; Anderson and Franks, 2001). However, to our knowledge, no formal test has been done. One approach would involve counting the number of intermediate-level part types in a variety of higher-level entities—say, at the colony level—and testing for a correlation with colony size. To support a generalization across levels, the same test would need to be done in multicellular organisms; a test at the eukaryotic cell level might even be possible.

Finally, an interesting possible test case may be available at an even higher level (i.e., one major level *above* the colony level). As described by Queller (2000), the Argentine ant *Linepithema humile* has invaded California recently, where much of its success has been attributed to the absence of fighting among colonies (unlike in their native Argentina). This peaceful neighborhood or "Pax Argentinica," as Queller has termed it, allows colonies to spend more time foraging and reproducing than they could if they had been aggressive. In addition, there is some flow of queens and workers among colonies. This raises the possibility that the group of colonies constitutes a new, functional higher-level entity, a supercolony in which the parts are colonies.

Our thesis here is that the two hypotheses apply at all major levels. Thus, if the new higher-level entity is stable (and Queller doubts that it is), and if selection acts on it to produce functional capabilities, we would expect some differentiation and specialization among its parts, with some colonies within the supercolony performing certain tasks far more frequently than others. This suggestion is not necessarily as improbable as it might at first seem. At a lower level, in honeybee colonies in which the queen is usually multiply mated, genetic studies have shown that different patrilines have different propensities to perform various tasks. That is, the offspring of father A are more likely to forage for pollen while the offspring of father B are more likely to forage for nectar, and so on (see Dreller and Page, 1999 and references therein). In the new higher-level entity, we might expect the emergence of colonies that are task-specific in the same way, and that lose the ability to perform a more generalist role, rather like the slave maker ants. Of course, the emergence of function in the California entity has not been demonstrated. Also, the entity is probably only a few decades old, and the evolutionary changes predicted by the hypotheses presumably have not yet occurred. We raise this case mainly to illustrate the hierarchical scope of the hypotheses, and to encourage a broad-minded and opportunistic approach to testing.

Concluding Remarks

The two hypotheses together predict that the emergence of a new functional higher-level entity will involve transformations at four levels: (1) the level below the lower-level organisms, at which parts are lost (hypothesis 1; figure 8.1); (2) the lower-level organisms themselves, which are transformed into parts (hypothesis 2; (figure 8.2); (3) the intermediate level, where parts emerge as size increases (hypothesis 2; figure 8.2); and (4) the higher level, where an entity emerges and becomes functional, driving the other three transformations. Figure 8.3 combines figures 8.1 and 8.2 to show the predictions of both hypotheses together. The net effect is a transfer of parts from the lower level to the higher—in other words, what might be called a "repartification" or, more euphonically, a "remodularization" of the organism as a new hierarchical levels arises. (We switch terminologies here and in the title of this chapter—from parts to modules—somewhat reluctantly and with a reminder to the reader that "module" here refers to a unit in the operation of an organism, not in its development.)

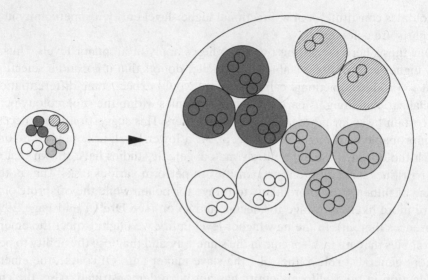

Figure 8.3
Remodularization. Figures 8.1 and 8.2 are combined to show the effect on parts organization suggested
by the two hypotheses together. Notice that the net effect of the advent of a new hierarchical level has
been a shift of parts from the level just below the lower-level organism to the intermediate level between
the lower-level organism and the higher-level entity. In this case, the original lower-level organism and
the larger higher-level entity have the same number of part types (five, including the outer membrane)
and the same parts organization (same pattern within the outer membrane on the left as within the larger
outer membrane on the right), but at different hierarchical levels. This was done for heuristic purposes,
to make the shift in parts organization vivid, and is not a necessary consequence of the hypotheses. See
text.

Notice that as figure 8.3 is drawn, it illustrates an extreme case in which the intermediate-level parts within the higher-level entity on the right are nearly the same as those within the lower-level organism on the left, and they are organized in the same way. This was done only to dramatize the point that the expected net effect of both hypotheses is a transfer of parts across levels. Importantly, however, neither is a necessary consequence of the hypotheses. Indeed, it seems likely that functional requirements would be different at different hierarchical levels (if only because of size differences; see Bonner's argument above), and therefore the part types would be different. On the other hand, it is not obvious that there will be systematic changes in the *number* of functions to be performed, and therefore in the number of part types, with increasing hierarchical level. In particular, it is not obvious that multicellular organisms ought to have more functional requirements and more part types than protists—or, more generally, that higher-level entities ought to require more functions and more part types than lower-level entities. If the functional demands of life at higher levels seem more numerous and more onerous, it may be mainly because they are more vivid and easier to imagine for higher-level entities like us.

A related issue here is complexity and how it changes in evolution. Complexity has a variety of senses, but in a narrow sense that has been advocated in biology (McShea, 1996), it has two principal components, hierarchical and nonhierarchical. Nonhierarchical or horizontal complexity (Sterelny, 1999) refers to the number of part types at a given hierarchical level in an organism. For example, the nonhierarchical complexity of an aquatic arthropod might be (partly) a function of the number of limb-pair types it has (Cisne, 1974). Hierarchical or vertical complexity (Sterelny, 1999) refers to the number of levels of nesting of parts within wholes. Thus, a colony is hierarchically more complex than a multicellular organism, which in turn is hierarchically more complex than a solitary protist, and so on.

It is clear that hierarchical complexity has increased a number of times over the history of life, with increases occurring in the transition from prokaryote to eukaryotic cell to multicellular eukaryote, and finally—by the early Phanerozoic—to individuated colony (McShea, 2001). For nonhierarchical complexity, the expectation might be that increases should also occur at every level, simply on the assumption that selection for greater functional capability will favor increased numbers of part types at all levels. What might seem to emerge from these considerations is a picture of evolution in which organisms arise that are ever deeper hierarchically (with the addition of new levels) and more diverse structurally, more richly ornamented (with the addition of new parts at each level).

However, the hypotheses offered here suggest a different picture. The suggestion is that while the origin of a new major level does produce gains in nonhierarchical complexity at one level (i.e., the transformation of lower-level organisms into parts and the gains in intermediate-level parts), it also produces a partly offsetting reduction in nonhierarchical complexity at a much lower level, a kind of hollowing out of the lower-level organisms. Of course, owing to constraints, the hollowing out is likely to be incomplete. And therefore, even if the number of functional demands were the same at all major levels, there would still be some net gain in nonhierarchical complexity for the organism as a whole, in the total number of parts summed over all levels. But the net gain could be considerably less than one would expect.

We stress that neither the two hypotheses nor the above view of complexity evolution have been demonstrated. Studies of evolutionary trends in nonhierarchical complexity have been done in certain groups (Cisne, 1974; McShea, 1993; Valentine et al., 1993), but all have targeted change at a single level (e.g., cell types within metazoans). An appropriate test would require investigating evolutionary changes in the number of part types at multiple levels simultaneously.

Acknowledgments

We thank the organizers and participants in the Fifth Altenberg Workshop.

References

Anderson C, Franks NR (2004) Teamwork in ants, robots and humans. Adv Stud Behav 33: 1–48.

Anderson C, McShea, DW (2001a) Individual versus social complexity, with particular reference to ant colonies. Biol Rev (Cambridge) 76: 211–237.

Anderson C, McShea DW (2001b) Intermediate-level parts in insect societies: Adaptive structures that ants build away from the nest. Insectes Sociaux 48: 291–301.

Arthur W. (1997) The Origin of Animal Body Plans. Cambridge: Cambridge University Press.

Banta WC, McKinney FK, Zimmer RL (1974) Bryozoan monticules: Excurrent water outlets? Science 185: 783–784.

Beklemishev WN (1969) Principles of Comparative Anatomy of Invertebrates. Vol. 1. Promorphology (Kabata Z, ed, MacLennan JM, trans). Chicago: University of Chicago Press.

Bell G, Mooers AO (1997) Size and complexity among multicellular organisms. Biol J Linn Soci 60: 345–363.

Boardman RS, Cheetham AH (1973) Degrees of colony dominance in stenolaemate and gymnolaemate bryozoa. In: Animal Colonies: Development and Function Through Time (Boardman RS, Cheetham AH, Oliver WA Jr, eds), 121–220. Stroudsburg, PA: Dowden, Hutchinson, and Ross.

Bonner JT (1988) The Evolution of Complexity by Means of Natural Selection. Princeton: Princeton University Press.

Brandon RN (1996) Concepts and Methods in Evolutionary Biology. Cambridge: Cambridge University Press.

Buss LW (1987) The Evolution of Individuality. Princeton: Princeton University Press.

Campbell DT (1958) Common fate, similarity, and other indices of the status of aggregates of persons as social entities. Behav Sci 3: 14–25.

Changizi MA (2001) Universal scaling laws for hierarchical complexity in languages, organisms, behaviors and other combinatorial systems. J Theor Biol 211: 277–295.

Cisne JL (1974) Evolution of the world fauna of aquatic free-living arthropods. Evolution 28: 337–366.

Corning PA (1983) The Synergism Hypothesis: A Theory of Progressive Evolution. New York: McGraw-Hill.

Detrain C, Pasteels JM (1992) Caste polyethism and collective defense in the ant *Pheidole pallidula*: The outcome of quantitative differences in recruitment. Behav Ecol Sociobiol 29: 405–412.

Dewel RA (2000) Colonial origin for Eumetazoa: Major morphological transitions and the origin of bilaterian complexity. J Morphol 243: 35–74.

Dreller C, Page RE Jr (1999) Genetic, developmental, and environmental determinants of honey bee foraging behavior. In: Information Processing in Social Insects (Detrain C, Deneubourg J-L, Pasteels JM, eds), 187–202. Basel: Birkhäuser.

Eldredge N., Salthe SN (1984) Hierarchy and evolution. Oxford Surv Evol Biol 1: 184–208.

Gould SJ (1985) A most ingenious paradox. In: Gould, The Flamingo's Smile, 78–95. New York: Norton.

Gould SJ (1996) Full House: The Spread of Excellence from Plato to Darwin. New York: Harmony Books.

Hanken J, Wake DB (1993) Miniaturization of body size: Organismal consequences and evolutionary significance. Ann Rev Ecol Syst 24: 501–519.

Hull DL (1980) Individuality and selection. Ann Rev Ecol Syst 11: 311–332.

Hyman LH (1940) The Invertebrates: Protozoa Through Ctenophora. New York: McGraw-Hill.

Kirk DL (1998) Volvox: Molecular-Genetic Origins of Multicellularity and Cellular Differentiation. Cambridge: Cambridge University Press.

Kozloff EN (1990) Invertebrates. Philadelphia: Saunders College Publishing.

Leigh EG Jr (1983) When does the good of the group override the advantage of the individual? Proc Natl Acad Sci USA 80: 2985–2989.

Leigh EG Jr (1991) Genes, bees and ecosystems: The evolution of a common interest among individuals. Trends Ecol Evol 6: 257–262.

Maynard Smith J, Szathmáry, E (1995) The Major Transitions in Evolution. Oxford: Freeman.

McKinney FK (1990) Feeding and associated colonial morphology in marine bryzoans. Rev Aquat Sci 2: 255–280.

McShea DW (1993) Evolutionary change in the morphological complexity of the mammalian vertebral column. Evolution 47: 730–740.

McShea DW (1996) Metazoan complexity and evolution: Is there a trend? Evolution 50: 477–492.

McShea DW (2000) Functional complexity in organisms: Parts as proxies. Biol Philos 15: 641–668.

McShea DW (2001) The hierarchical structure of organisms: A scale and documentation of a trend in the maximum. Paleobiology 27: 405–423.

McShea DW (2002) A complexity drain on cells in the evolution of multicellularity. Evolution 56: 441–452.

McShea DW, Venit EP (2001) What is a part? In: The Character Concept in Evolutionary Biology (Wagner GP, ed), 259–284. San Diego: Academic Press.

Michod RE (1999) Darwinian Dynamics: Evolutionary Transitions in Fitness and Individuality. Princeton: Princeton University Press.

Mishler BD, Brandon RN (1987) Individuality, pluralism, and the phylogenetic species concept. Biol Philos 2: 397–414.

Mittenthal JE, Baskin AB, Reinke RE (1992) Patterns of structure and their evolution in the organization of organisms: Modules, matching, and compaction. In: Principles of Organization in Organisms (Mittenthal JE, Baskin A, eds), 321–332. Reading, MA: Addison-Wesley.

Mori A, Le Moli F (1988) Behavioural plasticity and domestic degeneration in facultative and obligatory slave-making ant species (Hymenoptera, Formicidae). Monit Zool Ital n.s. 22: 271–285.

Noll FB (1999) Has colony size increased concomitantly with caste dimorphism some Epiponini wasps: A phylogenetic perspective. In: Social Insects at the Turn of the Millennium. Proceedings of the 13th International Congress of the International Union for the Study of Social Insects, IUSSI, Adelaide, 29 December 1998–3 January 1999 (Schwarz MP, Hogendoorn K, eds), 348. Adelaide: Flinders University Press.

Oster GF, Wilson EO (1978) Caste and Ecology in the Social Insects. Princeton: Princeton University Press.

Pettersson M (1996) Complexity and Evolution. Cambridge: Cambridge University Press.

Queller DC (2000) Pax Argentinica. Nature 405: 519–520.

Raff RA (1996) The Shape of Life. Chicago: University of Chicago Press.

Riedl R (1978) Order in Living Organisms (Jefferies RPS, trans). New York: Wiley.

Salthe SN (1985) Evolving Hierarchical Systems. New York: Columbia University Press.

Salthe SN (1993) Development and Evolution. Cambridge, MA: MIT Press.

Seeley TD (1989) The honey bee colony as a superorganism. Amer Sci 77: 546–553.

Simon HA (1962) The architecture of complexity. Proc Amer Phil Soc 106: 467–482.

Smith A (1776) The Wealth of Nations, books I–III. Reprinted. Skinner A (ed) Harmondsworth, UK: Penguin, 1986.

Sterelny K (1999) Bacteria at the high table. Biol Philos 14: 459–470.

Sudd JH (1963) How insects work in groups. Discovery (London) 25: 15–19.

Sudd JH (1965) The transport of prey by ants. Behaviour 25: 234–271.

Swedmark B (1964) The interstitial fauna of marine sand. Biol Rev (Cambridge) 39: 1–42.

Taylor PD (1999) Bryozoans. In: Functional Morphology of the Invertebrate Skeleton (Savazzi E, ed), 623–646. New York: Wiley.

Topoff H (1999) Slave-making queens. Sci Amer (November): 84–90.

Turner JS (2000) The Extended Organism. Cambridge, MA: Harvard University Press.

Valentine JW, Collins AG, Meyer CP (1993) Morphological complexity increase in metazoans. Paleobiology 20: 131–142.

Valentine JW, Erwin, DH (1987) Interpreting great developmental experiments: The fossil record. In: Development as an Evolutionary Process (Raff RA, Raff EC, eds), 71–107. New York: Liss.

Valentine JW, May CL (1986) Hierarchies in biology and paleontology. Paleobiology 22: 23–33.

Wagner GP, Altenberg L (1996) Perspective: Complex adaptations and evolution of evolvability. Evolution 50(3): 967–976.

Westheide W (1987) Progenesis as a principle in meiofauna evolution. J Nat Hist 21: 843–854.

Wheeler WM (1911) The ant colony as an organism. J Morphol 22: 307–325.

Wheeler WM (1928) The Social Insects: Their Origin and Evolution. London: Kegan Paul, Trench & Trubner.

Wilson DS, Sober E (1989) Reviving the superorganism. J Theor Biol 136: 337–356.

Wilson EO (1975) Slavery in ants. Sci Amer (June): 32–36.

Wimsatt WC (1974) Complexity and organization. In: PSA 1972 (Schaffner KF, Cohen RS, eds), 67–86. Dordrecht: Reidel.

Wimsatt WC (1986) Developmental constraints, generative entrenchment, and the innate–acquired distinction. In: Integrating Scientific Disciplines (Bechtel W, ed), 185–208. Dordrecht: Nijhoff.

Wimsatt WC (1994) The ontology of complex systems: Levels of organization, perspectives, and causal thickets. Can J Phil supp. 20: 207–274.

9 Modularity: Jumping Forms within Morphospace

Diego Rasskin-Gutman

. . . modularity, is an ineluctable feature of biological order. It is arguably the most crucial aspect of living organisms and their ontogenies, and is the attribute that most strongly facilitates evolution.
—Rudolph Raff (1996)

The design of multicellular organisms seems to rely almost uniquely on one single set of construction rules: make parts, repeat them, change them. The result of this iterative process is a segmentation of the body—sometimes concealed by an elaborated web of developmental processes—and a separation of tightly integrated repetitive patterns. This is what we call modularity. The potential generative competence of these embryonic developmental processes defines a theoretical space of possible morphologies that, during the course of the evolution of a lineage, is explored at length. However, as a result of the constraints imposed by developmental demands (e.g., genetic network dynamics, protein assemblies, cellular dynamics, metabolism, etc.), some regions of morphospaces will be filled while others will remain empty. Thus, in geologic time, evolving species jump from location to location within morphospace, following the logic of development. I would like to argue in the following sections that modularity, as the most conspicuous design element of organic form, defines those areas that are putative homes for future species. In order to build up this argument, I will discuss the nature and origins of morphological modularity and the significance of modularity regarding evolutionary change in the framework of morphospace theory.

The phenomenon of repetition was already considered by Bateson (1894) as the most important architectural pattern in the design of organisms (see also Weiss, 2002). Bateson was interested in the problem of how to analyze morphological variation—specifically, a phenomenon he identified as "meristic characters," those that would vary in number and that would be repeated in the body plan of a species, such as fingers or ribs. The recurrence of these kinds of characters in very different groups of organisms convinced Bateson that repeated parts generate a geometrical constraint with respect to what we would today called evolvability. More recently, Carroll (2001) also recognized that the complexity we see in living organisms is due to their modular design and the developmental independence for each embryo part that stems from their modularity.

The use of repetitive motifs is now known to occur at all levels of the biological organization (Duboule and Wilkins 1998; von Dassow and Munro, 1999; Klingenberg et al., 2001). Genes are made of four units that repeat themselves in patterns that are not yet well understood. Proteins are organized in domains, such as helixes and sheets, which occur again and again in all organisms. Cells are the paradigm of modular design as units of self-sufficient organization that form other modules, such as the alveoli in the lungs, the compartments in the heart or the limb pairs. Also, symmetry, as a design motif in multicellular organisms, can be viewed as repetition of parts (Bateson, 1894), that is, as an emergent pattern that arises as a product of modularity. Indeed, both radial and bilateral symmetry are among the most evident morphological patterns in fungi, plants, and animals, having been conserved since their first appearance, more than 500 million years ago (Buscalioni, 1999).

Although a single definition of modularity may be elusive, a good departure point is to recognize modularity as the nexus between morphological organization and functional integrity of an organic structure. Furthermore, modularity should be treated as a phenomenon that manifests itself to various degrees. Thus, organic structures would presumably exhibit different "degrees of modularity" relative to their organizational and functional properties. An important question that will be addressed is how to quantify modular variation. The problem of the origin of modularity in metazoan design (at a supracellular level) is, conceptually, a simpler one.

The repeated binary division of cells provides the simplest case of modular design in metazoans (see figure 9.1). During embryogenesis, this self-organizational binary dynamics generates pattern formation processes with functional properties such as near-decomposability (Simon, 1962) and preferential communication clustering of adjacent cells. The limb primordium in vertebrates is a clear case, in which there is a distinctive mass of mesenchymal cells that can even develop in other regions of the embryo after grafting and transplantation. At a higher level, modularity is

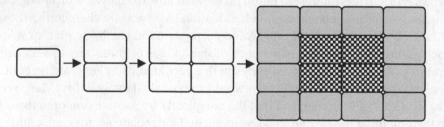

Figure 9.1
Repeated binary divisions self-organize modular structures in multicellular organisms. The first such manifestation is the inside-outside border in a blastula. Communication among the inside cells is favored by their proximity in detriment of their communication with the outside layer. Differential intercellular adhesive forces add up to generate the first step into the modularization of the organism.

produced by generic physical properties of tissues with differential adhesive surfaces, repeatedly generating structures such as tubes and sheets (Newman, 1992). However, the subsequent nested folding and inductive processes of the elaborate and complex embryonic dynamics conceal a clear distinction of their modular composition (see, for example, Bard, 1990; Slack 1991; Gilbert, 2000).

The evolutionary significance of modularity can be best appreciated within morphospace. Thus, a fundamental problem in morphospace and evolutionary theory is to elucidate how evolution moves from one region of morphospace into another. If life takes modularity as a set of construction rules, then evolution should necessarily proceed by moving around regions of modular design, consistently leaving gaps in nonmodular regions. In this metaphor, modular space can be viewed as an intermediary between the all-encompassing theoretical morphospace, which provides the mathematical and generative possibilities of form, and empirical morphospace, which gives a portion of the actual realization in nature. I will conclude by proposing the notion of modular space as the design pool from which lineages can jump in speciation events, generating evolutionary sequences.

Morphological Modularity

Living matter exhibits four distinct levels of morphological organization: proportions, orientations, connections, and articulations (Rasskin-Gutman and Buscalioni, 2001; Rasskin-Gutman, 2003). These four levels provide not only a way to generate a descriptive comparative framework but also a way to understand the logic of organic form. Each of these levels is a manifestation of a functional integrity among organic parts. In this context, function is understood as a successful interaction among parts, so that the whole "works" in the sense of maintaining the integrity of the organism. Thus the function of a heart is not to pump blood but to interact with other parts, such as blood, arteries, veins, muscles, and nerves. Function is understood as interaction without purposive elements. For any organic part to have a successful role in the life of an organism, it is necessary that the functional integrity be manifested at the four levels of morphological organization. Its proportions must be right, its orientation has to be right, the connections among its elements and with other elements of the organism have to be right, and the articulation or ability to change the orientation has to be right.

If modularity is an unavoidable design feature of organic life, then we should expect a series of phenomena to recur over and over in different lineages: seriality, redundancy, specialization, and integration. If it is not, we should not expect any of them (or at least we should not expect them to appear as a general pattern in distantly separated lineages). Ever since the appearance of life on Earth, evolution has been operating over units of change constituting variations on the same theme. At above-molecular levels, multicellularity is the first sign of the usage of modular design in evolution. At subcellular levels, modularity might also be an organizational standard. The segment polarity network, a genetic network present in the fruit fly, has clearly been shown to exhibit a modular nature (von Dassow et al., 2000).

These two instances exemplify the critical condition of modularity, its highly organized whole. Modules are parts of a system that have semi-independence in the sense that the ties within these parts are stronger than any other ties between other parts not belonging to the module. A hand is a module; its fingers have more relations among themselves than with other parts of the body (e.g., the toes). These relations are connectivity relations, in the sense that they are topologically related, with the resulting sharing of anatomical resources such as blood supply or innervation patterns.

Thus, a fundamental property of a modular system is its connectivity relations, on which all other properties are dependent. Moreover, connectivity provides the condition for the origin of a modular structure. It is precisely these connectivity relations that determine the modular properties characterizing the architecture of biological design. Modules can change semi-independently of the surroundings for each of the four morphological levels of organization, providing suitable raw materials for evolvability in a lineage (see figure 9.2). Furthermore, all other sorts of organismic modularity, such as developmental and functional modularity, are subordinated to the properties of morphological modules.

Modularity should be treated as a phenomenon that manifests itself to various degrees. Organic structures would presumably exhibit different "degrees of modularity" relative to their organizational and functional properties. The tetrapod pelvic girdle is a good example. Vertebrates have evolved this structure that initially appeared as a point of anchorage (connection) for the paired pelvic fins in fishes. The pelvis has three paired bones: the ilium, the ischium, and the pubis. In the evolution of tetrapods, the pelvis has been transformed into a ringlike structure with the ilia forming the dorsal half of the ring and the other two bones closing it ventrally. Archosaurs, a clade formed at present only by crocodiles and birds, but including pterosaurs, dinosaurs, and some other primitive reptiles, show five types of configurations (Rasskin-Gutman and Buscalioni, 2001). Figure 9.3 illustrates these five configurations that have evolved for all archosaur groups in the past 250 million years.

Since morphological modularity is characterized by the interconnection among elements in a system, networks can provide an effective way to represent and

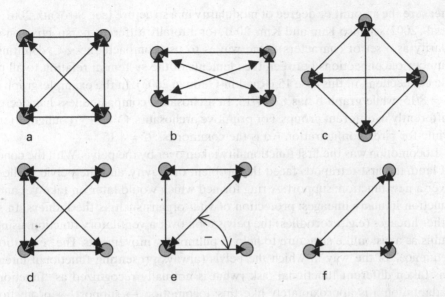

Figure 9.2
Modular organization. Modularity is a set of construction rules: make, repeat, change. For each individuated module there is a tight interconnection among elements. (*a*) initial module; (*b*) change in proportions; (*c*) change in orientation; (*d*) change in connections; (*e*) change in articulations; (*f*) loss of elements.

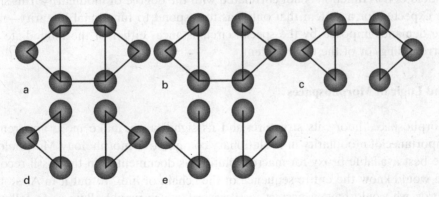

Figure 9.3
The pelvis as a graph with six nodes. Edges are physical connections between nodes. These five configurations are present in archosaurs. The total number of possible configurations is 156; the total number of possible connections is 15. (After Rasskin-Gutman and Buscalioni, 2001.)

measure the amount or degree of modularity in a structure (see Strogatz, 2001; Milo et al., 2002; see also Kim and Kim, 2001, for a totally different perspective on modularity as a set of characters). One way is to use compactedness, C, counting the number of connections between the elements of the system in relation to all possible connections, in this case 15 (Rasskin-Gutman 2003). In the example, graph a has $C = 8/15$, while graph b has $C = 7/15$. In archosaurs, compactedness has decreased differently in different groups. For primitive archosaurs, $C = 8/15$ (configuration (a), while for birds, configuration (e) is the commonest, $C = 4/15$.

Locomotion was the first functionality taken over by the pelvis. With the conquest of land, the first tetrapods faced the problem of gravity, and the pelvic girdle took over a new function: support. A ring formed which would later on take yet another function in many lineages: protection of vital organs such as the kidneys. In some other lineages (e.g., crocodiles) the pelvis took over a ventilatory function, using the pubis as a pistonlike structure to help in pulmonary movements. The "evolutionary sequence" of the way in which the pelvis (always presenting functional integrity) has taken different functional task (what is normally recognized as "function" or "adaptation") is approximately like this: locomotion → support → protection → ventilation. However, the degree of modular integration was reduced as the system took over different functions. For example, in a modern bird, three of these four functions are present, yet $C = 6/15$. The take-home message is that what is usually understood as function is not correlated with the degree of modularity. This should be expected for a system that only has to respond to functional integrity—not to the demands imposed by the special circumstances either of the immediate body surroundings or of the environment.

The Logic of Morphospaces

Morphospace theory, its structure, and its significance, make more apparent the importance of modularity in evolutionary comparative morphology. Morphology is the best available proxy for macroevolution as documented in the fossil record. If we would know the entire sequence of the "chain of life," to put it in Aristotelian words, we would have a perfect fossil record, which would allow us to follow the emergence of life and its diversification in all lineages. The formation of taxonomic groups or clades (kingdoms, phyla, orders, families, genera, and species) would hold no secrets. It would be like documenting the history of a civilization that left a perfect record of its whereabouts. Speciation events would be equivalent to the declaration of independence in a region of the globe. Thus, we would be able to tell, with dates included, when a change occurred that made a dromaeosaur a bird, or a therapsid a mammal. Unfortunately, we do not have this information; patterns and processes are all we can look at of the evolutionary past.

As cladists put it, patterns are logically prior to any hypothesis about an evolutionary process, because it does not make sense to postulate the way lineages can change before effectively knowing the lineage (with its kin relationships included) itself. Characters, the workhorses of comparative biology, are the solution. But how to break down the morphology of an organism so that the comparison makes sense and processes of change can be postulated? That is precisely Bateson's problem. The cladistic solution is to generate as much information as possible at any level of detail, usually in terms of presence/absence of a certain trait. The cluster of derived characters brings together lineages in so-called monophyletic groups in terms of congruence (i.e., the more derived characters are shared, the closer each taxon is to the others) (see chapter 7 in this volume). This, of course, has the problem that some derived characters will not be grouped together, originating the phenomenon of homoplasy in phylogenetic hypotheses. When patterns of phylogenetic relationship are hypothesized, it is necessary to have an explanation for both the trends seen and the multiple origins of certain traits, making the analysis of evolutionary processes very difficult.

All in all, this is the best we can do. Rieppel (1986) expressed this point by noting that cladistics is nothing but the recognition of our ignorance. Could things be different? Could we have a tool that is able to overcome this difficulty? Perhaps modularity is the answer. The key is to find a predictive tool for evolutionary change: a way to decompose the morphology of organisms so that some sort of "logic of change" can be inferred, rather than to be subjected to the ties of a phylogenetic pattern as offered by a cladistic analysis. The expectation of such a tool would be that morphological change is restrictive, constrained as opposed to the null expectancy in cladistics, where every change is possible, as a result of the congruence dictum. What we are looking for is a way to characterize constraints in their morphological materialization, an evolutionary arrow, a possibility that has been explored by many authors (e.g., Simpson, 1944; Needham, 1968; Riedl, 1978; Bonner, 1988; Atchley and Hall, 1991).

The prediction of the existence of constraints in the evolutionary process is that not all morphologies are possible and, hence, a certain "logic of change" is to be expected in phylogenetic patterns. Even adaptationist hard-liners will agree that constraints exist (Amundson, 1994), but they reduce constraints on form to constraint on adaptation, also claiming that if the adaptive force is strong enough, any possible constraint on form will be overcome. However, constraints on adaptation are dependent upon the fitness value of a given trait versus constraints on form, which are dependent upon a generative rule for a given trait. To reduce constraints on form to constraints on adaptation is to simplify the issue, because form may or may not restrict the adaptiveness of a species.

Morphospace and Module Space

Morphospaces have been used extensively and in a variety of scenarios as a tool in comparative morphology (see, for example, Raup and Michelson,1965; Rasskin-Gutman, 1995; Foote, 1997; McGhee 1999, Thomas et al., 2000; Chapman and Rasskin-Gutman, 2001). This space metaphor stems from three different sources: the adaptive landscape of Sewall Wright (Arnold et al., 2001); the theory of transformation envisaged by D'Arcy Thompson (1942; see also Stone (1997); and the phenetic school of comparative morphology (Sneath and Sokal, 1973). See Rasskin-Gutman and Izpisúa-Belmonte (2004) for a more detailed account of this relationship.

Morphospace is understood as a matrix of morphologies that is larger than the subset of realized morphologies in nature. The common denominator among these constructions is their usage of morphological information and/or the simulation of form dissociated from any other biological properties, with the exception of growth or developmental parameters.

Generative morphospaces use a parameter space and a set of rules that generate forms. Combinatorial morphospaces use variables as abstractions of characters, setting up a hyperspace of forms where each form can be mapped as a coordinated point (see figure 9.4). Both types of morphospaces can be theoretical or empirical. Depending on the usage or not of real data. Thus, a generative morphospace can use real data as an initial condition to simulate new forms (generative empirical) or can generate new data from scratch, using only parameters (generative theoretical). At the same time, combinatorial morphospaces can use real data as variables (combinatorial empirical) or can use fictitious variables to simulate new forms (combinatorial theoretical). Morphospace can be made with any kind of variable, any sort of abstraction that, at any level, depicts the morphological features of organisms. From landmarks to connectivity patterns, there is a variety of ways to break down the external- and internal-structures of organic parts. Morphospaces, though, are devoid of any functional assumption (and there reside their power of abstraction and generalization). They are abstractions to the last possible in biology: form, only form.

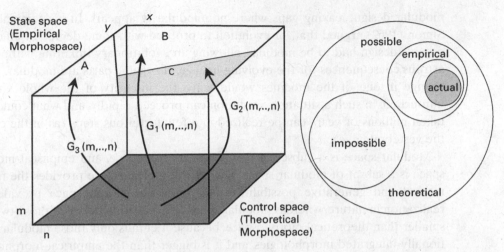

Figure 9.4
Relationship between theoretical morphospace and empirical morphospace. Parameters such as m, n control the outcome of the generative function G, which maps onto empirical morphospace of morphometric variables x, y. Right, theoretical morphospace encompasses all morphologies, impossible and possible. Empirical morphospace accounts only for the actual occurrence in nature that happens to enter the measured pool. Module space is the set of possible forms.

In theoretical morphospace, form can achieve any representation, as long as at some level of detail it resembles the appearance of the real structure it represents. Theoretical morphospace encompasses possible and impossible forms. The subset of the possible forms in turn encompasses realized and nonrealized forms. One of the goals of morphospace theory is to understand how evolution moves from one region of morphospace into another. To understand this dynamics is to understand the constraints imposed on evolution. What are the paths taken by the evolutionary process inside the morphospaces of possibilities and impossibilities? What are the paths of constraint in the evolutionary process? (For more discussion on developmental constraints, see Alberch, 1982; Wimsatt, 1986; Amundson, 1994; Beldade et al., 2002.)

Organic form is modular, be it as a result of self-organization of components through development (for example, you can expect that the organization of a structure formed by discrete components such as cells that reproduce by successive dichotomous divisions will have a modular architecture) and/or as a result of a loosely defined action by natural selection. The expectation is that life takes modularity as a construction principle. Thus, the answer to the question of how evolution proceeds within theoretical morphospace is simple: it moves around regions of modular design, leaving gaps where nonmodularity appears. In his seminal work Simon (1962) argued that for evolution to proceed with some degree of efficiency, organic design had to be modular, allowing an evolutionary tinkering without disastrous consequences for the evolving lineage. If organic parts are modular, then a change in one of the modules would leave the integrity of the whole virtually untouched. In such a situation, evolution can proceed rapidly, and what could have taken millions of years can be realized in a few fortuitous steps (as in the case of the vertebrate eye).

Modular space is a subset of theoretical morphospace, and empirical morphospace is a subset of modular space. Theoretical morphospace provides the mathematical and generative possibilities, and empirical morphospace provides the realization in nature, whereas modular space is a transition between those two. It is smaller than theoretical morphospace, because it entails only those modular, functionally integrated morphologies, and it is bigger than the empirical morphospace because it gives all the possible forms to which evolution has access, even those that have not been explored. Modular space is the real pool from which lineages can jump in speciation events. An evolutionary sequence then moves within theoretical morphospace, going from module to module. Whereas self-organization provokes the jump between species, natural selection provokes the jiggling of a given module, perhaps accelerating the likelihood of making a new jump, or just putting the species in a position to make the jump to another module state (see figure 9.5).

Conclusion

Modularity seems to have been taken almost exclusively as an expression of adaptation (Wagner, 1996; see also chapter 2 in this volume). In this view, modules are literally the biological architecture that allows complex "adaptations" to take place. This puts an excessive weight on a neo-Darwinian view of evolution, and an emphasis on a population dynamics for the mere definition of modularity, thus imposing a very particular view on its origin and possibilities for change. However, modularity is the mark of organization in living beings, and is also an excellent raw material for complexity. It has been identified at all levels of biological organization: gene sequences, protein motifs, cell types, generic tissue geometries, bone configuration, and brain structure, to name just a few. Modularity suggests that evolution proceeds as changes in the organization of living matter that allows for functional specialization. Thus, modularity provides a reason to state that function follows form in the most Geoffroyian sense of the concept.

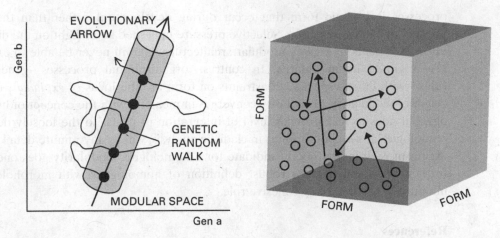

Figure 9.5
The exploration of genetic space is random (left), but morphospace is explored from regions of modularity to regions of modularity (right). Only those changes that "hit" the constraints of viable development are successful. As a result, an evolutionary arrow is created that is bounded by possible modular architectural design. The arrow on the left is equivalent to the arrows on the right.

Unfortunately for a convinced (rational) Platonist like myself, form is not dissociated from matter in nature. Luckily, this does not preclude the fact that they are logically independent and, hence, that they can be studied separately with different aims and, of course, different tools. Once again, for a Platonist like myself, it is only trivial that function follows form, and that any understanding of functional performance of a biological structure rests on a solid understanding of its form peculiarities. This being said, I believe that modularity provides the long-sought nexus between form and function (where function is a particular instantiation that is environmentally dependent of the ability of a form to interact with other forms).

The pervasiveness of modularity across phylogeny raises the question of its inevitability. Just as in the origin of complexity, one can argue that modularity might have evolved for no other reason than there not being other way to increase that same complexity. As Bateson recognized, cell division is the building rule of multicellular organisms, a design that only allows for modularity. If that is the case, then modularity is a cause for a discrete occupancy of theoretical morphospace. The problem of the discontinuity of form in morphospace is a problem that concerns developmental constraints, that is, those restrictions that developmental interactions in a modular design impose over the body plan of an organism.

Morphological modules, arising as a product of generative rules (make parts, repeat them, change them), are necessarily a product of self-organizational processes in organic form that occur during development rather than the end product of natural selection. Selective pressure may shed some light on the distribution of variants of a given modular architecture but will never be able to explain the origin of its modularity. In contrast, organizational processes—generative rules—with their respective constraints on form are the locus of *explanans* of the origin of modularity. Modularity converges in meaning with the concept of homology, but adds to it a stronger level of integration of parts. To the loosely defined homologies as synapomorphies in cladistic analysis, where any minute detail of an organism is already a good candidate for a homologue, modularity goes one step further, providing a more robust definition of homologues with morphological integration playing a key cohesive role.

References

Alberch P (1982) Developmental Constraints in Evolutionary Processes. In: Evolution and Development (Bonner JT, ed.) 313–332. Berlin: Springer-Verlag.

Amundson R (1994) Two concepts of constraint: Adaptationism and the challenge from developmental biology. Phil Sci 61: 556–578.

Arnold SJ, Pfrender ME, Jones AG (2001) The adaptive landscape as a conceptual bridge between micro and macroevolution. Genetica 112–113: 9–32.

Atchley WR, Hall BK (1991) A model for development and evolution of complex morphological structures. Biol Rev 66: 101–157.

Bard J (1990) Morphogenesis: The Cellular and Molecular Processes of Developmental Anatomy. Cambridge: Cambridge University Press.

Bateson W (1894/1992) Materials for the Study of Variation. Baltimore: Johns Hopkins University Press.

Beldade P, Koops K, Brakefield PM (2002) Developmental constraints versus flexibility in morphological evolution. Nature 416: 844–847.

Bonner JT (1988) The Evolution of Complexity by Means of Natural Selection. Princeton, NJ: Princeton University Press.

Buscalioni AD (1999) Animales Fantásticos. La creación de un reino hace mil millones de años. Madrid: Ediciones Libertarias.

Carroll SB (2001) Chance and necessity: The evolution of morphological complexity and diversity. Nature 409: 1102–1109.

Chapman RE, Rasskin-Gutman D (2001) Quantifying Morphology. In: Paleobiology, vol. II (Briggs D, Crowther P, eds), 489–492. Oxford: Blackwell.

Duboule D, Wilkins AS (1998) The evolution of "bricolage." Trends Genet 14(2): 54–59.

Foote M (1997) The evolution of morphological diversity. Ann Rev Ecol Syst 28: 129–152.

Gilbert SF (2000) Developmental Biology. Sunderland, MA: Sinauer.

Kim J, Kim M (2001) The mathematical structure of characters and modularity. In: The Character Concept in Evolutionary Biology (Wagner GP, ed), 215–236. San Diego: Academic Press.

Klingenberg CP, Badyaev AV, Sowry SM, Beckwith NJ (2001) Inferring developmental modularity from morphological integration: Analysis of individual variation and asymmetry in bumblebee wings. Amer Nat 157(1): 11–23.

McGhee GR Jr. (1999) Theoretical Morphology: The Concepts and Its Applications. New York: Columbia University Press.

Milo R, Shen-Orr S, Itzkovitz S, Kashtan N, Chklovskii D, Alon U (2002) Network motifs: Simple building blocks of complex networks. Nature 298: 824–827.

Needham J (1968) Order in Life. Cambridge, MA: MIT Press.

Newman SA (1992) Generic physical mechanisms of morphogenesis and pattern formation as determinants in the evolution of multicellular organization. In: The Principles of Organization in Organisms (Mittenthal JE, Baskin AB, eds) 241–267. Reading, MA: Addison Wesley.

Olson EC, Miller RL (1958) Morphological Integration. Chicago: University of Chicago Press.

Raff RA (1996) The Shape of Life: Genes, Development, and the Evolution of Animal Form. Chicago: University of Chicago Press.

Rasskin-Gutman D (1995) Modelos geométricos y topológicos en morfología. Exploración de los límites del morfoespacio afín. Aplicaciones en paleobiología. Ph.D. thesis, Autonomous University of Madrid.

Rasskin-Gutman D (2003). Boundary constraints for the emergence of form. In: The Origination of Organismal Form (Müller G, Newman S, eds), 305–322. Cambridge, MA: MIT Press.

Rasskin-Gutman D, Buscalioni AD (2001) Theoretical morphology of the archosaur (Reptilia: Diapsida) pelvic girdle. Paleobiology 27(1): 59–78.

Rasskin-Gutman D, Izpisúa-Belmonte JC (2004) Theoretical morphology of developmental asymmetries. Bioessays 26: 405–412.

Raup DM, Michelson A (1965) Theoretical morphology of the coiled shell. Science 147(3663): 1294–1295.

Riedl RJ (1978) Order in Living Organisms: A Systems Analysis of Evolution (Jefferies RPS, trans). London: Wiley.

Rieppel OC (1988) Fundamentals of Comparative Biology. Basel: Birkhäuser Verlag.

Simon HA (1962) The architecture of complexity. Proc Amer Phil Soc 106: 467–482.

Simpson GG (1944) Tempo and Mode in Evolution. New York: Columbia University Press.

Slack JMW (1991) From Egg to Embryo: Regional Specification on Early Development. Cambridge: Cambridge University Press.

Sneath P, Sokal R (1973) Numerical Taxonomy. San Francisco: W.H. Freeman.

Stone JR (1997) The spirit of D'Arcy Thompson dwells in empirical morphospace. Math Biosci 142: 13–30.

Strogatz SH (2001) Exploring complex networks. Nature 410: 268–276.

Thomas RDK, Shearman RM, Stewart GW (2000) Evolutionary exploitation of design options by the first animals with hard skeletons. Science 288: 1239–1242.

Thompson D'AW (1942) On Growth and Form, 2nd ed. Cambridge: Cambridge University Press. (1st ed., 1917.)

Von Dassow G, Mier E, Munro EM, Odell GM (2000) The segment polarity network is a robust developmental model. Nature 406: 188–192.

Von Dassow G, Munro E (1999) Modularity in animal development and evolution: Elements of a conceptual framework for EvoDevo. J Exp Zool (Mol Dev Evol) 285: 307–325.

Wagner GP (1996) Homologues, natural kinds, and the evolution of modularity. Amer Zool 36: 36–43.

Weiss K (2002) Good vibrations: The silent symphony of life. Evol Anthro 11: 176–182.

Wimsatt WC (1986) Developmental constraints, generative entrenchment, and the innate–acquired distinction. In: Integrating Scientific Disciplines (Bechtel W, ed), 185–208. Dordrecht: Martinus Nijhoff.

10 Morphological Modularity and Macroevolution: Conceptual and Empirical Aspects

Gunther J. Eble

A notion of morphological modularity is often implicit in systematics and paleontology. Indeed, the perception of morphological modularity is manifested in the very existence of anatomy, comparative anatomy, and taxonomy as disciplines, and provides a rational basis for treating organic diversity as a combinatorial problem in development and evolution. In practice, it corresponds to the recognition that phenotypic wholes can be decomposed into parts, or characters. This basic analytic stance has been present to varying degrees throughout the history of biology, was particularly important in Darwin's and Mendel's work, and persists to this day (Darden, 1992; Rieppel, 2001).

Yet the parts and characters routinely identified by the morphologist reflect hypotheses of modularity based on observational or quantitative criteria, without reference to the generative mechanisms or the theoretical contexts to which modules relate. In contrast, a notion of developmental modularity has recently been explicitly advanced in terms of mechanisms of genetic and epigenetic specification of units of phenotypic evolution (R. A. Raff, 1996; G. P. Wagner, 1996; G. P. Wagner and Altenberg, 1996; Bolker, 2000). Because morphological patterns of organization emerge in ontogeny, morphological modularity might thus be seen as an aspect of developmental modularity. Accordingly, a research program emerges: the validation of putative morphological modules as developmental modules. This is of particular interest, as it could help further approximate evolutionary morphology (including systematics and paleontology) and evolutionary developmental biology.

A complementary research program presents itself, however, once modularity is seen as a property that is differentially expressed across hierarchical levels (Simon, 1962; Bolker, 2000; R. A. Raff, 1996; E. C. Raff and Raff, 2000; R. A. Raff and Sly, 2000; G. P. Wagner, 2001). Descriptively, mechanistically, and theoretically, modules at different levels may demand level-specific characterizations and may reveal phenomena unique to particular spatial and temporal scales. Descriptively, morphological modules are objects defined in terms of geometry, topology, and statistical considerations. A standard of discreteness is usually present, and the amount of information they encapsulate can often lead to rigorous characterizations. This information may be biased by taxonomic practice and the history of comparative anatomy, but reliable identification and justification of characters is possible beyond their use as a means to the distinction of taxa.

Mechanistically, definitive morphological modules are established usually late in ontogeny, are subject to considerable epigenetic specification, and their variation will be mostly related to allometric growth. They define a unique, post-morphogenetic organizational level where module identity is maintained by morphostatic mechanisms partially decoupled from the developmental pathways of various module components (G. P. Wagner and Misof, 1993; G. P. Wagner, 1994).

Morphological modules and modularity are thus a legitimate level of causal explanation and study, to which generative mechanisms relate necessarily, as a source of precursors (but not sufficiently). Theoretically, morphological modules have unique roles at the organismal level and above, participating causally in the structuring of ecological and genealogical systems in microevolution and macroevolution. They therefore stand as process-based natural kinds (Quine, 1969; Boyd, 1991; G. P. Wagner, 1996, 2001).

All in all, the recognition of a legitimate phenomenological domain for morphological modules does not reduce the value of describing their microscopic structure, of expressing them as instances of developmental modules and understanding their developmental origins, or of treating them as causally inconsequential in some theoretical contexts. It simply recuperates the ontological semi-independence of morphology, along with the entities and processes it helps define. The complementary research program is then the characterization, mechanistic interpretation, and theoretical articulation of morphological modularity *at* the morphological level, but with explicit conceptualization of morphology as a multifactorial phenomenon connected to multiple levels and multiple scales in development and evolution.

The advantage of the reification of morphological modularity is that it can be more directly interpreted in terms of classification and systematization; it can be studied in fossil groups and nonmodel organisms, thus allowing a broader window into the evolution of modularity; and it can more readily allow exploration of macroevolutionary issues. The challenge to evolutionary and developmental morphologists is to devise protocols of study of morphological modules and modularity per se, and to develop intepretive schemes that are consistent with, but that at the same time enrich, evolutionary theory. On the theoretical side, theoretical morphology and theoretical morphospaces provide a way of directly modeling the range of possibilities specified by particular modular organizations. This is not dealt with here (see chapter 9 in this volume). Empirically, the proper study of morphological modularity demands rigor in the description and representation of form, as well as consistent criteria for the decomposition of wholes into parts and for the definition of classes of autonomous behavior. This chapter addresses some key empirical aspects of morphological modularity, including the identification of modules, the quantification of modularity, and the relationship between modularity and macroevolution, while developing an operational conceptual framework.

Identification of Morphological Modules

Observing or inferring the existence of particular modules presupposes some definition of what constitutes a module. Different definitions are possible, depending on which criteria are chosen and on whether descriptive, mechanistic, or theoretical individuation is sought. While a unified notion of module is highly desirable, it may not always be useful in the morphological domain, because morphological units are complex, multidimensional geometrical objects whose identity, generation, and role may vary differentially over scales of time and space and often be discordant. Still, a minimal notion can help in highlighting similar assumptions and goals across research programs.

Minimally defined, morphological modules are cohesive units of organismal integration. Module cohesion will usually arise from stronger interactions within than among modules (see Simon, 1962), and organismal integration will reflect differential interactions among modules. This perspective leaves open the question of what constitutes "interactions," which can, for example, be seen as structural relations (Riedl, 1978; McShea and Venit, 2001), pleiotropic effects (Bonner, 1988; G. P. Wagner, 1996; G. P. Wagner and Altenberg, 1996; Mezey et al., 2000), patterns of gene expression (Gilbert et al., 1996; R. A. Raff, 1996; R. A. Raff and Sly, 2000), or function (G. P. Wagner and Schwenk, 2000; McShea and Venit, 2001; Schwenk, 2001). This minimal definition of a morphological module is thus consistent with, but not equivalent to, general definitions of a developmental module (e.g., Bolker, 2000).

Within an organism or body plan considered in isolation, modules are *organizational* units. Among organisms, they are also *variational* units. Organizational morphological modules refer explicitly to the interactions postulated to be important in organismal construction or activity. They invite observation or description in terms of mechanistic relations, whether variation among organisms is present or not. As such, organizational modules are units of stability. Variational morphological modules reflect the strengths of interactions and their potential disruption. They can be inferred from the variation and covariation patterns of descriptive units, which may or may not be modules themselves. There is no necessary one-to-one relation between particular organizational modules and particular variational modules, because the nonlinear mapping from genotype to phenotype, from part to whole, and from structure to function may affect organization and variation differently in time and space. Methodologically, a match or mismatch will also hinge on what counts as organization and as variation.

Organizational Morphological Modules

Different kinds of interaction justify different notions and partitionings of organization. (1) Structural relations characterize an organization as a set of geometrical objects, each of them spatially individuated by discrete boundaries or by shape differences, and thus standing as a module (or part—see McShea and Venit, 2001). (2) In terms of pleiotropic interactions, the relevant organization is the genotype—phenotype map, and modules are clustered pleiotropic mappings (viewed as routes, not vehicles) that "align" genotypic and phenotypic space (G. P. Wagner, 1996; G. P. Wagner and Altenberg, 1996). (3) Developmental interactions have multiple material bases, and several types of organizational modules stem from them, such as fields of gene expression (Gilbert et al., 1996); genetically mediated spatiotemporal patterns of cell and tissue differentiation, proliferation, and movement (R. A. Raff, 1996); domains of epigenetic dynamics (Goodwin, 1984; Webster and Goodwin, 1996); and regions with localized allometric growth. (4) Functional cooperation of parts, in turn, make organization a matter of functional integration and performance, and modules the sets of functionally individualized units (even if spatially distributed) underlying organismal survival and reproduction (e.g., G. P. Wagner and Schwenk, 2000; Schwenk, 2001).

Clearly, substantial overlap must exist among these various kinds of interactions, and the modules they underlie, for logical and evolutionary reasons. It is also conceivable that some of them are reducible to others (e.g., cell types to patterns of gene expression, function to structure, pleiotropy to function), but chance, redundancy, and differences in dimensionality render complete reduction unlikely and mismatches inevitable. This *is* of fundamental interest in the dissection of hierarchies and multiple chains of causality. Heuristically, valid organizational morphological modules can be variously identified as structural units by an anatomist, as functional units by a functional morphologist, as pleiotropic clusters by a quantitative geneticist, or as developmental units by a developmental biologist. If module identification in each case is also couched on consistent methodological criteria, if it is refutable, and if it allows modules to be units in theories of process (see below), the choice of approach will be anything but arbitrary.

Of much interest, concomitantly, is the comparison of differently identified modules. If there are mismatches, how substantial are they? Do they reflect a difference in evolutionary history, in constraints, in ecological contexts, or in ontogenetic stages? Can they sometimes be ascribed to chance or to inferential error? If there is some common denominator for morphological modules, the comparative study of differently constructed morphospaces (e.g., Lauder, 1995, 1996; Eble, 1998, 2003) may yield unifying insights on the multifariousness of organizational morphological modules and their representation.

Variational Morphological Modules

While organizational morphological modules are mechanistic units of stability, variational morphological modules are units of actual or potential change. The notion of character is central here, because most characters are identified primarily as units of variation in related organisms (Fristrup, 1992, 2001). More precisely, valid characters are routinely perceived as units of *independent* variation (Darden, 1992). Independence is operational, not absolute, being equivalent to the notions of "quasi-independence" (Lewontin, 1978) and "near-decomposability" (Simon, 1962). Further, the degree of inferred character independence may depend on how variation itself is sampled and analyzed.

In many contexts, independent characters are inferred from the observation of correlations among units of description and quantification (Olson and Miller, 1958; Lewontin, 2001). The units, such as morphometric variables, need not correspond to modules. Modules will correspond to different directions of variation and to covariation clusters. In addition, the units considered may vary to any degree within and among species. Modules are implied by actual dissociability in collections of organisms treated as contemporaneous.

In contrast, the discrete morphological characters often used in phylogenetic analyses stand for stable units of evolutionary variation across species, assumed to be independent by virtue of corresponding to individual historical events and thus suggesting potential dissociability in evolutionary time. Character correlation or coevolution does not affect their status as separate entities.

These different notions of character independence codify different timescales and potentialities of variation. Variational morphological modules may therefore also be of different kinds, as with organizational morphological modules, but here the primary determinant is the dynamics of interactions among units and of actualization of instances of units, not the interactions themselves. Characters as variational units are not incompatible with their also being units of organization when the partitioning of variation follows the "lines of least resistance" defined by differential organizational discreteness within the organism. But given various kinds of organization and of variation, the relationship between organizational modules and variational modules may not be straightforward. Making sense of this relationship is most relevant to further understanding of the nature of modularity, and it can originally inform analogous issues, such as homology and homoplasy or, more generally, lineage stasis and change.

Morphological Modules as Causal Actors

Beyond their organizational and variational properties, modules can also be construed as having causal roles. Morphological modules are then instances of process-based natural kinds, that is, units that play a role in a process or set of processes (Quine, 1969; Boyd, 1991, 1999; G. P. Wagner, 1996, 2001; Webster and Goodwin, 1996; Griffiths, 1999). In this sense, the characterization of morphological modules presupposes the choice of a reference class of processes. Such processes may be developmental, as existing modules affect the ontogeny of other modules; physiological, as in homeostasis; ecological, such as predation or competition; or evolutionary, such as selection or speciation. Modules become entities endowed with theoretical significance, and their individuation a matter of identifying dispositional properties (propensities) determining their potential participation in the processes of interest (G. P. Wagner, 2001). These properties are causally inert (as with fitness understood as a propensity—Sober, 1984). They reflect expected behavior, not actual causes. The actual causes are to be found in the mechanics of organization and variation.

A one-to-one correspondence between particular processes and particular causes is not a sine qua non, and therefore the identification of causal morphological modules is not reducible to the study of organizational and variational modules. Process-based individuation may seem appropriate only when a general theory is available, which may be the case for evolution but not for development (Bolker, 2000), but causal roles can be consistently identified under any degree of generality, and can be usefully referred to even if the respective theories and postulated causes turn out to be inadequate. The identification of modules is a heuristic endeavor—process-based and pattern-based approaches are best seen as complementary.

Quantification of Morphological Modularity

From the fact that modules can be identified observationally or inferentially, it follows that modularity is present. Its quantification, however, need not always demand making the modules explicit. Further, it is highly context-dependent. And given that modules can be organizational or variational, modularity can accordingly also be seen as a constitutional property of individual organisms (and their parts) or as a relational property of sets of organisms. Organizational modularity is the state of being modular. Variational modularity is the state of varying in modular fashion.

In some contexts, it may be of interest to treat modularity as a nominal, presence–absence feature on the scale of the whole organism or its parts. Indeed, for dynamic modules such as morphogenetic fields, or for the terra incognita of the genotype-phenotype map, assessing the presence of modularity is a major goal. Further, because modules may often have a nested arrangement, the issue of whether they are themselves composed of modules at a structurally similar level of organization arises. This is not a trivial issue because internal cohesion may be specified in different ways. For example, the tetrapod limb is usually considered a module, but is it composed of modules whose origin and maintenance can be referred to the same hierarchical level(s) of organization? This may depend on whether a morphogenetic or a structural approach is used. For whole organisms, modularity is bound to be present at some level, and intuitively at more than one level, but it may not be ubiquitous or isomorphic across levels.

Organisms and their modules are modular to different degrees (Kim and Kim, 2001). Given that we do not know how much modularity can vary, numerical assessments of discrete and continuous variation should be attempted whenever possible if we are to achieve a complete understanding of the evolution of modularity. Statistics for modularity are thus needed. Importantly, the comparative study of modularity justifies, and may at times require, the use of proxy data and testable working assumptions as strategies in research. Below is an outline of possible approaches.

Number of Characters or Parts as Proxies for Modularity

Discrete morphological characters are standard data in organismal research and may be the most amenable to quantitative analyses of the evolution of modularity. They are similar to what McShea and Venit (2001) called "parts": operational units of the construction of an organism which can be expected to be a representative sample of the "true" underlying units, to be defined mechanistically or by theoretical role. In general, most characters defined on consistent topological and geometrical grounds as discrete units will correspond to such modules by proxy—hypothetical when described but cohesive and bounded enough to justify a strong assumption of individuation *in some context* (e.g., developmental, functional, evolutionary). If made explicit, this assumption can be tested on other grounds. An arbitrary character is hardly a module; but a comprehensive list of characters or parts that takes into account organismal integration (or disintegration) could be seen as a hypothetical list of modules. In practice, if error in characterizing morphological units as proxy modules is reasonably small or random, useful estimates of modularity can be produced and comparisons made (see McShea and Venit, 2001).

Counting characters is not equivalent to counting modules, but large differences in numbers of characters at a similar organizational level are likely to correspond to differences in number of modules. In well-circumscribed groups, counts of characters standing for modules by proxy are intuitively appropriate as measures of relative modularity. This will be especially true for groups studied by a single author or for which consensus on morphological terminology exists.

Even so, the kind of module of interest may influence how many morphological units are counted. For example, in a study of bryozoans McShea and Venit (2001) provide a number of protocols for counting part types assumed to be functional units. Repeated structures are excluded and treated as belonging to the same functional unit. If the focus is on structural or developmental organization, however, finer assessments of modularity may be possible, and repeated structures can provide useful data. This will often be the case for skeletal features, which allow greater taxonomic and temporal coverage. Sea urchins, for example, have skeletons composed almost entirely of calcite plates. Plate number and shape can vary substantially, but since plates arise sequentially from standard locations in the apical system, they could be viewed as repetitions of the same type of module. On this scale, modularity is essentially uniform.

But on other scales, variation in rate, timing, and location of skeletal growth individuates additional types of modules. In terms of numbers of plate columns, sea urchins are more modular in the Paleozoic than in the post-Paleozoic. In terms of regional plate differentiation, irregular sea urchins are more modular than regular sea urchins. At the limit, each plate is a module of localized and potentially dissociable growth with stronger internal integration, afforded by the continuity of stereom trabeculae, than external integration, mediated by collagen fibers at boundaries between plates. Thus, in terms of plate numbers, a sea urchin with 1000 plates is more modular than one with 100 plates. While a focus on a single aspect or scale is justifiable on theoretical grounds, consideration of multiple contexts provides a window into the scale-dependent manifestations of modularity.

Morphological Integration

Counts of discrete parts or characters may provide good proxy estimates of modularity for many comparative studies, but they do not take into account the full extent of differential integration within and among modules. The quantification of changes in within- and among-module integration can also be important in assessing the relative frequency of various mechanisms of module evolution, such as co-optation or parcellation (R. A. Raff, 1996; G. P. Wagner, 1996; G. P. Wagner and Altenberg, 1996). When modules are hypothesized a priori, based on mechanistic criteria such as function or developmental identity (e.g., Mezey et al., 2000; Eble, 2000; Klingenberg et al., 2001), patterns of morphological integration within and among sets of traits provide tests of the importance of postulated mechanisms. Alternatively, morphological modularity can be hypothesized a posteriori, from analysis of nested patterns of physical association and of covariation among traits, and later validated on mechanistic grounds.

A focus on integration within modules is of interest not only as a distinct measure of modularity but also when a complete inventory of parts or characters is not possible for preservation reasons, as happens with incompletely known fossil species or when the theoretical focus is on particular modules. Mezey et al. (2000) devised a statistic for within-module integration: the ratio of the total number of module traits affected by a set of quantitative trait loci to the maximum number of traits that this set could affect. Higher than average integration is considered significant. The reference standard in assessing significance may be a randomized distribution of interactions within a population (see Mezey et al., 2000).

A strictly morphological generalization of this statistic, immediately applicable whenever gene effects on traits are not available (as will often be the case in systematic and paleontological studies) is to quantify trait interactions. The statistic then becomes the ratio of the total number of trait interactions within a module to the maximum possible number of interactions such traits could allow. What counts as a trait interaction can be either physical contiguity, in which case shapes and positions matter, or inferred sign of covariation against a chosen standard (zero, average, random, etc.). The reference distribution of interactions may be based on individuals within populations, or on species within clades, if interspecific variation is being considered. Further, theoretical models of morphological transformation can provide an alternative to randomization as a basis for formulating null predictions, when the shape and local connectivity of morphological traits are available (e.g., Rasskin-Gutman, 2003).

When the number of characters is small or constant, or when characters vary substantially in shape and connectivity, measures of morphological integration among putative modules will be most informative. The degree of integration among modules is inversely related to their parcellation. A statistic for parcellation was suggested by Mezey et al. (2000): a chi-square comparison of the observed versus expected number of module traits affected by each quantitative trait locus. A morphological generalization of this statistic is also possible, measuring trait interactions among modules either in terms of neighboring relations (e.g., Rasskin-Gutman, 2003) or of the sign of covariation.

The *strength* of covariation may vary substantially across morphological units, and can be used to produce more precise estimates of modularity. Morphometric approaches are readily applicable in this context. On morphometric grounds, wings as putative modules were confirmed in *Bombus empatiens* and *Drosophila melanogaster* (Klingenberg and Zaklan, 2000; Klingenberg et al., 2001). Yet the greatest potential of morphometrics lies in the recasting of exploratory studies of morphological integration in terms of modularity. A number of studies have postulated and documented the existence of morphological covariation sets, reflected statistically in trait correlations and interpreted in terms of function, development or other factors (Olson and Miller, 1958; Zelditch et al., 1992; Zelditch and Fink, 1996; Eble, 2000, 2004). Covariation sets, viewed as putative variational modules, become important data for documenting the evolution of morphological modularity across the phylogenetic hierarchy.

Disparity

Disparity, the spread or spacing of forms in morphospace, is an aspect of biodiversity relating to phenotypic distinctness in a sample. It has become an important quantity in macroevolutionary studies (e.g., Gould, 1989, 1991; Foote, 1993, 1997; Wills et al., 1994; P. J. Wagner, 1995; Eble, 2000, 2004), and it holds promise in ecology (Roy and Foote, 1997) and evolutionary developmental biology (Eble, 2002, 2003).

Disparity is a general measure of variation, and as such no assumptions are made about its causes. Yet it can often be decomposed or scaled into contributions likely to reflect variational modularity. Variational modularity relates to spatially and theoretically contextualized variation, and therefore stands as a major aspect of variability (the potential to vary—see G. P. Wagner and Altenberg, 1996). Because modularity specifies opportunities for semi-independent variation, a correlation between extent of modularity (in terms of numbers of parts, within-module integration, and among-module integration) and extent of disparity should be found. Disparity is *not* formally equivalent to modularity, but for many problems in morphological evolution, trends in disparity can be a useful proxy for trends in variational modularity. This interpretation of disparity follows from the established use of patterns of natural variation as guides to the existence of constraint and differential variability (Alberch, 1983, 1989; Shubin and Alberch, 1986; Foote, 1995, 1999; G. P. Wagner, 1995; G. P. Wagner and Altenberg, 1996; Eble, 2000). The recognition of disparity as a large-scale proxy for modularity suggests new research directions in quantitative morphology and new perspectives for the interpretation of the causal role of disparity in macroevolution.

If morphological disparity is to be used to quantify morphological modularity, the nature of the assumed modularity-disparity connection should ideally be specified, to allow for additional tests. This may involve postulating what the hypothetical modules are; identifying developmental, functional, or other mechanisms thought to affect overall modularity and disparity in a similar way (e.g., mutation rates, developmental constraints, functional integration); or indicating common causal roles (e.g., in evolvability, in innovation production, in species and clade selection and sorting, in homoplastic evolution, etc.). Eble (2000) explored this connection in heart urchins, by focusing on a comparison of temporal disparity patterns between two sets of landmarks thought to reflect differential functional and developmental modularity. The connection was validated by the finding that disparity change and differentiation of the more integrated unit (set earlier in ontogeny) were more protracted over time, in contrast with the less integrated unit, in which most of the disparity was produced early in the history of the group.

Character or part counts, morphological integration, and disparity are statistical estimates of morphological modularity reflecting both organizational and variational aspects. As such, they encapsulate a variety of causes and roles for modules. Especially in macroevolution, causes and roles may change in importance across clades and time. Extensive quantification of patterns of morphological modularity will be needed if the preeminence of particular causes and roles is to be ultimately validated.

Morphological Modularity and Macroevolution

How does morphological modularity change in macroevolution? Can macroevolutionary phenomena significantly affect the temporal patterning of morphological modularity expected from microevolutionary theory? Is the impact of phylogenetic constraints on modularity potentially different in macroevolution? Addressing such issues will be needed to properly contextualize modularity in macroevolution. A step in this direction is to consider from an explicitly macroevolutionary perspective the relationship between modularity and various features of evolution, such as complexity, evolvability, innovation, stochasticity, and trends.

Complexity

Bonner (1988) suggested that as complexity increases, selection for localization of mutational effects would lead to increasing prevalence of gene network organization—in other words, modularity. Complexity, treated as the number of cell types within organisms and as number of species in communities, was suggested to broadly correlate with size. It is unclear that complexity actually increases in evolution (Gould, 1996; McShea, 1996), but these suggestions lead to the expectation that modularity should correlate evolutionarily with the number of cell types, with species diversity, and with size. The number of cell types can itself be seen as a measure of organismal modularity, and yet circularity is avoided if it is contrasted with morphological modularity at other levels of organization. A correlation between number of cell types and number of body plans is often reported (Kauffman, 1993; Valentine et al., 1994). Recasting body plans in terms of disparity should allow a broader range of inferences to be made.

Species diversity could be seen primarily as an aspect of the "modularity" of ecological communities or of clades, but a connection with organismal modularity is also possible. The latter was investigated in a recent study (Yang, 2001) which suggested that holometabolous insects have higher diversification rates than hemimetabolous insects because their more extensive metamorphosis specifies more modular juvenile and adult stages. How characteristic diversification rates might mechanistically relate to modules is a difficult issue, but if correlations can be consistently found across clades, modularity would stand as an important causal aspect of species and of clade sorting and selection. Some macroevolutionary trends in modularity (see below) could therefore be documented and modeled in connection with long-term trends in species diversity through time.

A relationship between modularity and size follows from the connection between dissociability and allometry. Assessment of degree of allometry (Hughes, 1990) provides a way of indirectly studying how size influences modularity. Large size may provide greater opportunities for morphological individuation, but since the shape of allometric trajectories will be the critical factor, exploration of the size spectrum may be more important in the macroevolution of modularity than maximization of size per se.

Evolvability and Innovation

G. P. Wagner and Altenberg (1996) suggested that the modularity of the genotype-phenotype map determines evolvability. Evolvability was defined as "the genome's ability to produce adaptive variants when acted upon by the genetic system" (p. 970), and also as "the ability of random variations to sometimes produce improvement" (p. 967). This latter definition is immediately applicable to the morphological level. Yet by focusing on evolvability as adaptability, it remains most relevant to microevolutionary selection scenarios, given the expectation of adaptation to changing environments. In macroevolution, improvement may readily occur when morphological change is anagenetic, but will often be absent in cladogenesis, since speciation is nonadaptive with respect to species persistence. Major innovations and clade founding also may often not represent improvement because they usually correspond to discrete events decoupled from the adaptive context (the fitness landscape) of the parental clade.

Because a notion of adaptive improvement is not always justified to contextualize interspecific variation and macroevolution, a further generalization of evolvability is possible: "the ability of variations to sometimes produce evolutionarily significant change." What counts as significant may differ depending on temporal scale, on hierarchical level, and on the degree of concordance between morphospace structure and fitness landscape structure. In studies of macroevolution, significant morphological change may be identified as an improvement in functional efficiency in some instances, but for operational reasons (the data of systematics and paleontology) and theoretical reasons (the centrality of novelty in macroevolution, regardless of the causes of sorting), significant morphological change can usefully stand simply for "substantial distinction." In this way, evolvability can be effectively quantified throughout the history of clades.

Variability is the potential to vary. Evolvability is the potential to vary in a relevant way. Rate of positive mutation is a possible measure of evolvability at the molecular level. Origination and innovation rates might analogously be used for morphological data. Disparity, in turn, often measures variation only, and is a proxy for modularity. However, the amount of disparity produced relative to time or diversity is likely to reflect evolvability. Similarly, the ratio of major morphological innovation to minor morphological innovation (Eble, 1998, 1999) or the frequency of homoplasy (P. J. Wagner, 2000) could be used. Another possibility, appropriate for both modularity and evolvability, is to use measures of stationarity of morphological variation through geological time (e.g., Foote, 1995) or of cumulative change through the history of clades.

Modularity as a By-Product of Stochastic Morphological Evolution
In stochastic simulations of the evolution of independent morphological characters, Raup and Gould (1974) found that statistically significant pairwise character correlation is common. They interpret this as the result of stochastic lineage sorting of character combinations of clade founders, and of the progressively smaller probability of return to average original states as dimensionality increases (which is a property of random walks). In stark contrast, Kim and Kim (2001) argue that trait associations and character modularity are highly unlikely in the space of possible combinations, and hence require special explanation.

The two views can be reconciled if the reference space of the possible itself evolves in the history of individual clades. Contingency produces directionality in the form of phylogenetic constraint, and at each stage in the evolution of a particular taxon, not all possible morphologies are equally likely (Raup and Gould, 1974). *Particular* character associations are highly unlikely relative to the total reference space, but it is likely that *some* associations, and perhaps many, will occur relative to a more limited set of possibilities expressed in phylogenetically circumscribed subregions of morphospace. Whether or not they do so in any given instance may depend on the dimensionality of character complexes and of underlying causes, which specify the frequency of phylogenetic constraint relative to phylogenetic inertia. As Kim and Kim (2001) suggest, modularity relates to higher decomposability relative to a reference group (see also Mezey et al., 2000). Where the reference group lies in the phylogenetic hierarchy will determine the size of the reference space and the imprint of contingency on the macroevolution of modularity.

Macroevolutionary Trends

Given the existence of macroevolutionary correlates of modularity, are there trends in the macroevolution of modularity? In the history of clades, is modularity more often increasing by parcellation of integrated phenotypes or decreasing by integration of parcellated phenotypes? G. P. Wagner and Altenberg (1996) suggest that in metazoans, parcellation is more common, because innovation through differentiation from more generalized ancestors is frequent. This agrees with evolutionary interpretations of von Baer's laws and with the notion that biological versatility, or morphogenetic semi-independence, seems to increase in evolution (Vermeij, 1973). At the same time, the potential for innovations seems to decrease in metazoan history (Erwin et al., 1987; Eble, 1998) as well as in the history of individual clades (Foote, 1997; Eble, 1999), suggesting that integration is an important trend as well.

Whether parcellation or integration is more frequent may in fact depend on temporal scale and hierarchical level (Jablonski, 2000). The origin of body plans during the Cambrian radiation, for example, can be interpreted as an increase in parcellation, since cell and tissue specialization would have accompanied divergence from generalized colonial protozoans (G. P. Wagner, 1996) or larvae (Davidson et al., 1995). Later increase in integration *of* body plans would follow, with formerly evolvable characters becoming developmentally entrenched as subclades appear and diversify and new characters accumulate in hierarchical fashion (see Eble, 1998).

More generally, novelty across the phylogenetic hierarchy and across scales of time may often involve the differentiation of existing elements (reduction of serial homology, reduction of degree of isometry), and hence some degree of parcellation. While later persistence of novelties may be a matter of selective advantage alone, the building up of the hierarchy of homology leads novelties to become more integrated and to be maintained by developmental constraint. To the extent that the evolution of modularity by parcellation leads to long-lived modules in macroevolution, later integration of such modules as homologies is likely.

Conclusion

Despite its scope, morphological modularity remains remarkably understudied. Because morphology provides basic data in embryology, systematics, quantitative genetics, functional morphology, and macroevolution, and because modularity is a seemingly pervasive aspect of organization and variation, the recognition of morphological modularity as a target of empirical and theoretical study can help in generating new research questions and a more interdisciplinary discourse within biology. Morphological modules are hypotheses of individuation that may find validation in separate mechanistic or theoretical contexts, but can also be justified on their own, in terms of the distinct evolutionary and developmental dynamics that morphology entails. Morphological modularity may be particularly important in macroevolution. Understanding it in this context will demand a shift in conceptual thinking, but the research protocols are already available. As macroevolution joins evolutionary developmental biology in the expansion of evolutionary theory, morphological modularity should become an important basis for interaction and cohesion.

Acknowledgments

I thank B. David, D. Erwin, D. Jablonski, L. McCall, D. McShea, D. Rasskin-Gutman, R. Thomas, and G. Wagner for comments and/or discussion. Research was supported in part by postdoctoral fellowships from the CNRS (France), the Santa Fe Institute, and the Smithsonian Institution.

References

Alberch P (1983) Morphological variation in the neotropical salamander genus *Bolitoglossa*. Evolution 37: 906–919.

Alberch P (1989) The logic of monsters: Evidence for internal constraint in development and evolution. Geobios, mém spec 12: 21–57.

Bolker JA (2000) Modularity in development and why it matters to Evo-Devo. Amer Zool 40: 770–776.

Bonner JT (1988) The Evolution of Complexity by Means of Natural Selection. Princeton, NJ: Princeton University Press.

Boyd R (1991) Realism, anti-foundationalism and the enthusiasm for natural kinds. Phil Stud 61: 127–148.

Boyd R (1999) Homeostasis, species, and higher taxa. In: Species: New Interdisciplinary Essays (Wilson RA, ed), 141–185. Cambridge, MA: MIT Press.

Darden L (1992) Character: Historical perspectives. In: Keywords in Evolutionary Biology (Keller EF, Lloyd EA, eds), 41–44. Cambridge, MA: Harvard University Press.

Davidson EH, Peterson KJ, Cameron RA (1995) Origin of bilaterian body plans: Evolution of developmental regulatory mechanisms. Science 270: 1319–1325.

Eble GJ (1998) The role of development in evolutionary radiations. In: Biodiversity Dynamics: Turnover of Populations, Taxa, and Communities (McKinney ML, Drake JA, eds), 132–161. New York: Columbia University Press.

Eble GJ (1999) Originations: Land and sea compared. Geobios 32: 223–234.

Eble GJ (2000) Contrasting evolutionary flexibility in sister groups: Disparity and diversity in Mesozoic atelostomate echinoids. Paleobiology 26: 56–79.

Eble GJ (2002) Multivariate approaches to development and evolution. In: Human Evolution Through Developmental Change (Minugh-Purvis N, McNamara K, eds), 51–78. Baltimore: Johns Hopkins University Press.

Eble GJ (2003) Developmental morphospaces and evolution. In: Evolutionary Dynamics: Exploring the Interplay of Selection, Neutrality, Accident, and Function (Crutchfield JP, Schuster P, eds), 35–65. Oxford: Oxford University Press.

Eble GJ (2004) The macroevolution of phenotypic integration. In: The Evolutionary Biology of Complex Phenotypes (Pigliucci P, Preston K, eds), 253–273. Oxford: Oxford University Press.

Erwin DH, Valentine JW, Sepkoski JJ, Jr (1987) A comparative study of diversification events: The early Paleozoic versus the Mesozoic. Evolution 41: 1177–1186.

Foote M (1993) Discordance and concordance between morphological and taxonomic diversity. Paleobiology 19: 185–204.

Foote M (1995) Morphological diversification of Paleozoic crinoids. Paleobiology 21: 273–299.

Foote M (1997) The evolution of morphological diversity. Ann Rev Ecol Syst 28: 129–152.

Foote M (1999) Morphological diversity in the evolutionary radiation of Paleozoic and post-Paleozoic crinoids. Paleobiol Mem, supp to Paleobiology 25(2): 1–115.

Fristrup K (1992) Character: Current usages. In: Keywords in Evolutionary Biology (Keller EF, Lloyd EA, eds), 45–51. Cambridge, MA: Harvard University Press.

Fristrup KM (2001) A history of character concepts in evolutionary biology. In: The Character Concept in Evolutionary Biology (Wagner GP, ed), 13–35. San Diego: Academic Press.

Gilbert SF, Opitz JM, Raff RA (1996) Resynthesizing evolutionary and developmental biology. Dev Biol 173: 357–372.

Goodwin BC (1984) Changing from an evolutionary to a generative paradigm in biology. In: Evolutionary Theory: Paths into the Future (Pollard JW, ed), 99–120. Chichester, UK: Wiley.

Gould SJ (1989) Wonderful Life. New York: Norton.

Gould SJ (1991) The disparity of the Burgess Shale arthropod fauna and the limits of cladistic analysis: Why we must strive to quantify morphospace. Paleobiology 17: 411–423.

Gould SJ (1996) Full House. New York: Harmony Books.

Griffiths PE (1999) Squaring the circle: Natural kinds with historical essences. In: Species: New Interdisciplinary Essays (Wilson RA, ed), 209–228. Cambridge, MA: MIT Press.

Hughes NC (1990) Morphological plasticity and genetic flexibility in a Cambrian trilobite. Geology 19: 913–916.

Jablonski D (2000) Micro- and macroevolution: Scale and hierarchy in evolutionary biology and paleobiology. Paleobiology 26 (supp): 15–52.

Kauffman SA (1993) The Origins of Order: Self-Organization and Selection in Evolution. Oxford: Oxford University Press.

Kim J, Kim M (2001) The mathematical structure of characters and modularity. In: The Character Concept in Evolutionary Biology (Wagner GP, ed), 215–236. San Diego: Academic Press.

Klingenberg CP, Badyaev AV, Sowry SM, Beckwith NJ (2001) Inferring developmental modularity from morphological integration: Analysis of individual variation and asymmetry in bumblebee wings. Amer Nat 157: 11–23.

Klingenberg CP, Zaklan SD (2000) Morphological integration between developmental compartments in the *Drosophila* wing. Evolution 54: 1273–1285.

Lauder GV (1995) On the inference of function from structure. In: Functional Morphology in Vertebrate Paleontology (Thomason JJ, ed), 1–18. Cambridge: Cambridge University Press.

Lauder GV (1996) The argument from design. In: Adaptation (Rose MR, Lauder GV, eds), 55–91. San Diego: Academic Press.

Lewontin RC (1978) Adaptation. Sci Amer 239: 156–169.

Lewontin RC (2001) Foreword. In: The Character Concept in Evolutionary Biology (Wagner GP, ed), xvii–xxiii. San Diego: Academic Press.

McShea DW (1996) Metazoan complexity and evolution: Is there a trend? Evolution 50: 477–492.

McShea DW, Venit EP (2001) What is a part? In: The Character Concept in Evolutionary Biology (Wagner GP, ed), 259–284. San Diego: Academic Press.

Mezey JG, Cheverud JM, Wagner GP (2000) Is the genotype-phenotype map modular? A statistical approach using mouse quantitative trait loci data. Genetics 156: 305–311.

Olson EC, Miller RL (1958) Morphological Integration. Chicago: University of Chicago Press.

Quine WV (1969) Ontological Relativity and Other Essays. New York: Columbia University Press.

Raff EC, Raff RA (2000) Dissociability, modularity, evolvability. Evol Dev 2(5): 235–237.

Raff RA (1996) The Shape of Life: Genes, Development, and the Evolution of Animal Form. Chicago: University of Chicago Press.

Raff RA, Sly BJ (2000) Modularity and dissociation in the evolution of gene expression territories in development. Evol Dev 2: 102–113.

Rasskin-Gutman D (2003) Boundary constraints for the emergence of form. In: Origination of Organismal Form (Müller GB, Newman SA, eds), 305–322. Cambridge, MA: MIT Press.

Raup DM, Gould SJ (1974) Stochastic simulation and evolution of morphology: Towards a nomothetic paleontology. Syst Zool 23: 305–322.

Riedl RJ (1978) Order in Living Organisms (Jefferies RPS, trans). Chichester, UK: Wiley.

Rieppel OC (2001) Preformationist and epigenetic biases in the history of the morphological character concept. In: The Character Concept in Evolutionary Biology (Wagner GP, ed), 57–75. San Diego, Academic Press.

Roy K, Foote M (1997) Morphological approaches to measuring biodiversity. Trends Ecol Evol 12: 277–281.

Schwenk K (2001) Functional units and their evolution. In: The Character Concept in Evolutionary Biology (Wagner GP, ed), 165–198. San Diego: Academic Press.

Shubin N, Alberch P (1986) A morphogenetic approach to the origin and basic organization of the tetrapod limb. Evol Biol 20: 319–387.

Simon HA (1962) The architecture of complexity. Proc Amer Phil Soc 106: 467–482.

Sober E (1984) The Nature of Selection. Cambridge, MA: MIT Press.

Valentine JW, Collins AG, Meyer CP (1994) Morphological complexity increase in metazoans. Paleobiology 20: 131–142.

Vermeij GJ (1973) Biological versatility and earth history. Proc Natl Acad Sci USA 70: 1936–1938.

Wagner GP (1994) Homology and the mechanisms of development. In: Homology: The Hierarchical Basis of Comparative Biology (Hall BK, ed), 273–299. San Diego: Academic Press.

Wagner GP (1995) The biological role of homologues: A building block hypothesis. N Jb Geol Paläont Abh 195: 279–288.

Wagner GP (1996) Homologues, natural kinds, and the evolution of modularity. Amer Zool 36: 36–43.

Wagner GP (2001) Characters, units and natural kinds: An introduction. In: The Character Concept in Evolutionary Biology (Wagner GP, ed), 1–10. San Diego: Academic Press.

Wagner GP, Altenberg L (1996) Perspective: Complex adaptations and the evolution of evolvability. Evolution 50(3): 967–976.

Wagner GP, Misof BY (1993) How can a character be developmentally constrained despite variation in developmental pathways? J Evol Biol 6: 449–455.

Wagner GP, Schwenk K (2000) Evolutionarily stable configurations: Functional integration and the evolution of phenotypic stability. Evol Biol 31: 155–217.

Wagner PJ (1995) Testing evolutionary constraint hypotheses with early Paleozoic gastropods. Paleobiology 21: 248–272.

Wagner PJ (2000) Exhaustion of morphological character states among fossil taxa. Evolution 54: 365–386.

Webster G, Goodwin BC (1996) Form and Transformation. Cambridge: Cambridge University Press.

Wills MA, Briggs DEG, Fortey RA (1994) Disparity as an evolutionary index: A comparison of Cambrian and recent arthropods. Paleobiology 20: 93–130.

Yang AS (2001) Modularity, evolvability, and adaptive radiations: A comparison of the hemi- and holometabolous insects. Evol Dev 3: 59–72.

Zelditch ML, Bookstein FL, Lundrigan BL (1992) Ontogeny of integrated skull growth in the cotton rat *Sigmodon fulviventer*. Evolution 46: 1164–1180.

Zelditch ML, Fink WL (1996) Heterochrony and heterotopy: Stability and innovation in the evolution of form. Paleobiology 22: 241–254.

11 Hierarchical Integration of Modular Structures in the Evolution of Animal Skeletons

Roger D. K. Thomas

Living organisms grow by the development of hierarchically organized, modular structures that have nonarbitrary dimensions prescribed by scaling considerations and rates of physical processes (Thompson, 1942; Vogel, 1988; Schank and Wimsatt, 2001). Historically, the evolutionary emergence of new levels of structural complexity has been triggered by the appearance of key adaptations, leading to the elimination of constraints that limit the realm in which prior modes of adaptation could occur. When such new modes of structural organization emerge, theoretical models predict a rapid, logistic pattern of increase in the variety of forms, up to a limit defined by the set of viable designs that can exist as transformations of the basic module.

Complexity can increase in a system that is far from equilibrium as its scope, the number of possible forms it can take, expands. Layzer (1975, 1978) has suggested a general model (figure 11.1) for the evolution of complexity in such systems. In this model, the macroscopic orderliness of any natural system, at a specified level of organization, is represented by the difference between the amount of information observed and the number of modules (Layzer uses the term "microstates") that would be represented if the system were maximally disordered. The maximum number of modules is defined as

$$H_{max} = \log A$$

where A is the number of potential modules. The actual state of the system at a given time is characterized by Shannon's formula

$$H_{obs} = -\sum (p_i \cdot \log p_i),$$

where p_i is the probability of occurrence of each distinct module. The value of H_{max} increases over time to the extent permitted by available materials and the flow of energy through the system. In the application of Layzer's model to evolving organisms, the lower limit approached by H_{obs} is set by material, generally geometric, constraints on the absolute number of discrete modules that are accessible at the hierarchical level of organization under consideration.

In this chapter, the initial "Cambrian explosion" of animal skeletons is assessed in the context of this model. The use of available design elements in the skeletons of the earliest metazoans is documented first. The results of this analysis are then assessed in relation to the pattern of exploitation of modular structures predicted by Layzer's evolutionary model. Finally, this model is extended to embrace patterns

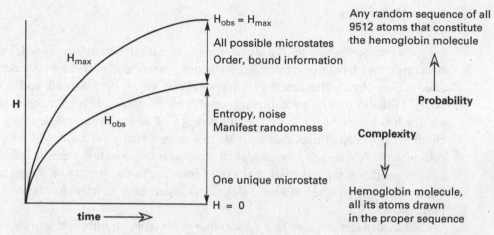

Figure 11.1
Model representing the simultaneous increase of order and entropy in an expanding open (with respect to energy inputs) system. H is an information function, a logarithmic measure of the entropy state of the system. This diagram, based on concepts set out by Layzer (1978), is due to Brooks and Wiley (1988), but my interpretation of its biological implications is different from that which they provided (see text).

of exploitation of structural options at successive hierarchical levels throughout the history of life.

Characterization of Skeletal Morphospace

Structural elements of the skeletons of extinct and living organisms are defined by the parameters of a theoretical morphospace which we call the skeleton space (Thomas and Reif, 1991, 1993). Seven fundamental characters of potential skeletal elements each have two, three, or four states (figure 11.2). The properties of an actual or potential design element are represented by a seven-letter formula. The horn of a unicorn, for example, is coded as BCTGLXR, because it is a single, external, rigid spike that grows in place by accretion and is sutured to the front of the skull. The skeleton space consists of 1536 potential character combinations like this one. However, we judge four pairings of characters to be illogical or inviable: a single element cannot be serial; imbricate solids are impractical; internal elements cannot be molted; and structures which are prefabricated before being moved into the position where they function cannot be remodeled later in development. If all character combinations that include these pairs are eliminated, 960 potentially viable

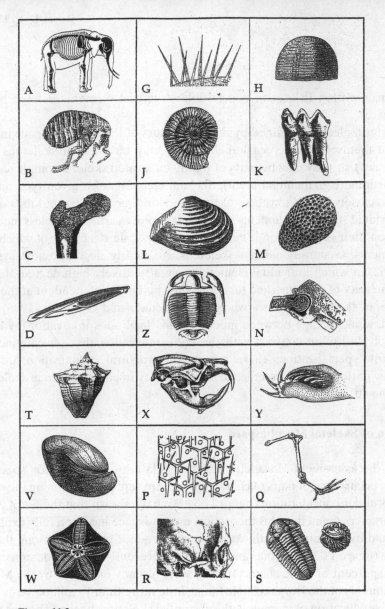

Figure 11.2
The skeleton space. The range of potential forms of animal skeletons or their subunits is defined in terms of seven essential properties, each with two to four possible states, yielding a total of 21 variables. The images illustrating each descriptor are drawn from a variety of sources.
(1) Situation: *A*, internal; *B*, external. (2) Material: *C*, rigid; *D*, flexible. (3) Number: *T*, one element; *V*, two elements; *W*, >two elements. (4) Shape: *G*, rods; *H*, plates; *J*, cones; *K*, solids. (5) Growth: *L*, accretionary; *M*, modular, serial units ± branching; *Z*, replacement/molting; *N*, remodeling. (6) Assembly: *X*, growth in place; *Y*, prefabrication. (7) Interplay of elements: *P*, no contact; *Q*, jointed; *R*, sutured or fused; *S*, imbricate.

designs remain within the set of conceivable structural elements defined by the skeleton space.

The structural elements defined by the parameters of the skeleton space may be skeletons in themselves (e.g., snail shells), they may be parts of skeletons (e.g., sponge spicules), or they may be parts of tightly integrated skeletal complexes with specific functions (e.g., mammal teeth). Skeletal elements of a given type may be more or less closely linked to others of the same kind, or to different kinds of elements. Individual structural elements and their complexes are more or less modular, depending on their organization in relation to larger-scale structures of which they are components. Generally, animal skeletons are nearly decomposable systems (Simon, 1962) in which individual elements have a relatively high degree of local, short-run integrity of structure and function while being interdependent at the level of operation of the organism as a whole. As Kirschner and Gerhardt (1998) have emphasized, weak linkage between modular units facilitates innovation by maintaining "evolvability," in contrast with the tight integration that typically accompanies narrow specialization. These aspects of structural organization are well illustrated by patterns in the emergence of complexity in early metazoan skeletons, as will be shown.

Exploitation of Skeletal Morphospace

To simplify the assessment of skeletal design in relation to the skeleton space, we consider the occurrence of pairs of characters that are represented among the skeletal elements under consideration (figure 11.3). We have shown that the set of viable pairs of characters defined in this theoretical morphospace has been fully exploited by extinct and living animals, with extensive convergence in design among the different phyla (figure 11.3D; Thomas and Reif, 1993). Recently, we have demonstrated that over 80 percent of these character pairs were already exploited by the Middle Cambrian animals of the Burgess Shale fauna, little more than 15 million years after the major diversification of animals with substantial skeletons began (Thomas et al., 2000).

Here, we document the early radiation of animal skeletons in more detail. Ediacaran faunas were dominated by organisms with hydrostatic skeletons. These organisms were prolific and worldwide in their distribution shortly before the emergence of animals with hard skeletons. Some of them grew to "astonishingly large sizes" (Glaessner, 1984, p. 102; he cites *Dickinsonia* up to 45 cm long, p. 60) in two dimensions, but in the third dimension the large forms were all relatively thin.

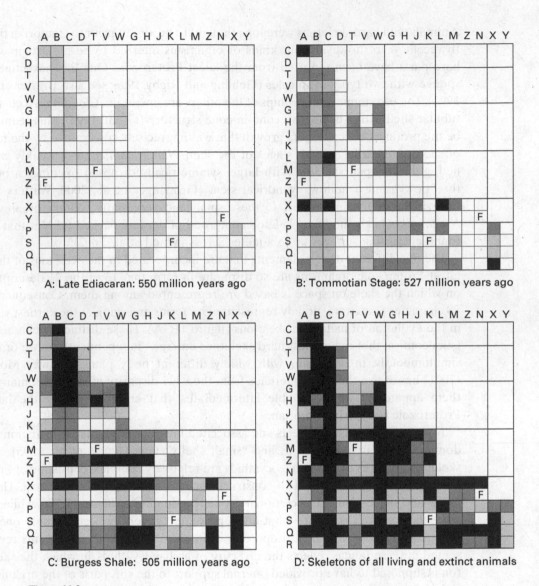

A: Late Ediacaran: 550 million years ago

B: Tommotian Stage: 527 million years ago

C: Burgess Shale: 505 million years ago

D: Skeletons of all living and extinct animals

Figure 11.3
Frequencies of occurrence of skeletal design elements, represented here in terms of pairs of characters. Each diagram represents the set of 186 pairs of potential attributes of animal skeletons or their components. Each structure is defined in terms of seven characteristics, such as shape, number of parts, and mode of growth. Co-occurring pairs of these characters constitute a simplified "design space" in which broad patterns in the evolution of animal skeletons can be charted. Character pairs marked F are illogical or inviable combinations. Characters A, B, C... are defined in figure 11.2. Relative frequency of occurrence of character pairs in real organisms: gray, present; cross-hatched, common; black, abundant.

Animals with hard skeletons were long thought to be absent prior to Cambrian time. In recent years, however, four kinds of organisms inferred to be metazoans with hard parts have been reported from the latest Proterozoic. One is a hexactinellid sponge with two types of spicules (Gehling and Rigby, 1996; see also Brasier et al., 1997). The taxonomic relationships of the others are uncertain. *Cloudina* is a curved, tubular shell with a distinctive cone-in-cone structure (Grant, 1990). This seems to be the product of accretionary growth that was imprecisely constrained by the form of successive earlier growth stages of the shell. *Namacalathus* has a weakly calcified, goblet-shaped skeleton with large, symmetrically disposed openings around the "bowl" and a hollow, cylindrical stem (Grotzinger et al., 2000; Watters and Grotzinger, 2001). In contrast to these small and apparently solitary organisms, *Namapoikia* has a modular skeleton composed of densely packed tubules that are found encrusting surfaces up to a meter across (Wood et al., 2002).

The five different design elements that are incorporated in the skeletons of these latest Proterozoic organisms are so disparate in form that 13 of the 21 descriptors on which the skeleton space is based are represented among them. Consequently, 58 character pairs were already represented by forms emerging at the earliest stage in the evolution of metazoan skeletons (figure 11.3A). These distinctive characteristics confirm that the earliest hard skeletons emerged independently, more or less simultaneously, in organisms with widely different body plans. Conway Morris (2001) has drawn the same inference from the wide disparity of skeletons, many of them apparently without viable intermediates, that emerged during the latest Proterozoic and Early Cambrian.

Tommotian faunas from reefs and associated facies of the Lena River region are dominated by archaeocyathids and "small shelly" taxa (Kruse et al., 1995). The skeletal structures of the archaeocyathids are relatively complex, consisting of fused plates and spines integrated to construct cones that branch in some taxa. These forms and single-element, accretionary cones predominate (figure 11.3B), while the few other types of structures that are represented generally occur in only one or two taxa. The relatively high proportion of internal, rigid skeletal elements represented in these faunas reflects the diversity of archaeocyathids, in which the skeleton is supposed to have provided internal support to the soft parts of the organism, as in other sponges.

The most notable feature of exploitation of the skeleton space by the Burgess Shale animals, apart from the large number of character pairs represented, is the preponderance of rigid or flexible external structures that are inferred to have grown by molting (figure 11.3C). This reflects the presence of a remarkable variety of arthropods and arthropod-like animals. Sponges are also very abundant, but they exhibit a limited range of skeletal elements, individually and collectively. Evidently, the skeletons of early Middle Cambrian animals were very likely to be spicular or segmental, and hence less highly integrated and differentiated than those of animals occurring in later faunas. It is particularly striking in regard to patterns of development that internal skeletal elements that may have grown by remodeling are represented in this fauna only by the notochord of the lancetlike animal *Pikaia*.

Striking differences in the patterns of employment of skeletal design elements between animals of the early Middle Cambrian and Tommotian faunas are expressed even more clearly in figures 11.4 and 11.5. These diagrams show the frequencies of occurrence of skeletal elements that grow in place, where they are used, as opposed to those that are prefabricated and subsequently moved into the position where they function. The overwhelming majority of design elements that have been recognized in extinct or living organisms fall in this part of the skeleton space. Each frequency represents a specific design element. The data are grouped here in such a way as to emphasize differences in the shapes of elements and their conjunction (relationship to adjacent parts), in relation to the four modes of skeletal growth.

The Ediacaran taxa already embodied some of the most distinctly different modes of skeletal construction. One consists of numerous internal spicules, each composed of three orthogonal rods, forming an internal scaffolding. Another is an external conical shell that enclosed the organism's soft parts. The third consists of two strongly curved plates enclosing a stem and the presumed body of the organism. The last is a modular structure composed of closely packed cylindrical tubes. The Tommotian fauna includes many small, conical and tubular shells. It lacks organisms with skeletal elements that grow by molting and it includes only one example of an element with jointed articulation. The apparent reduction in numbers of coiled shells in the Middle Cambrian is largely an effect of facies on the Burgess Shale fauna; "small shelly" organisms remained abundant in carbonate environments. However, the decline in forms that grew by budding of discrete units is real, reflecting a drastic decline in diversity of archaeocyathids. These were not replaced by other gregarious or colonial builders of modular, conical and tubular skeletons until the Ordovician.

The early Cambrian evolutionary radiation involved a rapid, accelerating increase in numbers and kinds of skeletal elements employed by individual organisms and by the diversifying fauna, collectively (table 11.1). Two Ediacaran organisms had a single shell and two others had skeletons composed of multiple, similar elements. Almost half the skeletal elements of the Lena River animals fall in each of these two categories. Only two symmetrically paired skeletal elements occur in the

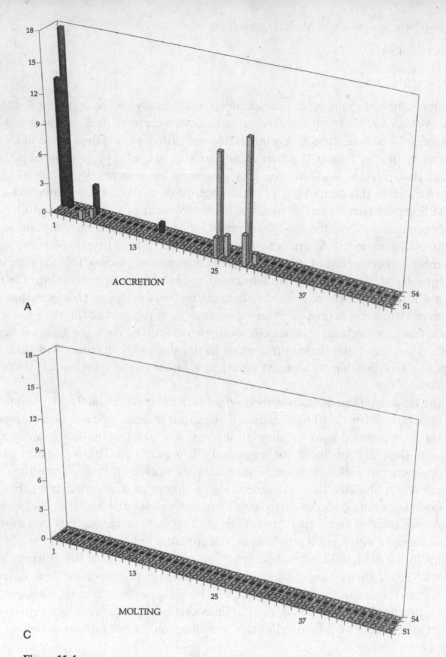

Figure 11.4
Exploitation by Tommotian animals of the Lena River region of potential designs for skeletons capable of supporting soft body tissues. That part of the skeleton space which represents structures that grow in place, where they function, is shown here. Few prefabricated elements have been recognized in the skeletons of Cambrian taxa, so nothing is omitted here (and only very few occurrences are omitted in figure 11.5). Within each group based on a particular growth pattern, rows 1–12 represent isolated elements; those in rows 13–24 are jointed; those in rows 25–36 are sutured or fused; and the elements in rows 37–48 have an imbricate, overlapping contact with one another.

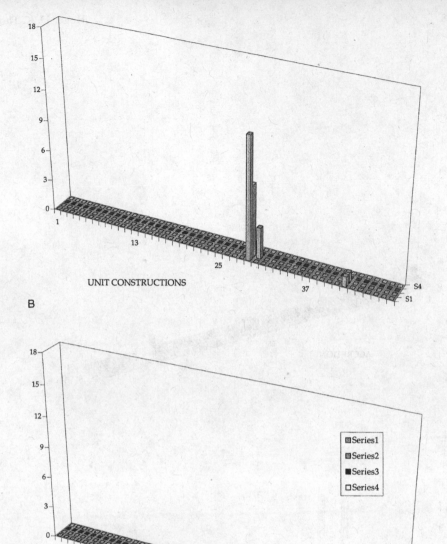

UNIT CONSTRUCTIONS

B

REMODELING

D

Series1
Series2
Series3
Series4

Figure 11.4 (continued)

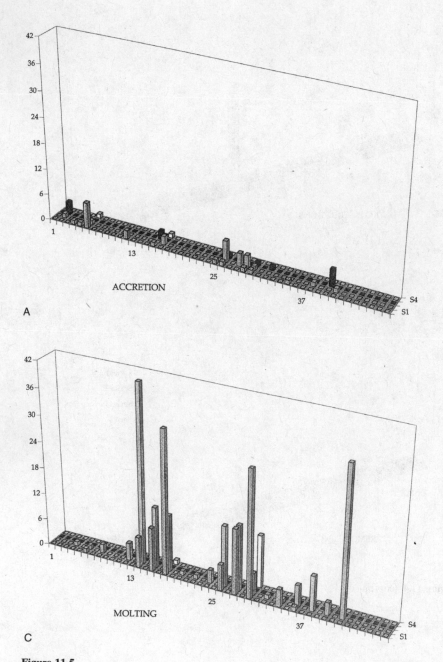

Figure 11.5
Exploitation by Middle Cambrian animals of the Burgess Shale of potential designs for skeletons capable of supporting soft body tissues. See figure 11.4 for further details.

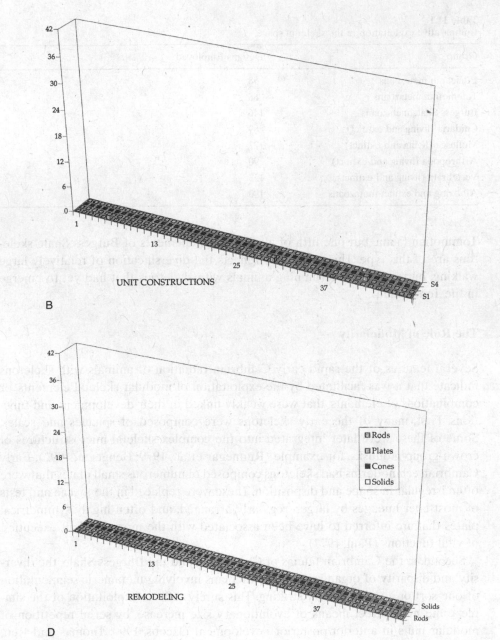

Figure 11.5 (continued)

Table 11.1
Comparative exploitation of the skeleton space

Group	Designs Employed
Ediacaran metazoans	58
Tommotian metazoans	88
Burgess Shale metazoans	146
Cnidaria (living and extinct)	89
Mollusca (living and extinct)	122
Arthropoda (living and extinct)	90
Vertebrata (living and extinct)	133
All living and extinct metazoans	180

Tommotian fauna, but one fifth of all the design elements of Burgess Shale skeletons are of this type. This rapid shift reflects the diversification of relatively large walking, burrowing, and swimming animals with skeletons that had yet to emerge in the Tommotian.

The Role of Modularity

Several features of the rapid early Cambrian radiation of animals with skeletons indicate that it was facilitated by the exploitation of modular skeletal elements, or combinations of elements, that were weakly linked in their development and functions. First, many of the early skeletons were composed of spicules and scales. Some of these were later integrated into the complex skeletal microstructures of crown-group mollusks, for example (Runnegar et al., 1979; Bengtson, 1992). Early Cambrian echinoderms had skeletons composed of numerous small plates that were often irregular in shape and disposition. These were replaced in the thecae and tests of most later lineages by larger, regularly arranged, and often highly symmetrical plates that are inferred to have been associated with the more efficient execution of vital functions (Paul, 1977).

Second, in the Cambrian faunas of Chengjiang and the Burgess Shale, the diversity and disparity of organisms with body plans involving metameric segmentation of one sort or another is very striking. This surely reflects exploitation of the simplest and most direct means of evolutionary size increase, by serial repetition of modular units in anterior/posterior development (Jacobs, 1990; Thomas and Reif, 1993). Subsequently, absolute numbers of segments tended to decrease while the number of kinds of segments specialized for different functions increased, as

Hughes et al. (1999) and Budd (2001) have shown for trilobites and arthropods, respectively.

Third, the skeletons of some Early and Middle Cambrian metazoans include elements so disparate in form that they were initially attributed to more than one organism. *Anomalocaris*, now believed to be a primitive arthropod, and *Halkieria*, which has been adopted alternatively as a stem-group mollusk or as a brachiopod, are notable on this regard. Reviewing this evidence, Conway Morris (1998) cautions that the seemingly bizarre attributes of the Burgess Shale animals may have more to do with their remoteness from our experience than with their actual adaptations. The skeletons of the Burgess Shale arthropods and arthropod-like animals do not incorporate more different kinds of skeletal elements than present-day crustaceans. On the other hand, some of these organisms have entire structural subsystems, such as the jaws of *Anomalocaris* and the proboscis of *Opabina*, with no counterparts among living arthropods. These observations are consistent with the hypothesis that weak developmental constraints (Valentine et al., 1996) allowed different elements of the earliest complex skeletons to evolve with considerable independence of one another.

Subsequent to the rapid, early exploitation of hard skeletons by metazoans, further change characteristically involved the integration of parts into more complex modules, the specialization of formerly similar parts to serve a variety of different functions, and in some cases the elimination of parts that became inappropriate or superfluous as the body plans of emerging higher taxa were refined. Reduction in the number of similar thoracic segments and the incorporation of larger numbers of segments into the pygidium have long been recognized as recurrent trends in the evolution of trilobites (Whittington, 1992; Hughes et al., 1999). Among crinoids, the number of circlets of plates that form the calyx was reduced from four cycles of similar plates in *Aethocrinus*, which is inferred to be the sister group of all other crinoids, to three and then two circlets of differentiated plates in more derived lineages (Simms, 1994).

Assessing crinoid morphology overall, on the basis of 75 discrete characters, Foote (1994) has shown that crinoids achieved their maximum disparity of form by the Middle Ordovician, notwithstanding considerable further taxonomic diversification that continued into the Devonian. The progressively increasing differentiation and specialization of crustacean appendages, well documented in terms of their levels of tagmosis (an information function) by Cisne (1974), stands in marked contrast to the low tagmosis of Cambrian trilobites, other arthropods, and lobopods (Budd, 2001). More often than not, the effect of these sorts of changes was to reduce the number of modules at one level of organization while at the same time introducing new hierarchical levels of structure into increasingly complex skeletons. So, for example, a series of similar appendages becomes a set of structures generated according to a common pattern, each modified to serve a somewhat different purpose.

The skeletons of many Cambrian taxa include components with no close analogues among their later relatives. Budd (2001) cites the "great appendage" of *Leanchoilia* in this regard. Other examples include the jaws of *Anomalocaris* noted above, the elaborate head shield of *Marella*, and the spines of armored lobopods such as *Hallucigenia*. The archaeocyathids and the lithistids employed a considerable variety of structural elements that have no counterparts among later sponges (Debrenne and Reitner, 2001). The elimination of these relatively specialized structures presumably represents an early stage in a long-continued macroevolutionary process of culling and refinement of structure.

In due course, this gave rise to mammals with less structurally disparate skeletons than the rhipidistian fish *Eusthenopteron* (Thomas and Reif, 1993) and advanced cephalopods in which the skeleton has been lost altogether. This reduction in complexity at a late stage in the history of exploitation of skeletal design options (see figure 11.7) occurs where emergent complexity at another hierarchical level—sophisticated behavior in the case of the examples just given—makes the skeleton or parts of it redundant. This parallels the "complexity drain" in number of parts, recently noted by McShea (2002), from the cells of free-living protists to cells of land plants and metazoans.

Logistic Increase in Skeletal Complexity

The numbers of design elements exploited at three early stages in the evolution of animal skeletons, documented in figure 11.3 and table 11.1, are consistent with a logistic pattern of diversification (figure 11.6A). Together with the essentially complete exploitation of design options that came later (Thomas and Reif, 1993), this implies a pattern of evolution occurring in a world of finite possibilities. There are only so many good ways in which to make skeletons. This is consistent with Layzer's evolutionary model. The full set of 1536 design elements prescribed by the skeleton space corresponds to the maximum value on which H_{max} converges. The number of design elements exploited in each fauna constitutes H_{obs}. The maximum value reached by H_{obs} has not been determined. It cannot be greater than 960, given the number of incompatible pairs of characters (see above). It is probably much lower still, on account of combinations of three or more characters that are likewise

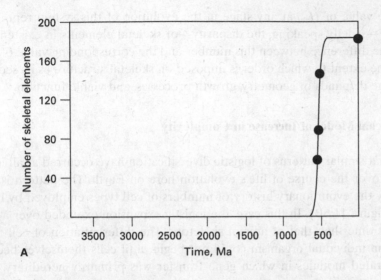

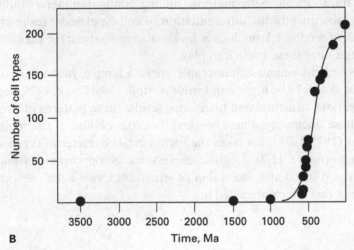

Figure 11.6
(*A*) Logistic curve fitted to a plot of numbers of skeletal elements employed against the geologic ages of the faunas in which they appear. The number of skeletal elements employed by all living animals has been arbitrarily set at 500, since this number has yet to be determined. The form of the curve is relatively insensitive to values of this number between 200 and 960, its maximum possible value. (*B*) Logistic curve fitted to a plot of minimum numbers of cell types against presumed times of origin of the higher taxa in which they occur. Data of Valentine et al. (1994) plus additional records for groups antecedent to the emergence of metazoans.

inviable. The value of H_{obs} at any stage in the evolution of this system represents the diversity—strictly speaking, the disparity—of skeletal elements in existence at that time. The difference between this number and the corresponding value of H_{max} represents the extent to which order is imposed on skeletal structure as its scope is limited by the demands of geometry, growth processes, and viable function.

A Hierarchichal Model of Increase in Complexity

It is likely that similar patterns of logistic diversification have occurred at all levels of structure over the course of life's evolution here on Earth. The pattern is well illustrated by the evolutionary history of numbers of cell types employed by living organisms (figure 11.6B). In this case, the logistic expansion extended over a very much longer time than that of animal skeletons. The differentiation of cell types comprising an individual organism could not begin until cells themselves became tightly integrated modules in which gene transfer was primarily hereditary, with limited lateral transfer of genetic material among contemporaneous individuals (Woese, 2002). Subsequently, the differentiation of cell types was presumably still impeded by lack of feedback from higher levels of organization, for their diversification accelerated once these came into play.

The fossil record documents innumerable rapid adaptive radiations, some of which have been shown to be logistic in form (Westoll, 1949; Cisne, 1974; Sepkoski, 1978, 1993), at various structural and taxonomic levels. These patterns of evolution, together with those documented here, suggest an extrapolation of the model proposed by Layzer (1975, 1978) that takes the hierarchical character of organic structure into account (figure 11.7). Logistic expansions of opportunities and their exploitation have occurred at a succession of structural levels, each one occurring more rapidly than that which preceded it.

Conclusion

The relatively complex skeletons that are characteristic of so many metazoans have necessarily evolved as modular, hierarchically integrated structures based on elements that are defined within the skeleton space. This mode of organization is unavoidable on functional, constructional, and historical grounds. At this level of complexity, only modular, hierarchical structures can be constructed economically, to operate effectively, within an acceptable (even geologically acceptable!) span of time (Simon, 1962).

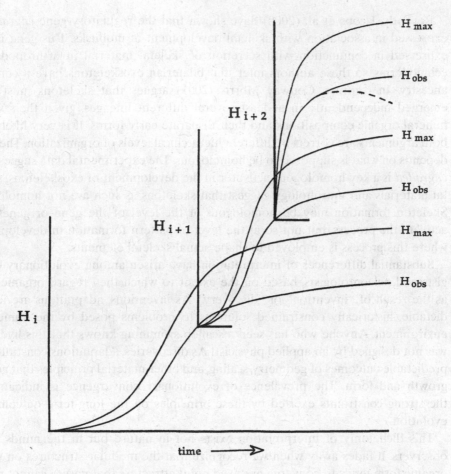

Figure 11.7
A hierarchical model of evolutionary change. The information functions represent the increasing numbers of modules available (H_{max}) and realized (H_{obs}) at successive structural levels. For example, i, $i+1$, and $i+2$ could be cells, tissues, and animal skeletons. The dotted curve for H_{obs} at the $i+2$ level is based on the observation that the skeletons of organisms which appear early in the evolutionary radiations of several higher taxa have more varied parts than the skeletons of animals in the same groups which emerge later on.

Recently, Jacobs et al. (2000) have shown that the regulatory gene *engrailed* is expressed in association with skeletal development in mollusks. This gene is also expressed in conjunction with secretion of skeletal material in arthropods and echinoderms, so these authors infer that bilaterian exoskeletons have a common ancestry. In contrast, Conway Morris (2001) argues that skeletons must have emerged independently in at least several different lineages, given their varied mineral/organic compositions and their disparate early forms. It is very likely that both arguments are correct at different hierarchical levels of organization. The issue depends on what is supposed to be homologous. The experimental data suggest that *engrailed* is a key homologous activator in the development of exoskeletons; skeletal materials and morphology suggest that skeletons as such are not homologous. Skeleton formation may be homologous at the level of the gene or genes that activate the process, but not so at the level of pattern formation in development, where this process is employed to shape actual skeletal elements.

Substantial differences of interpretation have arisen among evolutionary biologists and paleontologists, based on the extent to which they regard organic form as the result of "invention" or "discovery." As inventions, adaptations are unpredictable, historically constrained solutions to problems posed by the immediate environment. Anyone who has seen a scallop swimming knows that this hydrofoil was not designed by an applied physicist! As discoveries, adaptations constitute the predictable outcomes of geometry, scaling, and other material principles that govern growth and form. The prevalence of evolutionary convergence is indicative of the strong constraints exerted by these principles on the long-term outcomes of evolution.

This dichotomy of interpretation exists not in nature but in the minds of its observers. It fades away when we recognize that the modular structures on which organic form tends to converge are fixed point attractors that nature cannot avoid, as Kauffman (1989) has argued. Organisms in the whole, on the other hand, belong to that class of unpredictable, momentarily stable configurations of matter in motion that are now characterized as strange attractors. This view of organic form has strong implications for our understanding of macroevolution. To the extent that the skeletons of metazoans are defined by favorable combinations of fixed point attractors, they have properties which natural selection was bound to favor. Adaptation is locally opportunistic, in time and space, but the long-term outcomes in the evolution of animal skeletons represent differential exploitation of more, rather than less, felicitous designs.

Acknowledgments

I am very grateful to Werner Callebaut and to Diego Rasskin-Gutman for inviting me to participate in the stimulating workshop on which this book is based. This chapter would not have been completed without the encouragement and critical advice of Diego Rasskin-Gutman, to whom I am much indebted.

References

Bengtson S (1992) The advent of animal skeletons. In: Early Life on Earth (Bengtson S, ed), 412–425. New York: Columbia University Press.

Brasier MD, Green O, Shields G (1997) Ediacaran sponge spicule clusters from southwestern Mongolia and the origins of the Cambrian fauna. Geology 25: 303–306.

Brooks DR, Wiley EO (1988) Evolution as Entropy, 2nd ed. Chicago: University of Chicago Press.

Budd GE (2001) Ecology of nontrilobite arthropods and lobopods in the Cambrian. In: The Ecology of the Cambrian Radiation (Zhuravlev AYu, Riding R, eds), 404–427. New York: Columbia University Press.

Cisne JL (1974) Evolution of the world fauna of aquatic free-living arthropods. Evolution 28: 337–366.

Conway Morris S (1998) The Crucible of Creation. Oxford: Oxford University Press.

Conway Morris S (2001) Significance of early shells. In: Palaeobiology II (Briggs DEG, Crowther PR, eds), 31–40. Oxford: Blackwell.

Debrenne F, Reitner J (2001) Sponges, cnidarians, and ctenophores. In: The Ecology of the Cambrian Radiation (Zhuravlev AYu, Riding R, eds), 301–325. New York: Columbia University Press.

Foote M (1994) Morphological disparity in Ordovician–Devonian crinoids and the early saturation of morphological space. Paleobiology 20: 320–344.

Gehling JG, Rigby JK (1996) Long expected sponges from the Neoproterozoic Ediacara fauna of South Australia. J Paleontol 70: 185–195.

Glaessner MF (1984) The Dawn of Animal Life. Cambridge: Cambridge University Press.

Grant SWF (1990) Shell structure and distribution of Cloudina, a potential index fossil for the terminal Proterozoic. Amer J Sci 290A: 261–294.

Grotzinger JP, Watters WA, Knoll AH (2000) Calcified metazoans in thrombolite–stromatolite reefs of the terminal Proterozoic Nama Group, Namibia. Paleobiology 26: 334–359.

Hughes NC, Chapman RE, Adrain JM (1999) The stability of thoracic segmentation in trilobites: A case study in developmental and ecological constraints. Evol Dev 1: 24–35.

Jacobs DK (1990) Selector genes and the Cambrian radiation of Bilateria. Proc Natl Acad Sci USA 87: 4406–4410.

Jacobs DK, Wray CG, Wedeen CJ, Kostriken R, DeSalle R, Staton JL, Gates RD, Lindberg DR (2000) Molluscan engrailed expression, serial organization, and shell evolution. Evol Dev 2: 340–347.

Kauffman SA (1989) Origins of order: Self-organization and selection. In: Theoretical Biology: Epigenetic and Evolutionary Order from Complex Systems (Goodwin BC, Saunders P, eds), 67–88. Edinburgh: Edinburgh University Press.

Kirschner M, Gerhart J (1998) Evolvability. Proc Natl Acad Sci USA 95: 8420–8427.

Kruse PD, Zhuravlev AYu, James NP (1995) Primordial metazoan-calcimicrobial reefs: Tommotian (Early Cambrian) of the Siberian Platform. Palaios 10: 291–321.

Layzer D (1975) The arrow of time. Sci Amer 233(6): 56–69.

Layzer D (1978) Information in cosmology, physics, and the big bang. Int J Quant Chem 12(supp 1, for 1977): 185–195.

McShea DW (2002) A complexity drain on cells in the evolution of multicellularity. Evolution 56: 441–452.

Paul CRC (1977) Evolution of primitive echinoderms. In: Patterns of Evolution as Illustrated by the Fossil Record (Hallam A, ed), 123–158. Amsterdam: Elsevier.

Runnegar B, Pojeta J, Taylor ME, Collins D (1979) New species of the Cambrian and Ordovician chitons *Matthevia* and *Chelodes* from Wisconsin and Queensland: Evidence for the early history of polyplacophoran mollusks. J Paleontol 53: 1374–1394.

Schank JC, Wimsatt WC (2001) Evolvability, adaptation and modularity. In: Thinking About Evolution, vol. 2: Historical, Philosophical, and Political Perspectives (Singh RS, Krimbas CB, Paul DB, Beatty J, eds), 322–335. New York and Cambridge: Cambridge University Press.

Sepkoski JJ (1978) A kinetic model of Phanerozoic taxonomic diversity II. Early Phanerozoic families and multiple equilibria. Paleobiology 5: 222–251.

Sepkoski JJ (1993) Ten years in the library: New data confirm paleontological patterns. Paleobiology 19: 43–51.

Simms MJ (1994) Reinterpretation of the thecal plate homology and phylogeny of the class Crinoidea. Lethaia 26: 303–312.

Simon HA (1962) The architecture of complexity. Proc Amer Phil Soc 106: 467–482.

Thomas RDK, Reif W-E (1991) Design elements employed in the construction of animal skeletons. In: Constructional Morphology and Evolution (Schmidt-Kittler N, Vogel K, eds), 283–294. Berlin/Heidelberg: Springer.

Thomas RDK, Reif W-E (1993) The skeleton space: A finite set of organic designs. Evolution 47: 341–360.

Thomas RDK, Shearman RM, Stewart GW (2000) Evolutionary exploitation of design options by the first animals with hard skeletons. Science 288: 1239–1242.

Thompson D'A W (1942) On Growth and Form, 2nd ed. Cambridge: Cambridge University Press.

Valentine JW, Collins AG, Meyer CP (1994) Morphological complexity increase in metazoans. Paleobiology 20: 131–142.

Valentine JW, Erwin DH, Jablonski D (1996) Developmental evolution of metazoan body plans: The fossil evidence. Dev Biol 173: 373–381.

Vogel S (1988) Life's Devices: The Physical World of Animals and Plants. Princeton, NJ: Princeton University Press.

Watters WA, Grotzinger JP (2001) Digital reconstruction of calcified early metazoans, terminal Proterozoic Nama Group, Namibia. Paleobiology 27: 159–171.

Westoll TS (1949) On the evolution of the Dipnoi. In: Genetics, Paleontology and Evolution (Jepson GL, Simpson GC, Mayr E, eds), 121–184. Princeton, NJ: Princeton University Press.

Whittington HB (1992) Trilobites. Woodbridge, UK: Boydell Press.

Woese CR (2002) On the evolution of cells. Proc Natl Acad Sci USA 99: 8742–8747.

Wood RA, Grotzinger JP, Dickson JAD (2002) Proterozoic modular biomineralized metazoan from the Nama Group, Namibia. Science 296: 2383–2386.

12 Modularity in Art

Slavik V. Jablan

Modularity appears in different fields of science: in biology, physics, psychology, even in economic science, so it can be considered as a kind of unifying scientific concept. In art, modularity occurs when several basic elements (modules) are combined to create a large number of different (modular) structures. For example, different modules (e.g., bricks in architecture or in ornamental brickwork) occur as the basis of modular structures (Makovicky, 1989). In the natural sciences, modularity is represented by a search for fundamental units and basic elements (e.g., physical constants, elementary particles, prototiles for different geometric structures, etc.). In various fields of (discrete) mathematics, the search for modularity is the recognition of sets of basic elements, construction rules, and the exhaustive derivation of different generated structures.

In a general sense, modularity is a manifestation of the universal principle of economy in nature: the possibility for diversity and variability of structures resulting from some (finite and very restricted) set of basic elements by their recombination (figures 12.1 and 12.2). In all such cases, the most important step is the choice, by recognition or discovery, of the basic elements. This could be shown by examples from ornamental art where some elements originating from Paleolithic or Neolithic art are still present in ornamental art as "ornamental archetypes" (figure 12.3) (Christie, 1969; Washburn and Crowe, 1988).

In many cases, the derivation of discrete modular structures is based on symmetry. Using the theory of symmetry and its generalizations (simple and multiple antisymmetry, colored symmetry, etc.) for certain structures, it is possible to define exhaustive derivation algorithms, and even to obtain some combinatorial formula for their enumeration. Examples of modular structures that combine art and mathematics are (1) the set of modular elements for the derivation of possible and impossible objects, "space tiles," and the classification of structures obtained (figure 12.1b); (2) knot projections occurring in knotwork designs (Islamic, Celtic, etc.) derived from the regular and uniform plane tessellations by using a few basic elements, "knot tiles" (figure 12.2a); (3) antisymmetry ornaments and their derivation from a few prototiles—"op tiles"—as well as the algorithmic approach to their generation (figure 12.2b).

Things Are Not Always What They Seem

If we try to explain what we see as an object, we could conclude that we always make a choice between an infinite series of real three-dimensional objects having

(a)

(b)

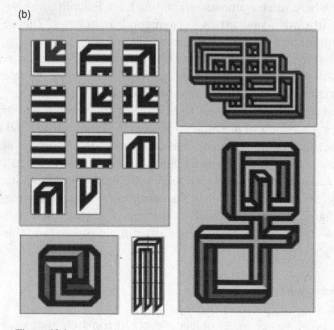

Figure 12.1
(*a*) Modular design variations on a singe theme: three letters T, created by the author. (*b*) Set of modular elements ("space tiles") and some examples of impossible objects from an infinite collection of possible and impossible objects that could be obtained.

(a)

(b)

Figure 12.2
(*a*) Set of modular elements ("knot tiles") for creation of different knots and links. (*b*) "Op tiles" and modular op art works from the same infinite gallery.

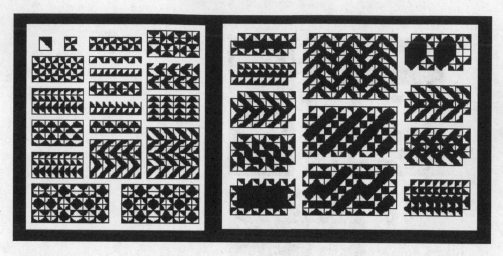

Figure 12.3
Black-and-white ornaments composed from a single element.

the same flat retinal projection (Gregory, 1970). From this infinite set of objects, our perception usually selects only one natural, most probable, or, sometimes, simplest interpretation. Our perception and interpretation of the normal 3-D world is strongly connected with the need to create a complete picture. But in the cases of isolated objects when no common reference system exists, ambiguity can occur, and it is impossible to determine a unique natural interpretation. The inability to form a natural interpretation increases when the object is an impossible 3-D object.

Research into mathematics, visual perception, and impossible 3-D objects has been stimulated by Penrose's impossible triangle (Penrose and Penrose, 1958) and by the works of M. C. Escher (Ernst, 1994). Their interpretations can be found in works by Piranesi and Albers, and in op art works (Teuber, 1986). If we look at the history of such objects, we can start with the mosaics from Antioch (Gombrich, 1979). These ancient tiles use Koffka cubes that create both a regular plane ornament and a 3-D structure, both of which are modular.

Let us next consider the Koffka cube (Koffka, 1935). It is multiply ambiguous: it could be interpreted as three rhombuses with a joint vertex, as a convex or concave trihedron, or as a cube. If we accept its "natural" 3-D interpretation as a cube, then a viewer can interpret it in any of three possible positions: upper, lower left, and lower right. Thus, for the corresponding three directions, a Koffka cube represents a turning point. Having multiple symmetries, it fully satisfies the conditions to be a basic modular element.

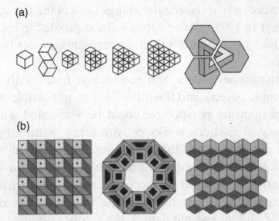

Figure 12.4
(*a*) Evolution of the Thiery figure, the tribar, and other impossible objects from the Koffka cube and hexagonal "op tiles." (*b*) Mosaics from Antioch.

With this in mind, we can analyze some well-known impossible objects (figure 12.4). Thiery figures (proposed at the end of the nineteenth century) consist of two Koffka cubes (Ernst, 1990). Objects created by Oscar Reutesward in 1934, the Penrose tribar, the Vasarely constructions, and Taniuchi's alphabet can all be shown to be modular structures made from Koffka cubes. For each of these objects, it is even possible to calculate its "volume." For example, the "volume" of a tribar with each bar formed by five cubes is twelve cubes. This leads to some interesting speculations.

According to the playwright Eugene Ionesco, we all know the sound of applause made by two hands, but what about the sound of applause made by only one hand? What will happen with the tribar as it gets smaller and smaller, with its edges consisting of three, two, or even only one unit? In the last case, the result is a singular Koffka cube. If it has two edges, the result is an impossible object, having as its envelope another interesting geometrical figure: a one-sided Moebius strip (Peterson, 1988).

Going in the opposite direction from the Koffka cubes, we could construct an infinite family of impossible figures. To create them, we could proceed in three directions (i.e., choosing from the six possible orientations). Our vision of some 3-D objects could be reduced to several possible cases as a result of different points of view. For example, a rectangular frame, when viewed from the corners, has three unique views (or four, if we distinguish enantiomorphic forms). Then, the drawings of its constitutive parts could be the basic elements (prototiles) of a modular plane tiling. Using the fact that there are only three regular tilings for a plane (by regular triangles, squares, and hexagons), in 1995 I derived the modular puzzle "space tiles" (figure 12.1b). They can be used to construct pictures representing possible and impossible objects in the plane.

For example, if we create the corner pieces of a rectangular frame, only three combinations will result in possible objects, and the others will be impossible (figure 12.5a). One such frame in the opposite perspective could be associated with the mosaic from Antioch, with some of Escher's works, or with many medieval paintings using counterperspective (Raushenbach, 1980). The basic elements of "space tiles" describe different possible vertex situations, and could be used for visual perception research and the criteria required to recognize a 3-D object from its drawing. If we introduce Archimedean (or uniform) plane tilings, we can obtain an infinite collection of possible and impossible figures, beginning with elementary ones and progressing to more sophisticated forms, as shown in B. Ernst's book *L'Aventure des figures impossibles* (1990) and the artistic creations of T. Farkas.

"Are impossible figures possible?" This question is the title of the paper by Kulpa (1983). The paper explains that the property of "impossibility" for a figure is not the property of the drawing alone, but the property of its spatial natural interpretation by a viewer (figure 12.6). We already noted that an object's visual interpretation is a selection from an infinite number of objects, some possible, some not. From the drawing of an impossible object, we can derive some of its objective realizations: (a) possible object(s) having the same retinal projection. In situations where an impossible object is simpler than the possible one(s), the eye and mind will accept it as the interpretation of a drawing. Beyond that, the conclusion will be that the object is impossible.

The next question concerning impossible objects is their "degree of impossibility" (figure 12.5b). This means our ability to recognize them as impossible (Huffman, 1971). For example, some frames in figure 12.5a are easily recognized as impossible, but this is not the case with some line arrangements in figure 12.5b, where it is much more difficult to determine if they are "impossible" or not. For the well-known Borromean rings (Adams, 1994), visual arguments are not sufficient, so we need a mathematical proof of their impossibility. (This means that they could not be realized by three flat rings.)

"Borromean rings" are three linked components (figure 12.5b), named after the Borromeas, an Italian family of the Renaissance who used them on their coat of arms (see Lindström and Zetterström, 1991). Whereas Borromean circles are impossible objects, Borromean triangles are possible. A hollow triangle is the planar region bounded by two homothetic and concentric equilateral triangles—flat

(a)

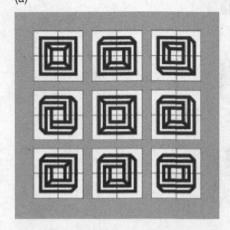

(b)

Figure 12.5
(a) Impossible and possible frames. (b) Borromean rings and line-segment figures.

Figure 12.6
Are impossible figures possible?

(a)

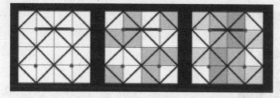

(b)

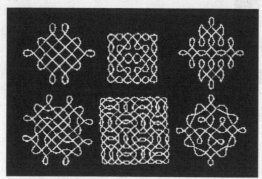

(c)

Figure 12.7
(*a*) Celtic knots with the corresponding mirror arrangements. (*b*) Knot, its mirror scheme and Lunda designs (*mod* 2 and *mod* 4). (*c*) Tamil and Tchokwe designs.

triangular rings. The Australian sculptor J. Robinson assembled three such triangular rings to form a structure (a sculpture titled *Intuition*), topologically equivalent to Borromean rings. Their cardboard flexible model will collapse to form a planar pattern. Peter Cromwel has found such an object in the details of a picture stone from Gotland (1995). That and other symmetrical combinations of three and four hollow triangles are also considered by Coxeter (1994).

Knots

Borromean rings are a simple example from the large field of knots and links that are the result of ancient human activities such as knotwork, weaving, and plaiting (Turner and van de Griend, 1996). From a mathematical point of view, a knot is a homeomorphic image of a circle—some placement of a circle in 3-D space—and a link is a placement of several circles (Conway, 1970; Rolfsen, 1976; Caudron, 1981; Burde and Zieschang, 1985; Kauffman, 1987; Livingston, 1993; Adams, 1994). Their 2-D representations are simple planar projections. In mathematics, the golden age of knot projections was the end of the nineteenth century. Most knots correspond to more than one projection. However, there is a de facto tradition that has been followed by mathematicians for decades: that descriptions of knots are given with only one projection for each knot. Their first choice may be attributed to Reidemeister (1932), and it is interesting to analyze his reasons for choosing each specific projection.

In art, knot designs have been present for thousands of years (Turner and van de Griend, 1996). Some of the most spectacular knots ever made are Celtic. They have been mathematically analyzed by Cromwel (figure 12.7*a*) (Bain, 1973; Cromwel, 1993). Their symmetry has also been analyzed by Gerdes (1989, 1990, 1996, 1997, 1999), who discussed the so-called mirror curves, knot-and-link Tchokwe sand drawings from the Lunda region (eastern Angola and northwestern Zambia), and Tamil designs (figure 12.7*c*).

Let us consider a polyomino (a collection of equal-size squares arranged edge-to-edge) in a regular plane tiling (Golomb, 1994), with a set of (two-sided) mirrors incident to its edges or perpendicular in their midpoints. The ray of light starting from such midpoint, after a series of reflections, will return to it, forming a closed path or a mirror curve. If the polyomino is completely covered by a singular curve, it always can be simply transformed (e.g., by alternation) into a knot projection. Otherwise, if it is exhausted by several components, it may give a link projection. A curve obtained in this manner will be symmetrical or asymmetrical, depending on

(a)

(b)

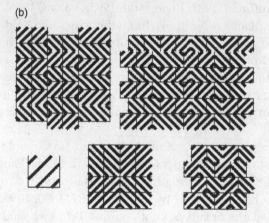

Figure 12.8
(*a*) Ornaments from Mezin (23,000–15,000 B.C., Ukraine). (*b*) Modularity of key patterns.

the shape of the polyomino and placement of internal mirrors, so symmetry is not necessarily a property of mirror curves. Mirror curves possess another remarkable property, that of modularity. Every mirror curve can be created covering a polyomino with only five basic elements (in the case of square tiling). I introduced such elements in 1994 and named them "knot tiles." The number of modular designs made by such tiles is unlimited, and different designs can be obtained by using topological variations of the prototiles (figure 12.2a).

Multiple variations can also be obtained by using different basic polyominoes resulting from Archimedean (uniform) plane tilings. It is interesting that the modularity of knotworks may have been discovered by Escher. He created several prototiles for their production (Schattschneider, 1990). Analyzing mirror curves, Gerdes discovered that if the successive small squares through which the curve passes are colored alternately (black and white), the result is a black-and-white mosaic, which he called a Lunda design. Lunda designs possess the local equilibrium property where the midpoint of each edge is equally surrounded by small black and white squares. Global equilibrium—the same number of black and white squares in every row and column—is the consequence of the local equilibrium. Thus, every square Lunda design is a modular black-and-white design, and it is formed by only three kinds of prototiles (two kinds of internal prototiles and one kind of border prototile). Many Lunda designs possess the amazing property of equality between the design and background. This means that they are antisymmetrical (figure 12.7b). If Gerdes was able to derive Lunda designs from mirror curves using black-and-white coloring, can we also find these designs in ancient ornamental art? Searching for the answer to this question, we will return to the very origins—Paleolithic and Neolithic ornamental art.

Ancient Art

"Key patterns" are a keylike ornaments that occur in Egyptian, Greek, Roman, Mayan, Chinese, and especially Celtic art. Because Celtic key patterns are so distinctive, they can be singled out (Bain, 1973). Most authors consider "keys" as common patterns. Grünbaum and Shephard (1997) hold this point of view in their book *Tilings and Patterns*. However, in their comment about the Chinese key pattern that opens the chapter on patterns, they note that it was inspired by a maze. Bain also mentions the connection between key patterns, spirals, meanders, mazes, and labyrinths.

The oldest examples of key patterns belong to the Paleolithic art from Mezin (Ukraine, about 23,000–15,000 B.C.) (figure 12.8a). If we compare Mezin patterns

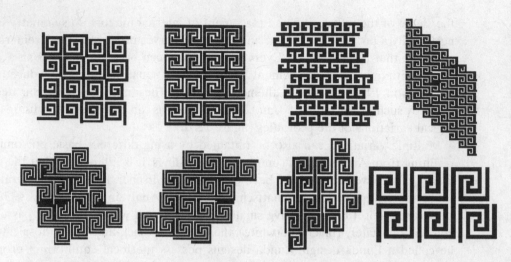

Figure 12.9
Neolithic key-pattern ornaments from Vincha (Yugoslavia), Dimini (Greece), and Podrinje (Yugoslavia).

with all other Paleolithic art ornaments, one can see that their patterns are radically different from the others and much more complex. All the ornaments from Mezin are systematically generated from the same basic prototile, which is a square with a set of parallel diagonal lines. Similar patterns (figure 12.8b) could be obtained by using only one or two such prototiles.

The next major occurrence of key-pattern ornaments are Celtic. In his book, Bain attempts to explain their construction, describing every key pattern by a series of numbers denoting the number of steps in a particular direction. Thus, from each key pattern it is possible to read its corresponding series. Unfortunately, it is not clear how to obtain such series. Key patterns make a very strong visual impression (figure 12.9), similar to the effects produced by op art works. Op art works are considered by Barrett (1970) as "interrupted systems," where "the pattern or system is broken or interrupted." This results in an extraordinary degree of flickering and dazzling. Similar structures are well known in the theory of visual perception.

Some key patterns produce the same strange and sometimes frightening visual impression. As a result, they have been used as charms against enemies or as symbols of a labyrinth. The maze pattern on the wall of the palace at Knossos with the motif of double-axe (*labris*), from which the word "labyrinth" may have originated, also can be constructed by the symmetrical repetition of a single prototile (figure 12.10).

Figure 12.10
Double axe from Dictean cave and the maze pattern on the wall of the palace at Knossos.

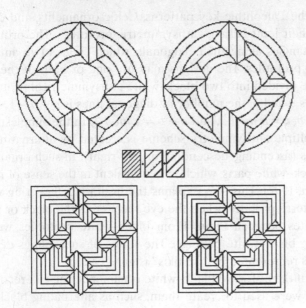

Figure 12.11
Roman mazes based on antisymmetry.

If we analyze Roman mazes (figure 12.11), we can recognize the same elementary meanders or meandering friezes that occur in key patterns (Phillips, 1992). Antisymmetry is probably the principle on which both are based. In the first example (taken from the Roman maze at Avenches, Switzerland), we can see a regular system formed by concentric circles, which is interrupted by four "dislocations": three rectangles with four, and one with five, diagonal lines. They are created by using a 6×4 rectangle with nine diagonal lines, from which two mutually antisymmetrical rectangles derive—one with four and other with five diagonal lines. In general, antisymmetry can be represented by the contrast of opposite colors (black and white) and by the principle of complementarity: two opposites creating a unity. In this case, the two opposites are two rectangles; in one of them, to each diagonal line there corresponds the empty field between two diagonal lines in the other, and vice versa. Superposed together, they create a unity: the original rectangle with nine diagonal lines. The other Roman mazes (and their reconstructions) are variations on the same idea of a regular system of concentric squares that is interrupted by several (regularly arranged) antisymmetric rectangles. The circular mazes are the simple topological equivalent of square mazes, and are derived directly from them.

If we consider again the Paleolithic key patterns, Celtic ornaments, and op art works, we can see that their joint basis is antisymmetric prototiles. The prototiles are created dividing rectangles along their diagonals, which creates two antisymmetric (complementary) prototiles. The designs may use one or both of the tiles. The rectangle can also be divided into two black–white antisymmetrical prototiles. For a rectangle with sides a and b, the number of diagonal lines is $a + b - 1$, so we distinguish the cases $a = b$, $a = b \bmod 2$, and $a + b = 1 \bmod 2$. In the simplest case, where $a = b = 2$, a 2-multiple antisymmetry scheme is created by alternating the direction of the diagonals (ascending–descending or left–right). In such ornaments we can recognize the black–white parts, which are equivalent in the sense of multiple antisymmetry (Jablan, 1992). This also explains the hesitating, flickering visual impression that such patterns produce, when the eye recognizes the black or white pattern and then oscillates between them. From black–white prototiles we can obtain the corresponding black–white patterns. The series of such tilings derived from the five prototiles is represented by "op tiles" (figure 12.2b).

After analyzing some of the historical black–white ornaments, one can recognize the simple methods that were used to create them, such as alternating black and white isohedral tilings. For others, however, it is much more difficult to understand both their conception and their construction. Some of these ornaments use multiple antisymmetry. There, a fundamental region is divided into several parts, and then produced as multiple-antisymmetrical tiles. Multiple antisymmetry does not require a very sophisticated rule. It is only a multiple 0–1 way of thinking in geometry, or a simple use of a binary numerical system (Boolean spaces). This idea, introduced in my paper "Periodic Antisymmetry Tilings" (Jablan, 1992), is used to obtain some nonstandard isohedral tilings by use of multiple antisymmetry. The Paleolithic key patterns mentioned represent probably the first use of antisymmetry in ornamental art. The Celtic key patterns, Roman mazes, and some op art works are based on the same principle, first discovered about 23000 B.C. Thus the question "Do you like Paleolithic op art?" is not as absurd as it first seems.

After discovering that many of the antisymmetric ornaments can be derived by combining a few basic "op tiles" as a modular structures (or simply patchworks), we may concentrate our research on the basic elements (modules) and their origins. Examples of such modules are a square with the set of diagonal lines, two anti-symmetrical squares, a black–white square (abundantly used in prehistoric or ethnic art, known also as the element of mosaics: Truchet tile) (figure 12.3). Their topo-logical equivalents are obtained by replacing straight diagonal lines with circle arcs (Smith, 1987).

After finding that key patterns are based on antisymmetry, we may next examine the antisymmetry in such ornaments. If you start from the simplest antisymmetrical squares with only one diagonal line, you can derive an infinite series of black-and-white ornaments. This includes many key patterns, Neolithic ornaments, and Kufic writing (Mamedov, 1986). The same prototile is also used in Renaissance and later European ornamental art as a basic element for the Persian scheme (Christie, 1969). Certainly, Kufic writings can be obtained from different basic elements (e.g., from a unit of black and white squares), but in this case in such writings may appear as 2×2 squares, and the proposed antisymmetric prototiles guarantee that the thick-ness of all the lines will be exactly 1 (figure 12.12).

In many cases the construction rules arising from the mathematical theory of symmetry differ from the construction rules used in ornamental art. For example, the mathematical concept of an asymmetrical figure, or an asymmetrical funda-mental region multiplied by symmetries, is not always used in ornamental art. Instead, artists use basic symmetrical or antisymmetrical figures: modules, rosettes, friezes, and others, by themselves or in combination (superposition, overlapping, etc.). Many Islamic ornaments are probably obtained by superposing very sim-ple patterns (figure 12.13) (Christie, 1969; Critchlow, 1976; El-Said and Parman, 1976).

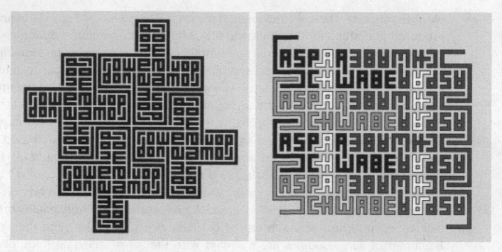

Figure 12.12
Kufic scripts designed by the author.

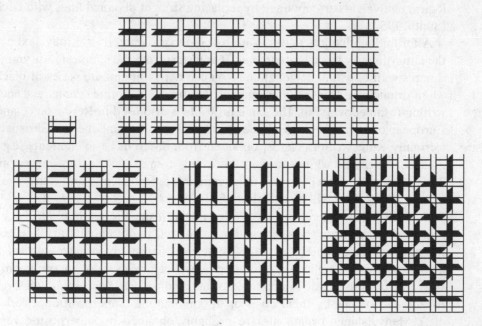

Figure 12.13
Design from a drawing in the Mirza Akbar collection, obtained by overlapping of elementary patterns.

Ornaments, Art, and Cultures

Mathematics answers the question "Which ornaments are derived?," but not "How are they derived?" or "Why are they derived?" Their origin could be from a manufacturing process (e.g., matting, plaiting, or basketry) or the application of simple rules for making different arrangements of modules. For example, similar or identical ornaments derived from basic modules occur in many different cultures (e.g., Neolithic black–white ornaments from Tell Halaf, or Cakaudrove patterns from Fiji) (Crowe and Nagy, 1992). After the Neolithic period, it is almost impossible to find a culture that has not used patterns derived from black–white squares. The question then becomes how many patterns have been derived (in the sense of exhaustive derivation according to symmetry rules) from the same prototile by different cultures (figure 12.3). When researching the history of intuitive or visual mathematics, it is important to follow the use of basic design elements in the same or different cultures, and their changes (geometrical and topological) over time, due to different cultural influences.

The path of such a basic prototile through different cultures can be traced in patterns from Mezin (Ukraine) (figure 12.8*a*), in Paleolithic artifacts from the Schela Cladovei culture (Romania), and its occurrence in all Neolithic cultures: Cucuteni (Ukraine, Moldavia, Romania), Gumelnitsa (Romania), Tisza (Hungary), Vincha (Yugoslavia), and Dimini (Greece) (figure 12.14). In his Ph.D. dissertation L.

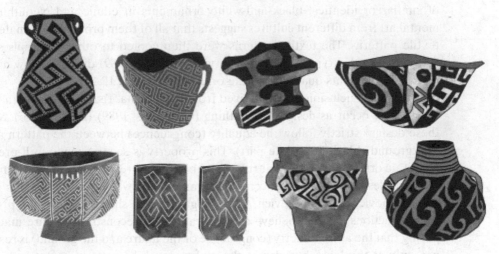

Figure 12.14
Neolithic ceramics: Tisza (Hungary), Cucuteni (Romania), Vincha (Yugoslavia), Dimini (Greece), Tisza and Miskolc (Hungary), Serra d'Alto (Italy), Rakhmani (Greece).

Figure 12.15
Neolithic textile ornaments.

Tchikalenko discussed for the first time the possibility that ornaments from Mezin are composed of one repeating module: a rectangle with parallel diagonal lines. Its black-and-white variant appears as the result of matting. Moreover, the occurrence of similar or identical black-and-white ornaments in ethnic and Neolithic ornamental art from different cultures suggests that all of them probably originated from textile patterns. The textile patterns were then copied to other materials such as ceramics, wood carvings, and stone carvings. Kalicz (1989) discusses how antisymmetrical ornaments qualify as textile ornaments (figure 12.15).

The same conclusion can be derived from the Vincha, Tisza, and Vadastra figures, where they occur as designs on clothing (Gimbutas, 1989) (figure 12.16). Most of these designs strictly follow the equality (congruence) between the pattern and the background (black-and-white part). This property is seen in many well-preserved Neolithic ornaments. Thus, that rule could be used to reconstruct the patterns (geometrical design) of Neolithic ceramic ornaments of which only fragments remain (V. Trbuhovich and Vasiljevich, 1983). Figure 12.17 shows one of our proposed reconstructions, which I believe are accurate. That reconstructions are made supposing that the antisymmetry (congruence of the figure and the ground) is respected not only at the local, but also at the global, level, because most patterns do not change appreciably over the area covered by the design.

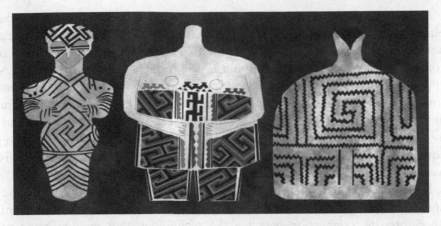

Figure 12.16
Neolithic artifacts from Vincha (Yugoslavia), Tisza (Hungary) and Vadastra (Romania).

Figure 12.17
Neolithic ornament from Podrinje (Yugoslavia) and its reconstruction.

One of the most interesting problems is to explain the origins of such "perfect" black–white ornaments. To some extent, we can try to do that by studying Escher's notebooks (Schattschneider, 1990), trying to analyze the present ethnic art and the construction methods it uses (Washburn, 1977; Washburn and Crowe, 1988; Ascher, 1991), or by trying to construct them independently. We have already mentioned some of the possible modular solutions: different kinds of op tiles, multiple-antisymmetry tiles, or Lunda designs. We need to note that black-and-white monohedral patterns are more general than Lunda designs. They satisfy the global equilibrium condition of congruence between their black and white parts, but usually they do not satisfy the local conditions for Lunda designs.

It is interesting that all of the monohedral Lunda designs and some of the plane antisymmetrical designs from Gerdes (1996) appeared in Neolithic ornamental art. Therefore, the connection between the technology that created basic patterns (matting, plaiting, knotwork, textiles, fabrics, etc.) and the Paleolithic and Neolithic ornaments as preserved on stones and bones or on ceramics, and the implicit mathematical knowledge that is their joint underlying basis, is an open field for research. In all such ornaments, we can see the domination of binary systems (black–white, left–right, above–below, etc.) based on simple and multiple antisymmetry (figure 12.15).

In art, we consider the specific properties of artistic expression: its style, the mode and the manner of execution, and construction rules characterizing a particular epoch. Over time, the concept of a pattern, inseparable from the idea of symmetry, has prevailed in the descriptive theories of ornamental styles. By classifying ornaments in terms of that concept or their underlying symmetry, it is possible to follow the consistencies and the variations of those parameters that create ornamental variations. Thus every epoch or ornamental style can be characterized by the geometrical and constructional problems solved in the ornamental art. "Style" is a term mainly used when analyzing artworks from modern epochs. You will not find the terms "Paleolithic style" or "Neolithic style" used to describe ancient art. But it is easy to recognize Paleolithic and Neolithic ornaments in mixed collections. Therefore, we need to find and explain exact geometrical-symmetrical criteria, closely connected with the theory of visual perception, that define the properties for a specific ornamental art (or style).

The use of symmetry to describe ornamental art is taken almost completely from mathematical crystallography. As we mentioned before, it gives us the answer to the question "Which ornaments are derived?," but not to the questions "How they are derived?" and "Why are they derived?" To explain the origins of ornamental art for each epoch, and to understand the anthropological, social, cognitive, and communicational senses of ornaments, together with the recognition of symmetry, we need to reconstruct the process by which the ornaments were made. This is especially important when analyzing the ornaments from the oldest periods, from which we only have archaeological artifacts. These ornaments can be reconstructed in two possible, mutually opposite, but complementary ways: by extending symmetry from local to global symmetry, or by using desymmetrizations ("symmetry breaking"), going from highly symmetrical structures to their symmetry subgroups. By a comparative analysis, we can follow the use of basic ornamental and geometrical elements and patterns (modules) in different cultures, their geometrical and topological change over time, and possible intercultural relationships.

References

Adams CC (1994) The Knot Book. New York: Freeman.

Ascher M (1991) Ethnomathematics: A Multicultural View of Mathematical Ideas. Pacific Grove, CA: Brooks and Cole.

Bain G (1973) Celtic Art: The Methods of Construction. New York: Dover.

Barrett C (1970) Op Art. London: Studio Vista.

Burchardt JJ (1988) Die Symmetrie der Kristalle. Basel: Birkhäuser.

Burde G, Zieschang R (1985) Knots. Berlin and New York: De Gruyter.

Caudron A (1981) Classification des Noeuds et des Enlancements. Orsay: Prépublications Université Paris-Sud.

Christie A (1969) Pattern Design. New York: Dover.

Conway JH (1970) An enumeration of knots and links and some of their algebraic properties, In: Computational Problems in Abstract Algebra (Leech J, ed), 329–358. New York: Pergamon Press.

Corbalis M (1990) Symmetry and asymmetry in psychology. Symmetry: Cult Sci 1: 183–194.

Coxeter HSM (1994) Symmetrical combinations of three or four hollow triangles. Math Intell 16(3): 25–30.

Critchlow K (1976) Islamic Patterns. London: Thames and Hudson.

Cromwel PR (1993) Celtic knotwork: Mathematical art. Math Intell 15(1): 36–47.

Cromwel PR (1995) Borromean triangles in Viking art. Math Intell 17(1): 3–4.

Crowe DW, Nagy D (1992) Cakaudrove Patterns. Ars Text 18: 119–155.

Dolbilin NP, Lagarias JC, Senechal M (1998) Multiregular point systems. Discr Comp Geom 20: 477–498.

El-Said I, Parman A (1976) Geometric Concepts in Islamic Art. London: World of Islam Festival Publishing.

Ernst B (1990) L'Aventure des figures impossibles. Berlin: Taschen.

Ernst B (1994) The Magic Mirror of M.C. Escher. Berlin: Taschen.

Gardner M (1989) Penrose Tiles to Trapdoor Ciphers. New York: Freeman.

Gerdes P (1989) Reconstruction and extension of lost symmetries: Examples from the Tamil of South India. In: Symmetry 2: Unifying Human Understanding (Hargittai I, ed), 791–813. Oxford and New York: Pergamon Press.

Gerdes P (1990) On ethnomathematical research and symmetry. Symmetry: Cult Sci 1: 154–170.

Gerdes P (1996) Lunda Geometry. Maputo, Mozambique: Universidade Pedagógica.

Gerdes P (1997) On mirror curves and Lunda designs. Comp Graph 21: 371–378.

Gerdes P (1999) Geometrical and Educational Explorations Inspired by African Cultural Activities. Washington, DC: Mathematical Association of America.

Gimbutas M (1989) The Language of the Goddess. San Francisco: Harper & Row.

Golomb SW (1994) Polyominoes, 2nd ed. Princeton, NJ: Princeton University Press.

Gombrich E (1979) Sense of Order. London: Phaidon Press.

Gregory RL (1970) The Intelligent Eye. London: Weidenfeld and Nicolson.

Grünbaum B, Shephard GC (1987) Tilings and Patterns. San Francisco: Freeman.

Hargittai I (ed) (1986) Symmetry: Unifying Human Understanding. Oxford and New York: Pergamon Press.

Hargittai I (ed) (1989) Symmetry 2: Unifying Human Understanding. Oxford and New York: Pergamon Press.

Huffman DA (1971) Impossible objects as nonsense sentences. In: Machine Intelligence 6 (Meltzer B, Michie D, eds), 295–323. Edinburgh: Edinburgh University Press.

Jablan SV (1992) Periodic antisymmetry tilings. Symmetry: Cult Sci 3: 281–291.

Jablan SV (1995) Theory of Symmetry and Ornament. Beograd: Matematicki Institut.

Jablan SV (1998) Modularity in Art. http://www.mi.sanu.ac.yu/~jablans/.

Jablan SV (1998) Mirror Curves. http://www.mi.sanu.ac.yu/~jablans/.

Kalicz N (1989) Die Gotter aus Ton. Budapest: Corvina.

Kauffman L (1987) On Knots. Princeton, NJ: Princeton University Press.

Koffka K (1935) Principles of Gestalt Psychology. New York: Harcourt, Brace and World.

Kulpa Z (1983) Are impossible figures possible? Signal Proc 5: 201–220.

Lindström B, Zetterström HO (1991) Borromean circles are impossible. Amer Math Month 98: 340–341.

Livingston C (1993) Knot Theory. Washington, DC: Mathematical Assocociation of America.

Locher P, Nodine C (1989) The perceptual value of symmetry. In: Symmetry 2: Unifying Human Understanding (Hargittai I, ed), 475–484. Oxford and New York: Pergamon Press.

Makovicky E (1989) Ornamental brickwork. In: Symmetry 2: Unifying Human Understanding (Hargittai I, ed), 4–6. Oxford and New York: Pergamon Press.

Mamedov KH (1986) Crystallographic patterns. In Symmetry: Unifying Human Understanding (Hargittai I, ed), 511–529. Oxford and New York: Pergamon Press.

Penrose LS, Penrose R (1958) Impossible objects: A special type of visual illusion. Brit J Psychol 49: 31–33.

Penrose R (1974) The role of aesthetics and applied mathematical research. Bull Inst Math Appl 10: 266–271.

Peterson I (1988) The Mathematical Tourist. New York: Freeman.

Phillips A (1992) The Topology of Roman Mazes. Leonardo 25: 321–329.

Raushenbach BV (1980) Prostranstvennye postroeniya v z'ivopisi [Space Constructions in Painting]. Moscow: Nauka.

Reidemeister K (1932) Knotentheorie. Berlin: Springer.

Rolfsen D (1976) Knots and Links. Berkeley, CA: Publish or Perish.

Schattschneider D (1990) M. C. Escher: Visions of Symmetry. New York: Freeman.

Schechtman D, Blech D, Gratias D, Cahn JW (1984) Metallic phase with long-range orientational order and no translational symmetry. Phys Rev Lett 53: 1951–1953.

Senechal M (1989) Symmetry revisited. In: Symmetry 2: Unifying Human Understanding (Hargittai I, ed), 1–12. Oxford and New York: Pergamon Press.

Senechal M (1995) Quasicrystals and Geometry. Cambridge: Cambridge University Press.

Smith CS (1987) The tiling patterns of Sebastien Truchet and the topology of structural hierarchy. Leonardo 20: 373–385.

Teuber ML (1986) Perceptual theory and ambiguity in the work of M. C. Escher against the background of 20th century art. In: M. C. Escher: Art and Science (Coxeter HSM, Emmer M, Penrose R, Teuber ML, eds), 159–178. Amsterdam: North-Holland.

Trbuhovich V, Vasiljevich V (1983) The Oldest Agriculture Cultures at Podrinje [in Serbo-Croatian]. Shabac: Narodni Muzej.

Turner JC, Van de Griend P (eds) (1996) History and Science of Knots. Singapore: World Scientific.

Washburn DK (1977) A Symmetry Analysis of Upper Gila Area Ceramic Design Decoration. Papers of the Peabody Museum of Archaeology and Ethnology, vol. 68. Cambridge: Harvard University.

Washburn DK, Crowe DW (1988) Symmetries of Culture. Seattle: University of Washington Press.

13 Modularity at the Boundary Between Art and Science

Angela D. Buscalioni, Alicia de la Iglesia, Rafael Delgado-Buscalioni, and Anne Dejoan

The Confines of Modularity

Is modularity an intrinsic property of natural beings (underlying processes such as organic growth), or is it a way of understanding and perceiving nature? Modularity seems to be a very intuitive idea, but why? We do not see it at first glance, in landscapes, animals, or physical processes. We believe that modularity stems from a conscious necessity to reduce the complexity of natural organization into a more comprehensible world. Artists and scientists struggle to build up a rational interpretation of natural and artificial objects, and only when they use their analytical "eyes" to try to solve questions such as "How can we enclose space?" "How can things be economically ordered or packed?" "Why do some entities become functionally integrated?" "What do we mean by regular growth?" "How do actions occur in time?" or even "How can we deal with the dichotomy between order and chaos?" do they come closer to or make recourse to the concept of modularity.

Nature is a complex universe with phenomena that we want to apprehend, understand, or solve (Simon, 1962). Nature is the common source and basis of knowledge for scientists and artists. Both groups endeavor to "understand what laws are in the bases of nature" (Klee, 1928). Whenever artists and scientists have explored how Nature is organized, they have discovered modularity. Initially, we may postulate that modularity is a model of organization. The "Mondrian tree" illustrates this idea (see figure 13.1). Throughout Mondrian's creative process, a natural object is revealed as being modularly arranged. Thus Mondrian's "tree sequence" starts by simplifying and partitioning a tree, in which the painter sets the organizing criteria that originate a new modular space. Mondrian's space is made up of orthogonal relationships generated by the tree branches in such a way that its modular composition is composed not only of the object (a branching tree) but also of the interstitial space (its spatial structure).

The looping of repeated elements or modules that may be connected, contiguous, accumulated, or mutually supported is a common feature of modular constructions. Modules build or delimit an organized whole. It is striking to see how modular domains might extend merely by introducing this simple rule. In addition, modular domains extend from a microscopic to a macroscopic organization, and may develop or be developed in space and time. Nature and humans may create modular arrangements by packing, ordering, or connecting identical or similar modules. A modular system may also be the product of enclosing or dividing a two- or

PHASE 1

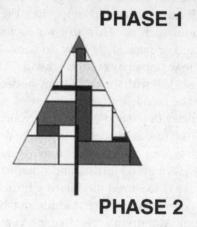

PHASE 2

Figure 13.1
Mondrian tree sequence used to illustrate how modularity is revealed during a creative process.

three-dimensional space. Modular repetition also occurs with the motion of an element in space and/or in time, such as the ubiquitous natural patterns generated by fractal and spiral growths. Any module in time generates a set of patterns that can be tackled analytically, such as in fluid dynamics, or in the kinetic, cubist, and futurist artistic vanguards.

As one may realize from the above examples, modular systems are developing systems. Thus, one may consider either the process of becoming modular or its final result, once a modular organization has been obtained.

Common characteristics of modular systems also give rise to common or universal properties, which are paramount for understanding how systems behave. It has been suggested that some properties of hierarchical structures in biology and paleontology (e.g., inclusivity, emergence, or asymmetry; Valentine and May, 1996) will form the basis of the predictive and explanatory power of a future theory of hierarchical systems.

The most immediate property of a developing modular organization is an ever-growing, self-similar system. In the simplest example constructed with a regular module and a stated relation ("adjacent modules"), we would have a continuous and infinite system, which is self-similar in the sense that there is only a single observed property at every scale of the construction. However, modular constructions also usually have compound properties. Think, for instance, of how the use of repetitions of a lemma or symbol (e.g., Babylonian mosaics, the popular propaganda of the Soviet Revolution, Buddhist mantras, and today's advertising) can give rise to complex social or religious messages. Emergent information may arise from a psychological (subjective) response induced by the perception of a modular system.

Certainly, the chance to observe the modular system from the top, such as a monk praying a mantra or a scientist who has finally understood a complex system, might provoke a feeling of transcendence. Conversely, if the observer stands at a functional or spatial unit, accessing only a local region of the system, s/he will surely feel the alienation induced by the weight of the whole (as in the chain of production parodied in Charlie Chaplin's film *Modern Times*). Other complex systems, such as those that are self-organized and hierarchical, are arranged in modular subsystems. These subsystems have their own properties and therefore interact with each other in complex ways. Simon (in the Foreword to this volume) considers the property of these hierarchical systems as being nearly decomposable, and one of us (Rafael Delgado Buscalioni) has set out, at the end of this chapter, the criteria needed for a modular description of self-organized systems.

Our aim in this chapter is to show that modularity may in fact be tackled as a single idea. What do paintings, mosaics, sculptures, animals, tissues, plants, music, or fluid dynamics have in common? To answer this was our challenge, and along with the present chapter, we have devised a visual and musical display concerning what we hold to be the domain of modularity (which can be visited at the KLI's Web page, http://kli.ac.at). Throughout this chapter we will follow the content and order of this Web page, extending and explaining in depth some modular systems according to a common group of criteria that constitute the bases for an eclectic view of modularity.

The Bases for an Eclectic Modularity

Obviously, the collection of words, ideas, and approaches to modularity is so vast that it is difficult to reduce its members to a single framework. References containing the keyword "modular" come from design (the Bauhaus [1919–1933]), architecture (Konrad Waschmann; Charles Le Corbusier; Eduardo Catalano, Frank Lloyd Wright, and Paolo Portoghesi), the arts (Paul Cézanne, László Moholy-Nagy, Oskar Schlemmer), visual communication (Gruppe μ, 1993), and the sciences (of which almost all contributions are summarized in this volume). In 1962, the German architect Konrad Waschmann proposed a classification of "building modules" according to twelve categories, including their nature, geometry, movement, elements, relationships, components, tolerance, and efficiency (Colectivo ETSAM, 1978).

Modularity in architecture tends to be quantitative, and is based on the standardization of the compound elements of a building. A modular organization is a skeleton that orders and coordinates the composition and dimensions of the building (Colectivo ETSAM, 1978). In the visual arts, on the other hand, modules are simplifications. For instance, three-dimensional geometric tiles, such as cones, bars, spirals, prisms, and circles, represent a reduction of multiplicity into essential units. In science, modularity has motivated the search for mechanisms and processes for understanding how modules (otherwise known as subsystems) evolve.

Modularity is one of the ways by which a developing system reaches an organized state. What is the opposite of modularity? One or even two elements will never be able to build a modular object. Nor will a completely unordered system with many elements in which there is never any correlation between its components be able to establish a modular organization. It is the repetition of modules and a set of relations or interactions that forms the basis for modularity. In fact, the smallest modular system would need at least three modules and two cloned relations between them.

We envision modularity as a triarchy: the modular whole, the module, and the model defined through the relations or interactions between the modules (see figure 13.2). Because of the need for multiple modules to build up a modular whole, a set of operations (the transformations) is required for the propagation of modules.

Transformations are at the boundary between the whole modular system and the individual modules, in part because transformations both affect and depend upon the nature of the modules (for instance, the "mitotic division" transformation may be applied to modules that possess the potential for duplication).

Properties are at the other edge of the triangle, between model and system. Properties belong to the model domain as long as they are the attributes emanating from the system within a particular model framework. Since a modular system may evolve or develop, properties may also change during the building process.

Due to the whole/unit dichotomy, modularity is like the two sides, creation and analysis, of the same coin. One may create a modular system by starting with the choice of a module. During this creative process, the module must be replicated in

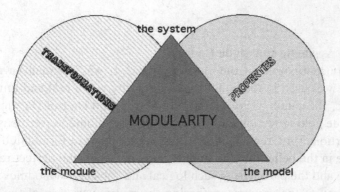

Figure 13.2
Modularity shown as a triarchy. A modular system is based on modules and models. The relations between the modules and the system, and the model and the system, define two domains: transformations and properties. (See text.)

order to make up the whole. This task must be carried out by using operators—in other words, transformations. However, the mere replication of modules does not necessarily give rise to a modular object unless we include a set of relations between them. These relations delimit the model of organization of the modular system, starting from its units. In this way the creator constructs a model that imposes a set of constraints on the whole, which accordingly furnishes a collection of attributes (or properties) of the system.

From the analytical perspective, the challenge is to cope with the whole or the system, which one tries to understand and reconstruct on a modular basis. Again, in this analytical assessment, a model outlining a sketch of the system has to be assumed. First, the model places the system within a preestablished framework (in other words, it constrains the properties of the whole), and second, it provides the basic, essential, or minimum principles of interaction between the modules. One of the aims of the analytical assessment is to discover (or deduce) the module, starting from the whole and projecting it into the model. Therefore, at a first glance, creation and analysis within a modular structure may be respectively understood as a bottom-up and a top-down process.

In any case, both creation and analysis usually involve successive iterations both from and to the module, with each iteration slightly modifying the model (or even its module conception) until the required degree of convergence is achieved. In what follows, we shall explore a variety of modules, transformations, and models. Three operative examples, taken from completely different domains, are analyzed in the light of the proposed triarchy on the Web page http://kli.ac.at.

The Module

The Module as an Aesthetic and Static Concept

Modularity was conceptualized in the classical world. It was originally a metric canon; the module was used in the belief that modularity was a real and universal property of Nature. The canon of Polycleitos and Myron, and later of Da Vinci and Dürer, as well as the golden section and the Le Corbusier modulor, were based on this idea of proportions (in Greek, *analoga*). These canons were formulated in art and in architecture in the belief that the human body possesses the perfect relation between the whole and the parts. Using such logical and "objective" modules would connect the harmony of the universe, gathering the metric variation of the world into an organic unity. Classic modules were and are both aesthetic and static. The module was an ideal metric unit to be adopted and applied to any construction, in order to maintain the similarity of the relationships between the whole and the parts.

The value of Φ (the golden section), for instance, arises from a geometric rule of construction: when dividing a segment into two unequal parts, the smaller fragment (a) is to the larger one (b) as (b) is to the sum of (a + b). As long as the value of Φ equals both ratios, $\Phi = b/a = (a + b)/b$. By repeating this rule, a continuous proportionality between parts will result within the object. The Blue and Red series of Le Corbusier are based on this same rule. A man's height of 183 cm is divided into two "golden" segments (113 cm and 70 cm), yielding the next golden segment (113 − 70 = 43). The difference between the latter two (70 − 43 = 27) makes up the minimal "golden unit" used by Le Corbusier for describing the ergonomic distances and positions of a human being. In this remarkable study he reduced architectural spaces to a human scale (figure 13.3).

Classic modules are used in economics by way of normalized systems (e.g., series A of the DIN regulations, ISO, ASA, etc.). Since a modular ratio preestablishes a modular construction, normalization and standardization have played a key role in improving productivity.

The Module as a Real Entity

Modern art "isms" have advanced in parallel with science in their attempt to cope with the task of modeling, partitioning, and integrating natural and artistic objects. Considering objects as systems that grow, develop, and evolve, driven by a set of natural, physical, or geometrical functions, has given rise to a richer definition of the idea of module. Modules can be real entities (here, the term "real" is opposed to "ideal"), bringing to the modular scenario the possibility of using an individual, a

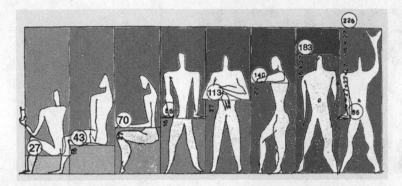

Figure 13.3
The Le Corbusier Blue and Red series describes the ergonomic distances and positions of a human being.

frequency, a neuron, a shape, and so on as a module. A module has structural and spatial properties; it may change; it delimits the space; and it interacts.

A serious dilemma, both in the creative and in the analytical process, is the particular choice (or guess) of the modular unit. This choice clearly affects, to a greater or lesser extent, the development and evolution of the system. We examine this critical dilemma in a simple system composed of a set of similar, adjacent, and aligned modules (a group of Mediterranean dancers). Of the two choices we can make to define the module, the key one is that which necessarily leaves "free appendages" at each end (figure 13.4).

Although the system apparently retains all its attributes, the final choice may constrain subsequent patterns in that system. For instance, the "free appendage series" may evolve by creating rolling circles through the joining of the two free ends. In the case of the other choice, however, the coarse modular series must always be limited to assembled movements. Evidently, this is a naïve example, but Nature has to resolve the dilemma of module choice from a small and limited number of possibilities (e.g., the different methods that Nature uses to fill the axial column in vertebrates).

The module is the terminal unit that cannot be (or is not) further decomposed, although, on the other hand, modules may aggregate. Such aggregation does not necessarily imply the creation of suprastructures or a hierarchical arrangement. In a simple system based on serial repetitions of modules (like the vertebral column or a nested Russian doll) the aggregates do not form suprastructures. The basic entities or modules are of the same type, and no distinct ranks can be established (see Valentine and May, 1996).

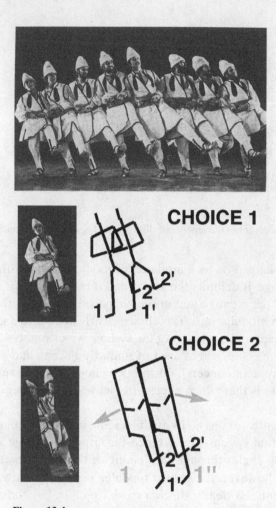

Figure 13.4
Dancers and the dilemma of the choice of a module. Compare the patterns that arises in the two systems. Choice 2 leaves elements (free appendages) out of the modular system, thus generating new functional possibilities.

A mosaic can summarize some patterns of aggregation into organized suprastructures. In this Escher mosaic (figure 13.5) we may perceive a set of "self-supporting" modules based on the coupling of two modules that form a background–figure combination of black and white silhouettes. In more elaborate mosaics, such as the Islamic ones (see the example on the supporting Web page by A. de la Iglesia) or in dynamic systems, modules generate suprastructures. Once the supramodular level is attained, there is a simplification of the number of transformations required to reproduce the aggregate. As discussed below, the genesis of suprastructures depends upon the kind of relations between modules.

Modularity as a Developmental Process

Transformations

Our definition of transformation depends on modularity being a developmental process, and thus a set of operators is needed that, when applied to a module, originates the elements of a modular system. Transformations are any kind of change in the spatial position or physical quality (shape, color, frequency, etc.) or any physical mechanism that induces a particular process within the system. Transformations are the mechanisms that generate and change modular systems.

An exhaustive definition of all possible transformations is not straightforward. Detailed classifications have been carried out to this end within a visual framework (see Gruppe μ, 1993 for the iconic representations of an object), but these classifications obviously did not include the transformations observed in Nature, which are somewhat complex. Below we summarize some transformations analyzed in the examples included on the Web page.

Geometric transformations are metric and projective. Metric transformations are simple: they consist of rotations, translations, and symmetries. The creation of mosaics uses combinations of these transformations (see Gardner, 1977, based on Penrose's work, and the excellent work of Jablan, chapter 12 in this volume). Nevertheless, the challenge of modern art has been the introduction of geometric transformations that are also projective transformations, which are homological or homothetic. In these transformations, certain properties of the module are conserved, but others, such as the length or the orientation, may change (see figure 13.6). Such a modular system will be composed of "similar-homologous" modules resulting from the stretching, projecting, increasing, or decreasing in size of a primary module.

The addition of geometric transformations to modules with other physical attributes, such as luminosity, saturation, and chromatic or complementary

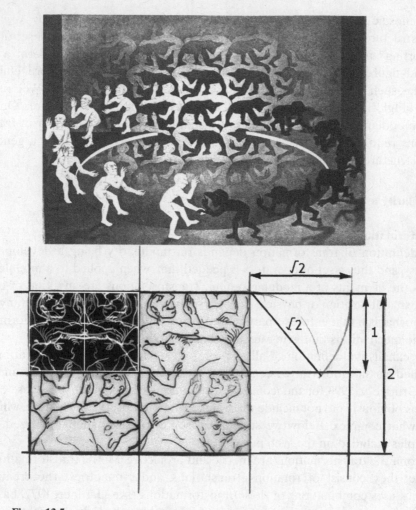

Figure 13.5
Escher mosaic. The module is one of the rectangles of the grid. The sides of the rectangle are $1/\sqrt{2}$. The module can be viewed as two self-supported modules (black and white characters). It builds up the mosaic by combining an axial symmetry and a translation. These two transformations generate a supramodule (a rectangle) that may be repeated successively by translations.

Figure 13.6
A Sophie Arp Tauber creation shows how a module is transformed by means of homothetic (stretching) operations.

dominance forms the basis of the optical transformations. The perception of optical phenomena in op art is based either on modular repetitions or on visual tricks, such as impossible structures, in which some of the aforementioned physical attributes are ordered according to particular geometric principles. We include cubist and futurist paintings as examples of kinetic transformations (those involving movement). They differ with respect to which subject is in motion; a moving observer defines cubism, while futurism depicts a sequence of movements of the represented object.

Transformations play a significant role in the study of complex modular systems. In simple systems with few modules, one can follow the path of each individual transformation. Surprisingly, large and complex systems, such as a turbulent fluid with many quasi-similar structures or modules, can be generated from a small set of "generative" transformations (e.g., instability mechanisms). Hence, in order to understand a turbulent fluid, for instance, the task is (or should be) to delimit these few fundamental transformations rather than to follow or deterministically describe each of the very large number of relations or "filiations" between the modules. An example of this is the Rayleigh–Taylor setup: a heavier fluid placed over a lighter one. Initial instability is due to buoyancy; the denser fluid falls and the lighter fluid rises. This leads to a situation in which the two fluids move in opposite directions.

Their contact surface is thus exposed to shear stresses that force the mixing of the fluids and leads to the formation of pairs of spiral rolls.

This mechanism (or transformation), known as the Kelvin–Helmholtz instability, is repeated time and again at smaller and smaller scales, finally leading to a system with a fractal appearance (see the images on the KLI Web page). The Kelvin–Helmholtz instability constitutes a fundamental kind of transformation in the transition to turbulence, appearing not only in the Rayleigh–Taylor setup but also in most conditions of turbulence. This was accurately described by Sharp (1984). It has been suggested that in fully developed three-dimensional turbulence, the geometrical structures formed have the properties of fractal sets, and that the formation of objects of shorter and shorter wavelengths is believed to be the result of repeated generations of the Kelvin–Helmoltz instability.

Having presented a cogent example of a type of "generative" transformation taken from fluid dynamics, we shall present two models that treat the transition to turbulence within a modular framework. (See "The route to turbulence", by Anne Dejoan at the supporting Web page.) As we shall see, these models in fact give direction to the development of the temporal and spatial evolution of the generative transformations.

The Model

We want to stress our conviction that it is impossible to conceive a modular system without a model. The model is the abstraction that summarizes the complexity of the modular structure. The model orders the system through the underlying relations and/or interactions between modules.

Nature uses a restricted number of models. It is not difficult, when comparing branching patterns (e.g., a river, a tree, or the circulatory system), to establish a number of analogies between them. Physical and spatial models constrain the shape of Nature, thus implying the existence of a large number of modular systems (see the supporting Web page for models and patterns used by artists and Nature, e.g., cube, cylinder, circle, sphere, and fan shape).

The use of models is also well known in morphology, elegantly developed in D'Arcy Thompson's book *On Growth and Form*. In his search for mathematical functions capable of describing biological change, Thompson used grids as models. The function (i.e., a transformation) deforms the grid, representing a way of quantifying phenotypic change. Modern art also makes use of deformed grids to make modules change. Op art, for instance, inscribes identical modules in a homological, deformed grid to give a sensation of volume.

In some instances, models can be easily recognized. One of the commonest is the trust used to create urban and architectural spaces. Trusts are scaffolds whose geometry organizes a developing space: its function, order, and construction (Colectivo ETSAM, 1978). These trusts, because they are models, are based on the principles of connectivity (figure 13.7). The choice of a set of connections introduces a logical order to the possibilities for linking the modules, establishing the boundaries between the compound elements of the system and disclosing the topological properties of the created space by means of their interconnections. The architect Eduardo Catalano explores trusts extensively in his book *La constante* (1995). Connectivity plays an important role in the construction of organisms. The repetition of simple connectivity rules makes it possible to form complex biological structures, such as muscles, in a body. By analogy, muscles and the sculptures of Naum Gabor are comparable: both can be generated merely by moving a "fiber" as a line from a "point" to a plane or from plane to plane, forming, by simply adding this movement, a warped surface resembling a muscular mass.

A mosaic, otherwise known as a tiled surface, or tessellation, is a modular system covering a planar surface. It is required to enclose the plane in a periodic manner. Mosaics use grids (rectangular, triangular, etc.) as models, and fulfill the condition that the angles between the intersecting vertices of the lines that make up the grid must add up to 360°. In mosaics, the module must be congruent with the grid, since

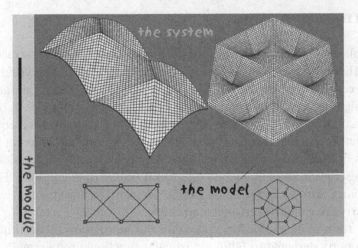

Figure 13.7
Connectivity models are used as a way of exploring the topological properties of a system. These examples were designed by the architect Eduardo Catalano as a modular tiled roof. (Modified from Catalano, 1995.)

Figure 13.8
Pentagons in a triangular grid. Note that the shape of the module is not a single pentagon.

otherwise it would not obey the condition of enclosing the plane. When a module does not fit into the model (e.g., pentagons in a triangular grid; the accumulation of pentagons does not add up to 360°), other forms must appear in the definition of the module (see figure 13.8).

Self-Organized Hierarchical Systems

An important class of modular systems is that of modules whose inner structure is also modular. The modules of a hierarchical system are therefore subsystems that may in turn be composed of smaller sub-subsystems (figure 13.9). This hierarchical organization might extend over a wide range of spatiotemporal scales, recurring commonly in natural systems as studied in biology, physics, sociology, and economics. The main feature of a hierarchical system is that the typical interaction of length and time scales between same-level subsystems decreases steeply as one descends each level of description (usually by one or several orders of magnitude). Following the terminology introduced by Simon (in the foreword to this volume), systems with this property are called "nearly decomposable."

The simplest form of hierarchical organization can be ruled by a constitutive or preimposed property designed to fulfill a particular functional requirement in the system as a whole. Think, for instance, of the hierarchical organization of the army or the interior of a computer: the commands that make the whole system run arise

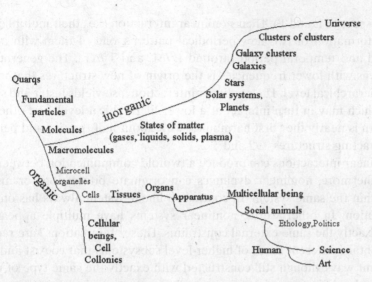

Figure 13.9
A proposal for a hierarchical universe

from a hierarchically structured set of sources. This kind of organization can, in a sense, be called "top-down" hierarchy. On the other hand, Nature usually selects a much more subtle type of hierarchy, which not only is imposed from the "top" but also can emerge from the bulk of the interactions between the smallest entities, and be set up by a kind of hierarchical process that is usually called "self-organization."

Self-organization is supported by a special type of interaction between the subsystems (or modules), which basically consists of either collaborative or competitive relations. The most significant characteristic of both collaboration and competition (which may in fact be considered as a "negative" collaboration) is that they are nonlinear interactions. The degree of collaborative "energy" (understood in a general sense) between two partners increases with the amount of energy they are capable of transmitting, multiplied by the amount of energy they can receive from each other. The multiplicative relation is the essence of the nonlinear interaction and supports two important requisites of self-organized hierarchical systems.

First, the nonlinear interaction of a pair of quasi-similar subsystems evolving in space and time produces a much larger and slower structure. For instance, consider two nearly similar periodical spatiotemporal patterns of the form $\cos(2\pi(x/L + t/T))$ (where x is position and t is time), whose spatial dimensions and time scales (periods) are respectively L, T, L + ΔL, and L, T, D + ΔT (ΔL and ΔT being very small with respect to L and T). Their nonlinear interaction (i.e., their multiplication) induces the formation of two new periodical patterns, one of them with a much larger spatial and temporal period (around $L^2/\Delta L$ and $T^2/\Delta T$). The generation of bigger patterns with lower frequencies is the origin of new structures that interact at a higher hierarchical level. The nonlinear interaction also yields faster and smaller structures, which may in turn interact at a lower hierarchical level. The other generated pattern is nearly the "first harmonic," with about half the size and period of the two interacting structures, L/2 and T/2.

Thus nonlinear interactions can produce a twofold communication between hierarchies. Furthermore, nonlinear dynamics can originate bifurcations or multiple solutions within the same system. In contrast, a linear system always has only one possible solution. In other words, nonlinear systems have multiple appearances exposed to exactly the same external constraints. These "bifurcations" are required in order to obtain different kinds of higher-level subsystems that coexist and interact in different ways, though still constructed with exactly the same type of underlying blocks: "the difference of the sameness."

Hierarchical Models of Description

As long as an emergent pattern arises from the nonlinear interactions within same-level subsystems that evolve over much longer scales of length and time, it is usually possible to analyze each hierarchical level with "nearly closed" models whose domain of description does not take into account the neighboring (upper and lower) hierarchies. Statistical mechanics and thermodynamics are a classic example from physics: both are independent theories describing the same kind of "object" on very different (microscopic and macroscopic) scales. In physics it is possible, by deductive means, to connect the several models of description to the lowest hierarchical level: from quantum mechanics to classical mechanics to statistical mechanics, and finally to thermodynamics. No similar type of deductive connection between models of different hierarchical levels has been found in other domains of knowledge (as, for instance, in the cases of complex or out-of-equilibrium physical matter or biological and economic systems). In those cases a "corpus of axioms" is needed to build a model for each hierarchy, in such a way that the subsequent models are completely or nearly isolated.

In view of this fact, a question arises: Is the disconnection between models for different hierarchies a temporary inability of current science, or is it instead a fundamental constraint on the whole system? In other words, is it always possible to use deduction to pass through the subsequent levels of description, starting from first principles, or does the very concept of "deduction" break down above some degree of complexity? Although neither we nor possibly anyone else can answer these questions, let us share some ideas that lead to three different categories of hierarchical models.

Before doing so, we need to include some kind of definition of "complexity" in order to continue with the following discussion. An initial and intuitive idea suggests that complexity increases with system size and the number of interacting entities. This first intuition clearly fails when one considers the connection between statistical mechanics and thermodynamics: any macroscopic thermodynamic system is composed of about 10^{24} individuals (particles), but their properties at equilibrium can nevertheless be quite well understood and predicted by the elegant theory of thermodynamics. The reason for this is that at thermodynamic equilibrium, the number of different kinds of interactions within the system is in fact very small compared to the number of interacting entities.

A better way of capturing the essence of complexity may be to consider that complexity grows with the ratio of the number of different kinds of interactions to the number of interacting objects. Henceforth, we will refer to this ratio qualitatively as the "complex dimension," whose value increases with the number of degrees of freedom of the interacting constituents. Thus the complexity of the simplest cell is probably much greater than that of the whole of intergalactic space. While a model for outer space has only to consider a few groups of fundamental interparticle forces (between photons and neutrinos), within a cell the number of different relations is greater than or equal to the number of cell constituents (macromolecules, organelles, etc.). Hence, complexity as defined above is also an emergent property of nonlinear interactions because, as stated above, nonlinearity originates multiple higher-level structures within the self-organized system.

Let us now proceed with a classification of models within the hierarchical structure. To illustrate this argument, we use the previously described "modular triarchy" (figure 13.10). As long as we are dealing with a hierarchical whole, the system comprising a certain hierarchical level, n (placed at the upper vertex of the corresponding triangle, S_n), is precisely the module, M_{n+1}, of the higher-level description $n + 1$ (therefore, $S_n = M_{n+1}$). Note that in figure 13.10, the side of consecutive triangles increases as a power law (in particular, as 2^n), as the size and time scales of subsequent hierarchical levels usually do. The line passing through subsequent systems ($S_0, S_1, \ldots$) determines the characteristic time scale and size of the emergent modules, whereas the "complex dimension" of the system advances along the line that passes through the models, on the opposite sides of the triangles.

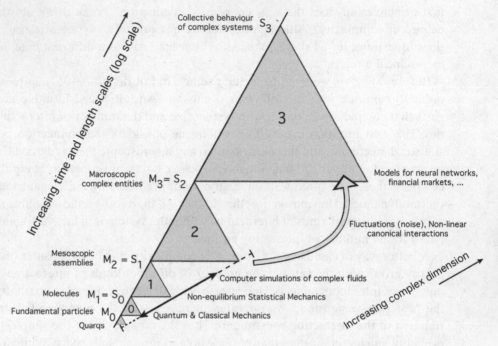

Figure 13.10
An illustration of a possible hierarchical organization. The black arrow indicates the expansion of deductive connection between models at different hierarchical levels. Dashed lines delimit a possible fundamental limit for deductive constructions. The gray arrow indicates two fundamental tools of the reductionistic models for the highest hierarchical levels: noise and nonlinearity. Both emerge from the ensemble of interactions at the lowest hierarchies.

Deduction and Construction: Models Based on First Principles

The model corresponding to the triangle that constitutes the lowest modular system (S_0) is constructed from very first principles. This model arises from an axiomatic corpus (e.g., in classical mechanics, this would be Newton's laws; in quantum mechanics, the Heisenberg Principle, Pauli's Principle of Exclusion, and so on). It is worth remembering that in science, any axiomatic corpus is open to alteration for as long as scientific endeavor continues to investigate smaller and smaller blocks. For instance, as illustrated in figure 13.10, some of the fundamental axioms of quantum and classical mechanics (at the 0th triangle) are currently under revision in the context of quark theory. Nevertheless, the models for the construction of the upper systems ($S_1, S_2, \ldots$) can be grounded on the fundamental principles while the complex dimension remains low enough. Some examples of models based on first principles are given in figure 13.10.

Complex Systems: Assumptions and Approximations, Quasi-heuristic Models

Usually, the degree of unpredictability grows quickly with the complex dimension of the system, leading to chaos. At this point the main tools of deduction fail because the concepts of cause and consequence merge. This may occur in such a way that any effect may arise from the coupling of a large number of causes, which may themselves be a consequence of the initially considered effect. In other words, at any given time, the memory of or correlation with previous states decreases with increasing complexity. This fact clearly imposes a barrier on any deductive effort that sets out to understand the whole system starting from first principles (i.e., from the lowest-level interactions). Nevertheless, this barrier can still be tackled in some contexts with the current tools of science. With the aid of computer simulation it is possible to reproduce, starting with Newton or quantum mechanics, very complex situations such as the formation of a cellular membrane, the self-assembly of proteins, or even a fully turbulent flow. In these simulations (or models) one commonly needs to introduce some heuristic reductions or simplifications due to a lack of theoretical, technical, or computational power. In fact, the simplifications used to investigate complex systems in the mesoscopic domain are gradually being accepted by the scientific community.

Highly Complex Systems: Fully Reductionistic Models

At the highest levels of the hierarchical system, the number of different possible interactions between individuals may be even larger than the number of individuals. One can think, for instance, of the neural network in the brain, or the Wall Street financial market, or even the World Wide Web: these systems are composed of many complex individuals interacting in many different ways, and it is not even clear that deductive reasoning may be of any use at all. It is certainly impossible to reconstruct the whole system with a set of deterministic rules of interaction between the units.

Therefore, in order to build models, modern science had to totally change its perspective in the 1980s. Instead of tracing all the individual interactions that compose the system, it searches for a new conceptual space where a very small number of new "paradigmatic" or "canonical" kinds of interactions can summarize the entire complex behavior. The modules of the new conceptual space are not the individuals anymore, but instead a few "overall behaviors" of the collective that are expected to be quasi-independent of each other. These new modules are often variables that interact by means of a set of equations (model). Although these variables may bear some kind of relation to the real system, very often they do not have an a priori "physical" interpretation. Instead, they are interpreted a posteriori—if they ever are.

The reduction of complexity is supported by two kinds of relations between and within the modules: nonlinear interactions and noise (i.e., fluctuations). We have already stated that nonlinearity is one of the requirements of self-organized systems. What about noise? The role of fluctuations may be to reproduce the integrated effect of high-frequency processes that emerge from the ensemble of interactions between individuals at the lower hierarchical levels. Alternatively, noise arises from the unpredictable short-range fluctuation of the external constraints of the system. Surprisingly, noise in these dynamical systems does not destroy their coherence (or correlation). Instead, as relatively recent studies on the effect of noise have shown, the fluctuations may be able to induce transitions to new states of the whole, or even increase the stability of the system by carrying it to spatially or temporally highly correlated states (see, e.g., Cabrera et al., 1999).

It is important to realize that the effect of low-amplitude fluctuations can become significant when they maintain a certain correlation (in time or space) that is of the same order as the typical scales of the system. In this way, a low-amplitude noise may produce resonant couplings to the overall forces that drive the macroscopic system (see Gammaitoni et al., 1998, for a review). The implications are indeed quite relevant and present throughout physics (e.g., fluids near the critical point—Sengers and Levelt Sengers, 1986), pattern formation, biology (e.g., neural behavior—Wiesenfeld and Moss, 1995; Gluckman et al., 1996; Kelso, 1995), ecology (Rohani et al., 1999), economics (Mantegna and Stanley, 1995), and social organizations (Stanley et al., 1996).

By employing nonlinear interaction and noise, modern mathematical and computational tools of stochastic dynamics have been able to classify several "canonical" models and discover several "universal" behaviors that may be found in a wide range of complex systems. Hence, one of the first objectives of a reductionistic model is to reproduce the qualitative behavior of a certain overall output extracted from the complex whole. If this task is successful, the system should be classified (at least partially) with the previously known (canonical) behaviors. The hope is that we may be able to use our knowledge of the canonical model to infer new properties of the complex system.

Conclusions

We firmly believe that modularity is a type of organization that exists in Nature. The vast and disparate range of examples we have identified in the arts and sciences leads us to the conclusion that there is common ground with respect to the meaning of modularity. For the sake of this broad panorama we have felt it necessary to stress the commonalities observed on any modular construction. In this chapter we offer an analytical approach to search for and generate modular systems. Since here modularity is within the domain of developing systems, the basic unit (the module) must propagate, and thus a set of operators (the transformations) is required. Because modularity is also part of the domain of organized and hierarchical systems, it is governed by a network of interactions or a set of relations among modules (or subsystems). These interactions delimit the model of organization in which the system is immersed. Therefore, modular systems may be modeled, thus opening a pathway for the exploration of universal behaviors and their properties. A revision and classification of models within the framework of scaled hierarchical systems is provided with the aim of showing the conflicts around the emergence of new properties in these systems. The discovery of such properties is a task in the future search for a theory of hierarchical modular systems.

References

Cabrera JL, Gorroñogoitia J, de la Rubia FI (1999) Noise-correlation-time-mediated localization in random nonlinear dynamical systems. Phys Rev Lett 82: 2816.

Catalano E (1995) La constante. Diálogos sobre estructura y espacio en arquitectura. Buenos Aires: Editorial Universitaria.

Chatté H, Manneville P (1987) Transition to turbulence via spatio-temporal intermittency. Phys Rev Lett 58: 112.

Chatté H, Manneville P (1988) Spatio-temporal intermittency in coupled map lattices. Physica D32: 409.

Colectivo para el Análisis de la Expresión Gráfica (1978) Dibujo técnico, Madrid: Bruño.

Feigenbaum M (1978) Quantitative universality for a class of nonlinear transformations. J Stat Phys 19: 25–52.

Gammaitoni L, Hanggi P, Jung P, Marchesoni F (1998) Stochastic resonance. Rev Mod Phys 70: 223–228.

Gardner M (1977) Extraordinary nonperiodic tiling that enriches the theory of tiles. Sci Amer 236(1): 110–121.

Gluckman BJ, Netoff TI, Neel EJ, Spano ML, Schiff, SJ (1996) Stochastic resonance in a neuronal network for mammalian brain. Phys Rev Lett 77: 4098.

Groupe μ (1993) Tratado del signo visual. Madrid: Cátedra.

Kelso SJA (1995) Dynamic pattern: The Self-Organization of Brain and Behavior. Cambridge, MA: MIT Press.

Klee P (1928) Exakte Versuche im Bereich der Kunst. Bauhaus, Zeitschrift für Gestaltung 2(2/3): 17.

Mantegna RN, Stanley HE (1995) Scaling behaviour in the dynamics of an economical index. Nature 376: 46–48.

Moholy-Nagy L (1932) The New Vision. New York: Wittenborn. (First German ed. 1928.)

Rohani P, Earn DJD, Grenfell BT (1999) Opposite pattern of synchrony in sympatric disease metapopulations. Science 286: 968–971.

Schlemmer O (1996) Católogo de la Exposición del Museo Nacional Centro de Arte Reina Sofia y Fundación Caixa. Madrid: Ministerio de Educación y Cultura.

Sengers JV, Levelt Sengers JMH (1986) Thermodynamic behavior of fluids near the critical point. Ann Rev Phys Chem 37: 189–222.

Sharp DH (1984) An overview of Rayleigh–Taylor instability. Physica D12: 3–18.

Simon HA (1962) The architecture of complexity. Proc Amer Phil Soc 106: 467–482.

Stanley MHR, Amaral LAN, Buldyrev SV, Havlin S, Leschhorn H, Mass P, Salinger MA, Stanley HE (1996) Scaling behaviour in the growth of companies. Nature 379: 804–806.

Thompson, D'AW (1942) On Growth and Form, 2nd ed. Cambridge University Press. (1st ed., 1917.)

Valentine JW, May CL (1996) Hierarchies in biology and paleontology. Paleobiology 22: 23–33.

Wiesenfeld K, Moss F (1995) Stochastic resonance and the benefits of noise: From ice age to crayfish and squids. Nature 373: 33–36.

IV MODULARITY OF MIND AND CULTURE

There is a wide consensus in the scientific community today that humans, like all other mammals, are a product of biological evolution. On the day we are writing this, the first hit in a Google search for *"Homo sapiens"* is a page of the National Center for Biotechnology Information of the U.S. National Institutes of Health, which states: "Fifty years after the double helix, the reference DNA sequence of *Homo sapiens* is now available for downloading." At least on a simplistic view of the genotype-phenotype map ("a gene for each trait"), some kind of modularity would seem to be vindicated as far as the human body, including its brain and maybe even behavioral traits of individuals, is concerned.

But what about the human mind? How, specifically, did our minds evolve? And what can we learn of it—namely, can evolutionary considerations illuminate the issue of the basic architecture of the human mind/brain? As soon as these questions are posed, theoretical opinions start to diverge widely, and relevant hard evidence is scarce to find or apply. The battles between nativism and constructivism—or nature versus nurture, in one guise—rage as vehemently as ever, it would seem; and those who have overcome these and other dichotomies remain a minority. Quite a few cognitive scientists, developmental psychologists in particular, are loath to attribute any innate predispositions to human infants, but—as Karmiloff-Smith (1992, p. 1) remarks—would not hesitate to do so with respect to the ant, the spider, the bee, or the chimpanzee. "Why would Nature have endowed every species except the human with some domain-specific predispositions?"

Most researchers who *do* attribute various innate predispositions to the human mind (at least that of the neonate) also accept the important roles of the physical and sociocultural environments, and continue to embrace "the deep-seated conviction that we are special—creative, cognitively flexible, and capable of conscious reflection, novel invention, and occasional inordinate stupidity" (Karmiloff-Smith, 1992, p. 1).

In the midst of these and other debates, such as those over parallelism versus seriality (in part occasioned by the evolution of computer technology), or one-shot learning versus gradual learning, modularity has provoked a vast amount of often heated debate in cognitive science and the philosophy of mind. Fodor's book *The Modularity of Mind* (1983), which intimately linked nativism with the ideas of domain specificity and modularity, is often cited as having had the most impact—both positively and negatively—on recent theorizing on these matters. An entire new and booming discipline, evolutionary psychology, rests on a massively modular conception of cognitive architecture; it views the mind (if not necessarily the brain) as composed largely or even entirely of innate, special-purpose computational modules, understood here as mechanisms.

Several of the chapters in this section deal with modularity of mind issues. Fodor's view in several respects echoed Chomsky's influential conception of an innately specified language "organ." However, with two exceptions—Chomskyan linguistics on the one hand, and abstract systems-theoretical approaches (often inspired by Simon's work) on the other—there has been little or no consideration of the possibility of modular structures and functions in the subject domains of the social sciences and humanities, at least under that name (but see chapter 17).

The chapters in this section all address aspects of the issue of the modularity of minds and cultures directly. They are, overall, critical of currently dominant views, and suggest constructive and more balanced alternatives. Some of them also begin to address questions, and in part suggest answers, concerning the Big Issue—*in pluribus unum*? If Simon (in the foreword to this volume) is right, it should eventually be possible to articulate a unified picture of modularity encompassing inorganic matter, life, mind, and culture.

Raffaele Calabretta and Domenico Parisi are concerned with brain/mind modularity as a contentious issue in cognitive science (chapter 14). Many cognitivists conceive of the mind as a set of distinct specialized modules and believe that this rich modularity is basically innate. Connectionism, on the other hand, which the authors endorse, views the mind as a more homogeneous system that, basically, genetically inherits only a general capacity to learn from experience. If there are modules, they are the result of development and learning. On their view, connectionist modules are anatomically separated and/or functionally specialized parts of a neural network, and may be the result of a process of evolution in a population of neural networks.

Their evolutionary connectionism allows simulation of how genetically inherited information can spontaneously emerge in populations of neural networks, instead of being arbitrarily hardwired in the neural networks by the researcher. It also makes it possible, they argue, to explore all sorts of interactions between evolution at the population level and learning at the level of the individual that determine the actual phenotype. Evolutionary connectionism differs from evolutionary psychology in that it (1) uses neural networks rather than cognitive models for interpreting human behavior, (2) adopts computer simulations for testing evolutionary scenarios, and (3) rejects the panadaptationist view of evolution.

Fernand Gobet (chapter 15), after surveying how the concept of modularity is used in different guises in psychology, applies several of its meanings to the (symbolic) CHREST architecture, which provides a computational definition of knowledge modules ("chunks"). Using data from chess results, he shows how relations and latencies between pieces offer converging evidence for the psychological reality of chunks. He suggests that such results indicate that CHREST captures a key feature of human cognition, and that its redundancy ensures that CHREST is not brittle or oversensitive to details. It proves incorrect to equate symbolic processing with top-down processing, as is often done. Gobet also discusses the relationship between functional and knowledge modularity as they interact within CHREST or other symbolic systems. At least in CHREST, given its limited capacity of short-term memory, complex objects can be represented only as a hierarchy built up from smaller objects.

Boris Velichkovsky (chapter 16) presents experimental data from investigations of human eye movements, perception, and memory demonstrating that modularity, in the sense of domain specificity, is only one of the aspects of human cognitive architecture. The effects he discusses are embedded into gradient-like or hierarchical mechanisms that can be preliminarily identified as related to their evolutionary origins. His focus is on mechanisms below perceptual and above semantic processing. He argues that even the mild version of modularity of mind conception, which stresses differences between perceptual input systems and semantic central systems (e.g., Fodor, 1983), must be incomplete as a model of human cognition, especially because of its ignorance of the "vertical" dimension of mental functioning. He therefore pleads for, and offers empirical data in support of, a psychological and neurophysiological analysis based on Bernstein's motor control theory, and aimed at developmental stratification of the evolving mind-and-brain mechanisms as an alternative to the current search for isolated modules of processing.

The economists Luigi Marengo, Corrado Pasquali, and Marco Valente (chapter 17) argue that economic theory has at its very core a strong, albeit implicit, idea of modularity grounded in the concepts of *division of labor*—creation of modules— and *market selection*—coordination of modules. Contemporary economic theory has little to say about the former, and even less about the connections and interdependencies between the two mechanisms: the existence and "granularity" of modules have always been assumed to be exogenously given and irrelevant for the outcome of the coordination process. They engagingly argue that the "morphogenetic" process of construction of modules determines the effectiveness of the market coordination mechanism. Only when the economic system is fully decomposable does the market mechanism achieve optimal coordination, irrespective of the underlying modules into which labor has been divided. In nearly decomposable systems—the typical case—there is a trade-off between the scope of market selection and its efficiency. Economic organizations and institutions are solutions of this dilemma, and their genesis and evolution can be fully understood in terms of this delicate balance. The authors also discuss the "representation problem" in evolutionary computation and biology: adaptation and evolution can work effectively only if variations are not too often very disruptive and not too rarely favorable. By acting on the representation, one can make simple adaptive mechanisms such as mutation and recombination either very effective or totally ineffective. Modularity and near decomposability, they show, can be a way to solve the representation problem.

In the final chapter, Kimbrough Oller raises the issue of the logical distinction between thought and language in the context of the discussion of modularity. Extrapolating from his earlier work on the emergence of the speech capacity, his fundamental contention is that a *natural logic* of (potential) communicative systems is a critical domain for the understanding of language structure (cf. Rasskin-Gutman's discussion of "theoretical morphospace" in chapter 9). Oller presents natural logic as a "third way" beyond the nature–nurture dichotomy. It includes a series of distinctions that constitute modular elements of communicative capability. Infrastructural natural logic, in his view, offers the opportunity to (1) understand at a very general level how language is structured in the mind in terms of modules of function; (2) relate the mature structure of language to its primitive beginnings in the human infant; (3) compare the nature of human vocal communication with that of extant nonhumans; (4) provide the basis for fruitful speculations about the evolution of language in our species; and (5) lay the groundwork for a theory of possible communicative evolution in any species: human, nonhuman, or—even more daringly—extraterrestrial.

References

Fodor J (1983) The Modularity of Mind: An Essay on Faculty Psychology. Cambridge, MA: MIT Press.
Karmiloff-Smith A (1992) Beyond Modularity: A Developmental Perspective on Cognitive Science. Cambridge, MA: MIT Press.

Raffaele Calabretta and Domenico Parisi

Connectionism Is Not Necessarily Antimodularist or Antinativist

In a very general and abstract sense modular systems can be defined as systems made up of structurally and/or functionally distinct parts. While nonmodular systems are internally homogeneous, modular systems are segmented into modules (i.e., portions of a system having a structure and/or function different from the structure or function of other portions of the system). Modular systems can be found at many different levels in the organization of organisms (for example, at the genetic, neural, and behavioral/cognitive levels). An important research question is how modules at one level are related to modules at another level.

In cognitive science, the interdisciplinary research field that studies the human mind, modularity is a very contentious issue. There exist two kinds of cognitive science, *computational* cognitive science and *neural* cognitive science. Computational cognitive science is the more ancient theoretical paradigm. It is based on an analogy between the mind and computer software, and it views mind as symbol manipulation taking place in a computational system (Newell and Simon, 1976). More recently a different kind of cognitive science, connectionism, has arisen, which rejects the mind/computer analogy and interprets behavior and cognitive capacities using theoretical models which are directly inspired by the physical structure and way of functioning of the nervous system. These models are called neural networks—large sets of neuronlike units interacting locally through connections resembling synapses between neurons. For connectionism, mind is not symbol manipulation. Mind is not a computational system, but the global result of the many interactions taking place in a network of neurons modeled with an artificial neural network. It consists entirely of quantitative processes in which physicochemical causes produce physicochemical effects. This new type of cognitive science can be called *neural cognitive science* (Rumelhart and McClelland, 1986).

Computational cognitive science tends to be strongly modularistic. The computational mind is made up of distinct modules which specialize in processing distinct types of information, have specialized functions, and are closed to interference from other types of information and functions (Fodor, 1983). Computational cognitive models are schematized as "boxes-and-arrows" diagrams (for an example, see figure 14.1). Each box is a module with a specific function. The arrows connecting boxes indicate that information processed by some particular module is then passed on to another module for further processing. In contrast, connectionism tends to be anti-modularistic. In neural networks information is represented by distributed patterns

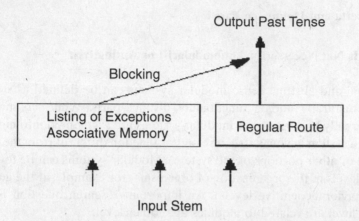

Figure 14.1
An example of a "boxes-and-arrows" model: the dual-route model for the English past tense (Pinker and Prince, 1988). "The model involves a symbolic regular route that is insensitive to the phonological form of the stem and a route for exceptions that is capable of blocking the output from the regular route" (modified from Plunkett, 1996).

of activation in potentially large sets of units, and neural networks function by transforming activation patterns into other activation patterns through the connection weights linking the network's units. Most neural network models are not divided into any kind of modules except for the distinction between input units, output units, and one or more layers of intermediate (hidden or internal) units (for an example, see figure 14.4).

One cannot really understand the contrast between modularism and antimodularism in cognitive science, however, if one does not consider another contrast: that between computational cognitive science (cognitivism) and neural cognitive science (connectionism). This is the contrast between innatism and anti-innatism. Cognitivists tend to be nativists. Modules are assumed to be specified in the inherited genetic endowment of the species and of each individual. For evolutionary psychologists, who tend to be cognitivists, the modular structure of the mind is the result of evolutionary pressures. Evolutionary psychologists are convinced that it is possible to identify the particular evolutionary pressures behind each module. Hence, evolutionary psychologists (Cosmides and Tooby, 1994) embrace a strong form of adaptationism. They not only think that modules are already there in the genetic material; they think that modules are in the genes because in the evolutionary past individuals with a particular module in their genes generated more offspring than individuals without that module.

This panadaptationism is not shared by all cognitivists, however. For example, the linguist Noam Chomsky believes that the mind is computational and that there is a specific mental module specialized for language (or syntax); but he does not believe that language in humans has emerged under some specific evolutionary pressure (cf. Fodor, 2000). As some evolutionary biologists, in particular Gould (1997), have repeatedly stressed, what is genetically inherited is not necessarily the result of specific evolutionary pressures. Nor is it necessarily adaptive: it can also be the result of chance, it can be the adaptively neutral accompaniment of some other adaptive trait, or it can be an exaptation, the use for some new function of a trait which has evolved for another function (Gould and Vrba, 1982). More recently, the contrast between Steven Pinker and Jerry Fodor, who are both well-known cognivists and nativists, has revealed how the adaptive nature of inherited traits can divide computational cognitive scientists. Pinker (1999) has argued for a strong form of adaptive modularism, while Fodor is in favor of a strong form of nonadaptive modularism (Fodor, 1998).

In contrast to cognitivists, connectionists tend to be antinativists. Connectionism is generally associated with an empiricist position that considers all of the mind to be the result of learning and experience during life. What is genetically inherited, in humans, is only a general ability to learn. This general ability to learn, when it is applied to various areas of experience, produces the diverse set of capacities exhibited by humans.

The matter is further complicated if one considers development. Development is the mapping of the genetic information onto the adult phenotype. This mapping is not instantaneous but is a process that takes time to complete. In fact, development consists of a temporal succession of phenotypical forms. When one recognizes that the genotype/phenotype mapping is a temporal process, the door is opened for learning and experience to influence the phenotype. Therefore, cognitivists tend to be not only nativists but also antidevelopmentalists.

Cognitivist developmental psychologists (e.g., Spelke et al., 1994; Wynn, 1992) tend to think that modules are already there in the phenotype from the first stages of development, and that there is not much of real importance that actually changes during life. Furthermore, as nativists, they think that even if something changes during development, this is due not to learning and experience but to some temporal scheduling encoded in the genetically inherited information, like sexual maturity, which is not present at birth but is genetically scheduled to emerge at some later time during life.

Developmental psychologists, on the contrary, who are closer to connectionism (e.g., Karmiloff-Smith, 2000) tend to think that modules are not present in the phenotype from birth (in newborns or infants), but develop later in life. Furthermore, they believe that modules are only very partially encoded in the genotype, and are the result of complex interactions between genetically encoded information and learning and experience.

In the present chapter we want to argue for a form of connectionism that is neither antimodularist nor antinativist. Connectionism is not necessarily antinativist. Even if many neural network simulations use some form of learning algorithm to find the connection weights that make it possible for a neural network to accomplish some particular task, connectionism is perfectly compatible with the recognition that some aspects of a neural network are not the result of learning but are genetically inherited. For example, since most simulations start from a fixed neural network architecture, one could argue that this network architecture is genetically given, and that the role of learning is restricted to finding the appropriate weights for the architecture.

In fact, Elman et al. (1996) have argued that connectionist networks allow the researcher to go beyond cognitivism (which simply affirms that this or that is innate), and explore in detail what can be innate and what can be learned by showing how phenotypical capacities can result from an interaction between what is innate and what is learned. These authors distinguish among different things that can be innate in a neural network: the connection weights (and therefore the neural representations as patterns of activation across sets of network units), architectural constraints (at various levels: unit, local, and global), and chronotopic constraints (which determine when things happen during development).

One could also add that the connection weights may be learned during life, but that there may be genetically inherited constraints on them. For example, their maximum value or "sign" (for excitatory or inhibitory connections) may be genetically specified, or the genotype may encode the value of learning parameters such as the learning rate and momentum (Belew et al., 1992). As we will show later in this chapter, modularity can emerge in neural networks as a function of genetically inherited architectural constraints and chronotopic constraints.

However, to argue that something is innate in a neural network, it is not sufficient that some of the properties of the neural network are hardwired by the researcher in the neural network; it is necessary to actually simulate the evolutionary process that results in these genetically inherited properties or constraints. Artificial life simulations differ from the usual connectionist simulations in that artificial life uses genetic algorithms (Holland, 1992) to simulate the evolutionary process and to evolve the genetically inherited properties of neural networks (Parisi et al., 1990; Calabretta et al., 1996).

Unlike traditional connectionist simulations, artificial life simulations involve not an individual network that on the basis of its individual experience learns some particular capacity, but an entire population of neural networks made up of a succession of generations of individuals, each of which is born with a genotype inherited from its parents. Using a genetic algorithm, the simulation shows how the information encoded in the inherited genotypes changes across the successive generations because reproduction is selective and new variants of genotypes are constantly added to the genetic pool of the population through mutations and sexual recombination. At the end of the simulation the inherited genotypes can be shown to encode the desired neural network properties that represent innate constraints on development and behavior. We call this type of connectionism "evolutionary connectionism."

We can summarize the three options that are currently available to study the behavior of organisms with table 14.1.

Evolutionary connectionist simulations not only allow us to study how genetically inherited information can spontaneously emerge in populations of neural networks, instead of being arbitrarily hardwired in the neural networks by the researcher, but also make it possible to explore all sorts of interactions between evolution at the population level and learning at the level of the individual that determine the actual phenotype.

In this chapter we describe two evolutionary connectionist simulations that show how modular architectures can emerge in evolving populations of neural networks. In the first simulation, every network property is genetically inherited (i.e., both the

Table 14.1
Three options for studying behavior and mind

Computational cognitive science or Cognitivism	Mind as symbol manipulation taking place in a computerlike system	Nativist	Modularist
Neural cognitive science or Connectionism	Mind as the global result of the many physicochemical interactions taking place in a network of neurons	Anti-nativist	Anti-modularist
Evolutionary connectionism	Mind as the global result of the many physicochemical interactions taking place in a network of neurons	Interaction between evolution and learning	Modularist

network architecture and the connection weights are inherited) and modular architectures result from genetically inherited chronotopic constraints and growing
instructions for units' axons. In the second simulation, the network architecture is
genetically inherited but the connection weights are learned during life. Therefore,
adaptation is the result of an interaction between what is innate and what is learned.

Cognitive versus Neural Modules

Neural networks are theoretical models explicitly inspired by the physical structure
and way of functioning of the nervous system. Therefore, given the highly modular
structure of the nervous system, it is surprising that so many neural network architectures that are used in connectionist simulations have internally homogeneous
architectures and do not contain separate modules. Brains are not internally homogeneous systems but are made up of anatomically distinct parts, and distinct portions of the brain are clearly more involved in some functions than in others. Since
it is very plausible that human brains are able to exhibit so many complex capacities not only because they are made up of 100 billion neurons but also because these
100 billion neurons are organized as a richly modular system, future connectionist
research should be aimed at reproducing the rich modular organization of the brain
in neural networks.

However, even if, as we will show by the two simulations described in this chapter,
connectionist simulations can address the problem of the evolution of modular
network architectures, it is important to keep in mind that the notion of a module
is very different for cognitivists and for connectionists. Cognitivistic modularism is
different from neural modularism.

For cognitivists, modules tend to be components of theories in terms of which
empirical phenomena are interpreted and accounted for. A theory or model of some
particular phenomenon hypothesizes the existence of separate modules with different structures and/or functions which, by working together, explain the phenomenon of interest. Therefore, cognitivist modules are *postulated* rather than
observed entities. For example, in formal linguistics of the Chomskyan variety, syntax
is considered as an autonomous module of linguistic competence in that empirical
linguistic data (the linguistic judgments of the native speaker) are interpreted as
requiring this assumption. Or, in psycholinguistics, the observed linguistic behavior
of adults and children is interpreted as requiring two distinct modules, one supporting the ability to produce the past tense of regular English verbs (e.g., *worked*)
and the second one underlying the ability to produce the past tense of irregular
verbs (e.g., *brought*) (Pinker and Prince, 1988; see figure 14.1). This purely theoreti-

cal notion of a module is explicitly defended and precisely defined in Fodor's book *The Modularity of Mind* (Fodor, 1983), one of the foundational books of computational cognitive science.

The same is true for evolutionary psychology, which, as we have said, has a cognitivist orientation. Evolutionary psychology's conception of the mind as a "Swiss Army knife" (i.e., a collection of specialized and genetically inherited adaptive modules) is based on a notion of module according to which modules are theoretical entities whose existence is suggested by the observed human behavior.

Neuroscientists also have a modular conception of the brain. For example, Mountcastle (cited in Restak, 1995, p. 34) maintains that "the large areas of the brain are themselves composed of replicated local neural circuits, modules, which vary in cell number, intrinsic connections, and processing mode from one brain area to another but are basically similar within any area." However, the neuroscientists' conception of the brain is based on empirical observations of the anatomy and physiology of the brain rather than on theory (see figure 14.2). The brain obviously is divided into a variety of "modules" such as distinct cortical areas, different subcortical structures,

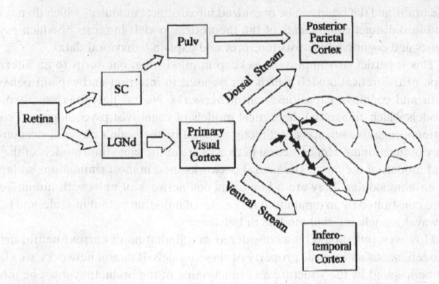

Figure 14.2
"The major routes of visual input into the dorsal and ventral streams. The diagram of the macaque brain [. . .] shows the approximate routes of the cortico-cortical projections from the primary visual cortex to the posterior parietal cortex and the inferotemporal cortex, respectively" (modified from Milner and Goodale, 1998). According to Ungerleider and Mishkin (1982), the ventral stream plays a critical role in the identification and recognition of objects (i.e., the "what" task), while the dorsal stream mediates the localization of those same objects (i.e., the "where" task).

interconnected subsystems such as the retina-geniculate-visual cortex for vision or the basal ganglia-frontal cortex subsystem for attention. This rich modularity of the brain, both structural (anatomical and cytoarchitectonic) and functional (physiological), is evidenced by direct (instrumental) observation, by data on localization of lesions in various behavioral/mental pathologies and on neuropsychological dissociations, and more recently and increasingly, by neuroimaging data.

One can look for correspondences between the two types of modules, the theoretical modules of computational cognitive science and the observed "modules" of the brain. This is what cognitive neuropsychologists are supposed to do. They interpret the behavioral deficits of patients using the "box-and-arrows" theoretical models of cognitive psychology, where boxes are modules and arrows indicate the relationship between modules (see figure 14.1), and then they try to match this modular analysis with observations and measurements on localization of lesions and other physical data on patients' brain. However, one cannot assume that the modular theoretical models of computational cognitive science necessarily correspond to the observed modular structure and functioning of the brain. Cognitive modules may not match the physical (neural) structural or functional "modules" of the brain, and the brain can be organized into distinct "modules" which do not translate into distinct components of the theoretical models in terms of which psychologists and cognitive scientists interpret and explain behavioral data.

This is particularly important to keep in mind when one turns to an alternative type of theoretical models which can be used to interpret and explain behavioral data and cognitive capacities: neural networks. Neural networks are theoretical models which, unlike the theoretical models of cognitivist psychology and computational cognitive science, are directly inspired by the brain's physical structure and way of functioning. Hence, neural networks are at the same time models of the brain and models of the mind. The neural networks used in most simulations so far have been nonmodular. They are a homogeneous network of units with minimal structure constituted by an input module (i.e., set of units), an output module, and (almost always) a single internal module in between.

However, this should be a considered as a limitation of current neural network models, not as an intrinsic property of these models. If neural networks are claimed to be inspired by the structure and functioning of the brain, they must be modular because the brain is modular. Notice, however, that the modules of neural networks will be more similar to the modules of the brain than to the theoretical "boxes" of the "boxes-and-arrows" models of computational cognitive science. A module in a modular neural network is a (simulated) physical module, not a postulated theoretical construct. A neural module can be a subset of network units with more internal connections linking the units of the module among themselves than external connections linking the units of the module with units outside the module.

Or, more functionally, a neural network module can be an observed correlated activity of a subset of the network's units, even without "anatomical" isolation of that subset of units. If the modular structure of a neural network is hardwired by the researcher, the researcher should be inspired by the actual modular structure of the brain rather than by theoretical considerations based on cognitive models. If, more in the spirit of neural cognitive science, the network architecture is not hard-wired by the researcher but is a result of evolution, development, and/or learning, the researcher should be interested in ascertaining if the emerging modular structure matches the actual modularity of the brain.

As we have said, connectionist research tends to be considered antimodularist, in contrast to the strongly modular cognitive models. This is factually correct because most neural network architectures actually used in connectionist simulations are nonmodular and because connectionism tends to underscore the role of general learning mechanisms rather than that of genetically inherited specific modules in shaping the behavior of organisms. However, as we have also said, neural network research need not be antimodularist and need not downplay the role of genetically inherited information. The real contrast between neural network models and cognitive models does not concern modularity in itself, but rather the nature of modules and the question of what theoretical models are appropriate to explain behavior and cognition.

Consider the cognitivist hypothesis that English speakers produce the past tense of verbs using two distinct modules, one for regular verbs and the other for irregular verbs (Pinker and Prince, 1988; see figure 14.1). There appears to be some empirical evidence that these two modules might reside in physically separate parts of the brain. Patients with lesions in the anterior portion of the brain tend to fail to produce regular past-tense forms while their ability to produce irregular past-tense forms appears to be preserved. In contrast, patients with lesions in the posterior portion of the brain tend to show the opposite pattern. They find it difficult to produce irregular past tenses, whereas they are able to produce regular ones. This may indicate that two distinct neural modules actually underlie past-tense production. This is completely acceptable for a connectionist (at least for the variety of connectionism represented by the authors of this chapter), who will try to simulate the behavior of producing the past tense of verbs using a modular network with two distinct modules, one for regular verbs and another for irregular verbs. (These two modules could be either structural or functional, in the sense defined above.)

What distinguishes the cognitive and the neural approaches to the treatment of past tense is the nature of the modules. Cognitivists claim that the regular past-tense module is a rule-based module. When producing the past of the verb *to work*, the brain is applying the rule "Add the suffix *-ed* to the verb root." In contrast, the irregular past-tense module is an association-based module containing a finite list of verb roots, each associated with its irregular past-tense form. The brain just consults this list of associations, finds the appropriate verb root (for example, *bring*), and produces the corresponding past-tense form (*brought*).

This theoretical interpretation of past-tense behavior is rejected by a connectionist simply because his or her theoretical tools (i.e., neural network models) do not allow for this interpretation. Neural network models are inspired by the brain, and brains are physical systems made up of physical entities and processes in which all that can ever happen is the production of physicochemical effects by physicochemical causes. Hence, in principle a neural network cannot appeal to a rule as an explanation of any type of behavior and cognitive ability. A connectionist can accept that separate and distinct portions of the brain, and of the neural network that simulates the brain, may be responsible for the production of regular and of irregular past-tense forms. However, both neural modules must function in the same basic way: units are activated by excitations and inhibitions arriving from other connected units. This does not rule out the possibility that one can discover differences in the organization and functioning of the two different neural modules for regular and irregular English verbs, and of course this requires an explanation of why the brain has found it useful to have two separate modules for controlling verb past-tense behavior instead than only one. This poses the question of the origin of modules, to which we turn in the next section.

Evolutionary Connectionist Simulations: An Evolutionary and Developmental Approach to the Study of Neural Modularity

In this section we describe two evolutionary connectionist simulations in which modular network architectures evolve spontaneously in populations of biologically reproducing neural networks. The two simulations address only some of the many different problems and phenomena that may arise as a result of the complex interactions between the adaptive process at the population level (evolution) and the adaptive process at the individual level (learning) and that may be studied using evolutionary connectionist simulations. In the first simulation a modular architecture emerges as part of a process of development taking place during the life of the individual which is shaped by evolution, but does not take experiential and environmental factors during development into consideration. Furthermore, the connection weights for this network architecture are also genetically inherited. In the second simulation, evolution actually interacts with learning because the network architecture evolves and is genetically inherited while the connection weights for this architecture are learned during life. (For other simulations on the evolution of modular network architectures, see Murre, 1992.)

Evolution and Maturation of Modules

Cecconi and Parisi (1993) have described some simulations of organisms which live in an environment containing food and water and which, to survive, have to ingest food when they are hungry and water when they are thirsty. The behavior of these organisms is influenced not only by the external environment (the current location of food and water) but also by the motivational state of the organism (hunger or thirst) which is currently driving its behavior. The body is hungry until a given number of food elements have been eaten, and then it becomes thirsty; similarly, thirst becomes hunger after a given number of water elements have been drunk. At any given time the motivational state of the organism is encoded in a special set of "motivational" units representing an internal input (coming from inside the body) which, together with the external input encoding sensory information about the location of food and water, sends activation to the network's hidden units and therefore determines the network's output. The network's output encodes the displacements of the organism in the environment to reach food or water.

In Cecconi and Parisi's simulations the network architecture is fixed, hardwired by the researcher, and nonmodular. By using a genetic algorithm for evolving the connection weights, the authors demonstrate that the organisms evolve the appropriate weights for the connections linking the motivational units to the hidden units in such a way that the current motivational state appropriately controls the organisms' behavior. When the organisms are hungry, they look for food and ignore water. When they are thirsty, they look for water and ignore food.

But what happens if, instead of hardwiring it, we try to evolve the architecture by means of a genetic algorithm? Is the evolved architecture modular or nonmodular?

To answer this question, Cangelosi et al. (1994) added a model of neural development to the simulation of Cecconi and Parisi (1993). In the new model the network architecture, instead of being hardwired by the researcher, is the eventual result of a process of cell division and migration, and of axonal growth and branching which takes place during the life of the individual organism. Unlike most simulations using genetic algorithms to evolve the architecture of neural networks (Yao,

1993), in Cangelosi et al.'s model the genotype does not directly encode the connectivity pattern of the network. What is specified in the genotype is the initial spatial location (in bidimensional space) of a set of simulated neurons (network units), the rules that control the migration of each neuron within the bidimensional space, and the growth parameters of each neuron's axon after the neuron has reached its final location.

When a new individual is born, a process of neural development takes place. First, each of the individual's neurons is placed in the bidimensional space of the nervous system according to the x and y coordinates specified in the genotype for that neuron. Second, each neuron displaces itself in neural space according to other genetically specified information until it reaches its final location. Third, after reaching its final location, the neuron grows its axon according to growth instructions (orientation and length of axonal branches), also specified in the genotype. When the axonal branch of a neuron reaches another neuron, a connection between the two neurons is established and is given a connection weight which is also specified in the genotype.

A genetic algorithm controls the evolution of the population of organisms. Starting from an initial population with randomly generated genotypes, the best individuals (i.e., those that are best able to eat when hungry and drink when thirsty) are selected for reproduction and the offspring's genotypes are slightly modified by some random genetic mutations. The result is that after a certain number of generations the organisms are able to reach for food when they are hungry, while ignoring water, and to reach for water when they are thirsty, while ignoring food.

Notice that in the genotype neurons are not specified as being input neurons, output neurons, or hidden neurons. The total bidimensional space of the brain is divided into three areas: a lower area that will contain input units (both external sensory units and internal motivational units), an intermediate area that will contain hidden units, and a higher area that will contain motor output units. If during development a neuron ends up in one of these three areas, it takes the function (input, hidden, or output) specified by the area.

Furthermore, if a neuron ends up in the input area, it can be either a sensory neuron encoding environmental information on location of food and water or a motivational neuron encoding internal (bodily) information on whether the organism needs food (is hungry) or water (is thirsty). Individual organisms can be born with a variety of defective neural networks (no input units for food or water or for hunger/thirst, no motor output units, no appropriate connectivity pattern), but these individuals do not have offspring and their defective genotypes are eliminated from the population's genetic pool.

What network architectures emerge evolutionarily? Are they modular?

Evolved network architectures contain two distinct neural pathways or modules: one for food and the other for water. When the motivational state is "hunger," only some of the hidden units have activation states that vary with variations in input information about food location; variations in input information about water location do not affect these hidden units (see figure 14.3, left). This is the food module. Conversely, when the motivational state is "thirst," water input information controls the activation level of the remaining hidden units, which are insensitive to sensory information about food. This is the water module. All successful architectures contain motivational units that send their connections to both the food module and the water module and, on the basis of their activation (hunger or thirst), give control of the organism's behavior to either food or water.

This shows that—unlike the network architecture hardwired by Cecconi and Parisi (1993), which was nonmodular—if we allow evolution to select the best adapted network architectures, the evolved architectures are modular. The neural network prefers to elaborate information about food and information about water in dedicated subnetworks that we can call modules.

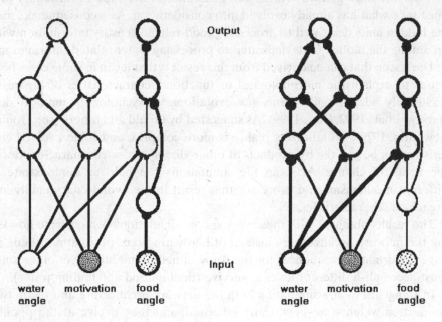

Figure 14.3
Food pathway and water pathway are shown in bold on the left and right side, respectively. (Modified from Cangelosi et al., 1994.)

However, as real brain modules as contrasted with cognitive "boxes-and-arrows" models and even hardwired modular architectures, evolved neural modules are not completely isolated or insulated modules. In the evolved architectures of Cangelosi et al., the water pathway includes some units which are specialized for processing information about water and some units which are also used to process information about food. In other words, while information about water is blocked by the network's connection weights when the organism is hungry and it is trying to approach food, information about food has some role even when the organism is thirsty and it is trying to approach water (see figure 14.3, right). Interestingly, the asymmetry between the two neural pathways or modules appears to be related to the history of the evolution of the abilities to find food and to find water, since the ability to find food begins to emerge evolutionarily in this population earlier than the ability to find water.

The fact that the water module includes some units that are also part of the food module, together with the historically contingent fact that the food module emerges earlier than the water module, demonstrates the role of historical contingency in evolved systems. Since for chance reasons the water pathway emerges evolutionarily after the food pathway (i.e., some generations later), the evolutionary process must take what has already evolved into consideration. As a consequence, some of the hidden units dedicated to processing food-related sensory information will end up among the hidden units dedicated to processing water-related information.

The lesson that can be derived from this result is that it can in some cases be erroneous to explain the morphological or functional characteristics of organisms in exclusively adaptationist terms (as evolutionary psychologists tend to do; see Barkow et al., 1992; Buss, 1999). As suggested by Gould and others (see Gould and Lewontin, 1979), evolutionary reality is more complex, and some evolved characteristics can be just the by-products of other, directly selected, characteristics or be the result of chance. Artificial life simulations can help us demonstrate these different mechanisms and processes that result in the evolutionary emergence of organismic characteristics.

The results obtained with these very simple simulations demonstrate how evolving the network architectures, instead of hardwiring (i.e., postulating) them, might have important consequences for the study of neural modularity in organisms that must accomplish different tasks to survive (finding food and finding water).

In Cangelosi et al.'s simulation both the network architecture and the network's connection weights are genetically inherited, and they evolve at the population level. The particular experience of the individual in its environment has no role in determining the individual's phenotype. It is true that the individual develops, in that the adult neural network is the result of a succession of developmental stages (the displacements of the network's units in bidimensional space and the process of axonal growth), but those changes should be called maturation rather than development because the environment and the individual's experience have no role in determining them. Rather, it is evolution that selects, at the population level, the most appropriate maturational sequence.

In the next section we describe another simulation in which evolution and individual learning during life both contribute to shaping the individual's phenotype. More specifically, evolution creates modular architectures as the most appropriate ones for the particular tasks the individual faces during life, and learning identifies the connection weights for these architectures.

Evolution and Learning in the Emergence of Modular Architectures

Ungerleider and Mishkin (1982) proposed the existence in primates of two visual cortical pathways, the occipitotemporal ventral pathway and the occipitoparietal dorsal pathway, which were respectively involved in the recognition of the identity ("what") and location ("where") of objects (see figure 14.2). (More recently, what was interpreted as the representation of the location of an object has been reinterpreted as representing what the organism has to do with respect to the object ["how"].) (See Milner and Goodale, 1998.)

This work has been very influential in both neuroscience and cognitive science, and Rueckl et al. (1989) used a neural network model for exploring the computational properties of this "two-systems" design. In their model, neural networks with different fixed architectures were trained in the "what" and "where" task by using the back-propagation procedure (Rumelhart and McClelland, 1986) and their performances were compared. The results of the simulations show that modular architectures perform better than nonmodular ones and construct a better internal representation of the task.

One way of explaining the better results obtained on the "what" and "where" tasks with modular networks than with nonmodular ones is to point out that in nonmodular architectures, one and the same connection weight may be involved in two or more tasks. But in these circumstances one task may require that the connection weight's value be increased, whereas the other task may require that it be decreased (see figure 14.4, left). This conflict may affect the neural network's performance by giving rise to a sort of neural interference. On the contrary, in modular architectures, modules are sets of "proprietary" connections that are used to accomplish only a single task, and therefore the problem of neural interference does not arise (see figure 14.4, right). Rueckl et al. (1989) hypothesize that this might be one of

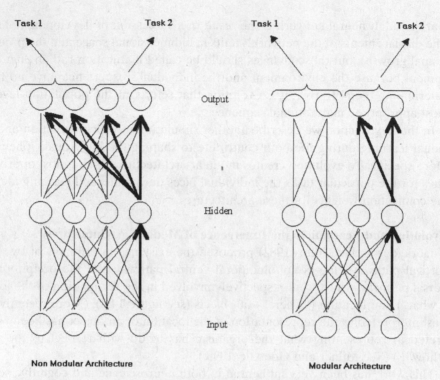

Figure 14.4
Neural interference in neural networks that have to learn to perform two tasks. Consider in the non-modular architecture (left) the highlighted connection that links one of the input units with one of the hidden units. Since the hidden unit sends its connections (highlighted) to both of the output units involved in task 1 and those involved in task 2, a modification of the connection's weight value would affect both tasks. In this kind of architecture, neural interference may arise because optimization of one task may require that this connection weight's value be increased, whereas optimization of the other task may require that it be decreased. In the modular architecture (right) the problem of neural interference does not arise because the connection goes to a hidden unit which sends its connections (highlighted) only to the output units involved in task 2, and therefore its value can be changed to satisfy the requirements of task 2 only.

the reasons for the evolutionary emergence of the two distinct neural pathways in real organisms.

To test this hypothesis Di Ferdinando et al. (2001) repeated the experiment of Rueckl et al. (1989) by allowing the evolution of the network architecture. In Rueckl et al.'s simulations the network architectures are hardwired by the researcher and the authors are able to find the best possible architecture (which is a modular architecture with more hidden units assigned to the more difficult "what" task and fewer hidden units assigned to the easier "where" task) by trying many different hardwired architectures and testing them.

Jacobs and Jordan (1992) used a developmental model in which the network architecture emerges as a result of a process of development in the individual. The individual network starts as a set of units, each placed in a particular location of a bidimensional physical space; then pairs of units may establish connections based on a principle of "short connections," according to which two units are more likely to become connected the closer they are in space. This is an interesting proposal based on a principle that favors short connections and is likely to play a role in neural development.

However, the resulting network architecture is not really self-organizing because it is the researchers who decide the location of units in physical space and therefore in a sense hardwire the network architecture. (In the simulations described in the section "Evolution and Maturation of Modules" there is also development of the connectivity pattern as in Jacobs and Jordan's simulations, but both the location of the network's units in space and the rules controlling the growth of connections are genetically inherited and are the result of a self-organizing evolutionary process.) In the simulations described in this section, although there is no development, the network architectures are the spontaneous outcome of a process of evolution which is independent of the researcher.

In a first set of simulations Di Ferdinando et al. (2001) used a genetic algorithm for evolving both the architecture and the connection weights of the neural networks. The results showed that the genetic algorithm was unable to evolve both the architecture and the weights. Furthermore, the network architecture that tended to evolve was different from the best architecture of Rueckl et al. in that it assigned more resources (hidden units) to the easier "where" task than to the more difficult "what" task. In other words, the evolutionary algorithm was not able to allocate the appropriate resources to the two tasks.

The failure of the genetic algorithm to find the best architecture for the "what" and "where" tasks, and therefore to reach appropriate levels of performance when both the network architecture and the connection weight are genetically inherited, appears to be due not only to the fact that a mutation affecting the architecture can suddenly make a set of weights evolved for the preceding architecture inappropriate for the new architecture, but also to a phenomenon analogous to genetic linkage. In simulations in which the architecture is fixed and is the best modular architecture (more units allocated to the "what" task than to the "where" task), the genetic algorithm appears to be unable to evolve the appropriate connection weights because a favorable mutation falling on the weights of one module can be accompanied by an unfavorable mutation in the weights of the other module. This interference at the genetic level appears to be unexpected according to models of population genetics (Wagner, personal communication).

Further analyses of the simulation results reveal other interesting phenomena that are due to the coevolution of architecture and weights—for example, freezing of the architecture at low mutation rates and oscillation of the evolved architecture at high rates.

The best results (i.e., the appropriate modular architecture and high levels of performance) are obtained in simulations in which evolution cooperates with learning. More specifically, the best solution, as suggested by Elman et al. (1996), is to have evolution take care of the architecture and learning of the connection weights. With this solution evolution is free to zero in on the best network architectures without fear that inherited weights that were appropriate for previous architectures may turn out to be inappropriate for mutated architectures (genetic linkage) and learning during life is free to find out the best connection weights for each inherited architecture. These simulation results clearly show that evolution and learning are not dichotomous, as empiricists and nativists sometimes seem to believe, but that their cooperation is necessary if organisms must be able to acquire complex capacities.

As a final observation we note that, as in the simulations described in the previous section, the evolved neural modules are not completely isolated and the modular architecture is not as clean as a "boxes-and-arrows" model. While most connections are proprietary of the two modules, the "what" module and the "where" module, there are some connections that are shared by the two modules.

Conclusions

In this chapter we have described a new approach to studying brain/mind modularity which takes into consideration the phylogenetic history of an organism's brain modules. This approach, artificial life, allows us to simulate in the same model an organism at the genetic, neural, and behavioral levels, and may help us to reveal how modules at one level may be related to modules at another level.

Brain/mind modularity is a contentious issue in current cognitive science. Cognitivists tend to conceive of the mind as a set of distinct specialized modules, and they believe that this rich modularity is basically innate. Evolutionary psychologists even think that each module is adaptive, in that it has been biologically selected as a result of specific evolutionary pressures. However, other cognitivists, such as Chomsky and Fodor, believe that modules are innate but not necessarily adaptive (Fodor, 2000). On the other hand, connectionists tend to think that the mind is a more homogeneous system that basically genetically inherits only a general capacity to learn from experience, and that if there are modules, they are the result of development and learning rather than being innate.

We have maintained that connectionism is not necessarily antimodularist and antinativist. On the contrary, since neural network models are said to be inspired by the brain, they must be modular (even if most network architectures used in connectionist simulations are nonmodular) because the brain is a rich structure of specialized modules. Viewing neural networks in the perspective of artificial life allows us to develop an appropriately modular and nativist connectionism, evolutionary connectionism. Artificial life simulations simulate evolving populations of organisms that inherit a genotype from their parents which, together with experience and learning, determines the individual phenotype. The way is open then for simulations that explore whether modular or nonmodular network architectures emerge for particular tasks, and how evolution and learning can cooperate to shape the individual phenotype.

In any case, even if connectionism can be modularistic, this does not imply that when connectionists talk about modules, they mean the same thing as cognitivists. Cognitive modules are theoretical entities which are postulated in "boxes-and-arrows" models used to explain behavioral data. Connectionist modules are anatomically separated and/or functionally specialized parts of the brain. There may be only partial coextensiveness between the two types of modules, and in any case research on neural modules is very differently oriented than research on cognitive modules and considers different types of empirical evidence.

Evolutionary connectionism shares the main goal of evolutionary psychology: to develop a "psychology informed by the fact that the inherited architecture of human mind is the product of the evolutionary process" (Barkow et al., 1992), but it differs from evolutionary psychology in three main ways: (1) it uses neural networks rather than cognitive models for interpreting human behavior; (2) it adopts computer simulations for testing evolutionary scenarios; (3) it has a less panadaptionist view of evolution and is more interested in the rich interplay between genetically inherited and experiential information. The simulation of evolutionary scenarios allows us to take chance and other nonadaptive evolutionary factors into consideration, and therefore prevents us from explaining all the morphological or functional characteristics of organisms in exclusively adaptationist terms.

We have presented two artificial life simulations in which the genetic algorithm actually selects for modular architectures for neural networks. In one simulation both the network architecture and the network weights are genetically inherited and they evolve, but evolution selects for appropriate maturational sequences. In the other simulation, evolution and learning cooperate in that evolution selects for the network architecture and learning finds the weights appropriate for the inherited architecture. These simulations weaken Marcus's criticism when he says that "none of [. . . connectionist] models learn to divide themselves into new modules" (Marcus, 1998, p. 163).

The first of the two simulations described in this chapter also shows that modules can be inherited (innate), but their exact structure is not necessarily adaptive and the result of specific evolutionary pressures, but it can be the result of other evolutionary forces, such as chance and preadaptation.

More artificial life simulations are needed to explore how modular architectures evolve or develop during life, and how selective pressures at the population level or experience during life may shape the existing modules. But the artificial life perspective allows us to explore other research directions that involve the interactions of modules at the genetic, neural, and behavioral levels (Calabretta et al., 1998). For example, using neural networks in an artificial life perspective, one can explore if genetic duplication would facilitate the evolution of specialized modules. This would represent an important confirmation of the general hypothesis that gene duplication facilitates the evolution of functional specialization, which was originally proposed by Ohno (1970) and modified by Hughes (1994). We have already shown with artificial life simulations that this might actually be the case (Calabretta et al., 2000; see also chapter 2 in this volume). Another research direction would be testing of whether sexual reproduction might decrease the kind of genetic interference (linkage) postulated by Di Ferdinando et al. (2001); this would strengthen one of the several hypotheses formulated about the role of sexual reproduction in evolution (Michod and Levin, 1988).

References

Barkow J-H, Cosmides L, Tooby J (eds) (1992) The Adapted Mind: Evolutionary Psychology and the Generation of Culture. New York: Oxford University Press.

Belew RK, McInerney J, Schraudolph NN (1992) Evolving networks: Using the genetic algorithm with connectionist learning. In: Artificial Life II (Langton CG, Taylor C, Farmer JD, Rasmussen S, eds), 511–547. Redwood City, CA: Addison-Wesley.

Buss DM (1999) Evolutionary Psychology: The New Science of the Mind. Boston: Allyn and Bacon.

Calabretta R, Galbiati R, Nolfi S, Parisi D (1996) Two is better than one: A diploid genotype for neural networks. Neural Proc Lett 4: 149–155.

Calabretta R, Nolfi S, Parisi D, Wagner GP (1998) A case study of the evolution of modularity: Towards a bridge between evolutionary biology, artificial life, neuro- and cognitive science. In: Proceedings of the Sixth International Conference on Artificial Life (Adami C, Belew R-H, Kitano H, Taylor C, eds), 275–284. Cambridge, MA: MIT Press.

Calabretta R, Nolfi S, Parisi D, Wagner GP (2000) Duplication of modules facilitates the evolution of functional specialization. Art Life 6: 69–84.

Cangelosi A, Parisi D, Nolfi S (1994) Cell division and migration in a "genotype" for neural networks. Network 5: 497–515.

Cecconi F, Parisi D (1993) Neural networks with motivational units. In: From Animals to Animats 2. Proceedings of the 2nd International Conference on Simulation of Adaptive Behavior (Meyer J-A, Roitblat HL, Wilson SW, eds), 346–355. Cambridge, MA: MIT Press.

Cosmides L, Tooby J (1994) Origins of domain specificity: The evolution of functional organization. In: Mapping the Mind: Domain Specificity in Cognition and Culture (Hirschfeld LA, Gelman SA, eds), 85–116. Cambridge, MA: MIT Press.

Cosmides L, Tooby J, Barkow JH (1992) Introduction: Evolutionary psychology and conceptual integration. In: The Adapted Mind (Barkow J-H, Cosmides L, Tooby J, eds), 3–15. New York: Oxford University Press.

Di Ferdinando A, Calabretta R, Parisi D (2001) Evolving modular architectures for neural networks. In: Connectionist Models of Learning, Development, and Evolution (French RM, Sougné JP, eds), 253–262. Berlin: Springer.

Elman JL, Bates EA, Johnson MH, Karmiloff-Smith A, Parisi D, Plunkett K (1996) Rethinking Innateness: A Connectionist Perspective on Development. Cambridge, MA: MIT Press.

Fodor J (1983) The Modularity of Mind. Cambridge, MA: MIT Press.

Fodor J (1998) The trouble with psychological Darwinism. London Rev Books 20(2): 11–13.

Fodor J (2000) The Mind Doesn't Work That Way: The Scope and Limits of Computational Psychology. Cambridge, MA: MIT Press.

Gould SJ (1997) Evolution: The pleasures of pluralism. NY Rev Books, June 26, pp. 47–52.

Gould SJ, Lewontin R (1979) The spandrels of San Marco and the Panglossian paradigm: A critique of the adaptationist programme. Proc Roy Soc London B205: 581–598.

Gould, SJ, Vrba ES (1982) Exaptation: A missing term in the science of form. Paleobiology 8: 4–15.

Holland JH (1992) Adaptation in Natural and Artificial Systems: An Introductory Analysis with Applications to Biology, Control, and Artificial Intelligence. Cambridge, MA: MIT Press.

Hughes AL (1994) The evolution of functionally novel proteins after gene duplication. Proc Roy Soc London B256: 119–124.

Jacobs RA, Jordan MI (1992) Computational consequences of a bias toward short connections. J Cog Neurosci 4: 323–335.

Karmiloff-Smith A (2000) Why babies' brains are not Swiss army knives. In: Alas, Poor Darwin (Rose H, Rose S, eds), 144–156. London: Jonathan Cape.

Marcus GF (1998) Can connectionism save constructivism? Cognition 66: 153–182.

Michod RE, Levin BR (1988) Evolution of Sex: An Examination of Current Ideas. Sunderland, MA: Sinauer.

Milner AD, Goodale MA (1998) Precis of The Visual Brain in Action. Psyche 4(12). http://psyche.cs.monesh.edu.au/va/psyche-4-12-milner.html

Murre JMJ (1992) Learning and Categorization in Modular Neural Networks. New York: Harvester.

Newell A, Simon HA (1976) Computer science as empirical inquiry: Symbols and search. Commun Assoc Comp Mach 19: 113–126.

Ohno S (1970) Evolution by Gene Duplication. New York: Springer.

Parisi D, Cecconi F, Nolfi S (1990) Econets: Neural networks that learn in an environment. Network 1: 149–168.

Pinker S (1999) Words and Rules: The Ingredients of Language. New York: Weidenfeld and Nicolson.

Pinker S, Prince A (1988) On language and connectionism: Analysis of a parallel distributed processing model of language acquisition. Cognition 28: 73–193.

Plunkett K (1996) Development in a connectionist framework: Rethinking the nature–nurture debate. CRL Newslet, Univ Calif, San Diego. *http://crl.ucsd.edu/newsletter/10-4/*.

Restak RM (1995) The Modular Brain. New York: Simon and Schuster.

Rueckl JG, Cave KR, Kosslyn SM (1989) Why are "what" and "where" processed by separate cortical visual systems? A computational investigation. J Cog Neurosci 1: 171–186.

Rumelhart D, McClelland J (1986) Parallel Distributed Processing: Explorations in the Microstructure of Cognition. Cambridge, MA: MIT Press.

Spelke ES, Katz G, Purcell SE, Ehrlich SM, Breinlinger K (1994) Early knowledge of object motion: Continuity and inertia. Cognition 51: 131–176.

Ungerleider LG, Mishkin M (1982) Two cortical visual systems. In: The Analysis of Visual Behavior (Ingle DJ, Goodale MA, Mansfield RJW, eds), 549–586. Cambridge, MA: MIT Press.

Wynn K (1992) Addition and subtraction by human infants. Nature 358: 749–750.

Yao X (1993) A review of evolutionary artificial neural networks. Int J Intell Syst 8: 539–567.

15 Modularity and Chunking

Fernand Gobet

Introduction

Like the questions of nature versus nurture, parallelism versus seriality, one-shot learning versus gradual learning—to mention only a few—the concept of modularity, in its many guises, has provoked a vast amount of debate and controversy in psychology. In its biological connotation, the concept goes back at least to Gall's phrenology, which proposed that the shape of the cranial bones was a good indication of the mental faculties residing within. Used as a way to characterize human knowledge, the concept goes even farther back in time; for example, in the seventeenth century, John Locke proposed a "mental chemistry" of the human mind, explaining how knowledge is built up hierarchically from simple ideas.

In recent years, the question of modularity has been made central to psychology for a number of reasons. These include developments in computer science and artificial intelligence, where a modular knowledge representation is often presented as a desirable feature; progress in neuroscience, where new anatomical, imaging, and experimental data have identified a number of brain modules at various levels of granularity (e.g., Churchland and Sejnowski, 1992); and a resurgence of the nature versus nurture debate, where modularity has often been seen as conceptually supportive of a nativist position (e.g., Fodor, 1983).

It is obviously not possible to cover all of these strands in a single chapter. Given the recent controversy about the role of modularity, if any, in "classical" information-processing models of cognition, I thought it interesting to examine to what extent such models can be modular. Rather than reviewing several examples superficially, I have preferred to analyze in detail a single architecture, and to assess the empirical evidence supporting some notion of modularity. I have therefore chosen to focus on CHREST (Chunk Hierarchy and REtrieval STructures), the computational architecture I have developed in recent years to simulate various aspects of human cognition. Thus, this chapter can be seen as a case study of the extent to which modularity can be meaningfully applied to a symbolic, computational model.

The chapter is organized as follows. First, I discuss several meanings of the concept of modularity in psychology. In the second section, I attempt to use these to characterize the CHREST architecture. Emphasis is given to one of these meanings (modularity of knowledge), and to how it fits the chunking mechanisms inherent in CHREST. The third section discusses empirical data that attempt to validate the notion of chunking, and the fourth section presents simulations of these data using CHREST. The different meanings of modularity typically used in psychology are revisited in the conclusion.

Hierarchical Levels in Science

In an interdisciplinary book such as this, it may be useful to make explicit the type of philosophy of science that one uses in everyday research, since it is likely to vary across fields. It is probably accurate to describe most work in psychology, and particularly research using computational modeling, as what Simon (1969) called "pragmatic holism" (i.e., the belief that complex systems, including the nature of the mind, can be characterized as a hierarchy of levels and each level can be studied in relative isolation from the levels below and above). While, in principle, explanations of behavior could be formulated in terms of neurons, molecules, atoms, or even of entities situated lower in this hierarchy, it is assumed that explanations using high-level concepts such as symbols, schemas, stimulus-response bonds, and so on are more likely to offer tractable research questions. Simon's characterization of science as a hierarchy of levels, which can be distinguished from the sort of strong reductionism currently fashionable in neuroscience (e.g., Churchland and Sejnowski, 1992), obviously derives from his analysis of near-decomposability in nature (Simon, 1969).

The Concept of Modularity in Psychology

One difficulty with the concept of modularity—perhaps one reason behind this book—is that this concept has various meanings across different fields; worse, it has different meanings within a single field such as psychology. Consider the case of neuroscience and psychology, which are both concerned with the study of the human mind. At first blush, one might expect each of these two sciences to use the concept of modularity with a clear meaning: a biological structure with specific neuronal and connectivity properties, on the one hand, and a relatively well circumscribed mental faculty, on the other. However, there are actually a variety of definitions within each of these fields, definitions that are sometimes clearly incompatible pairwise. I will focus upon psychology, say a little about neuroscience, and refer the reader to other chapters of this book for an overview of the ways modularity is used in other fields.

Meanings of Modularity in Psychology

The concept of modularity is omnipresent in psychology, although its relevance has sometimes been disputed (see Lashley, 1950). Simplifying the matter a bit, one can categorize the various uses of the term into three classes of meanings: biological, functional, and knowledge-related.

Biological Meaning There is a large overlap between the meanings under this heading and those used in neuroscience. While neuroscience is mostly interested in modules at the cell level—including their physiology, connectivity, and chemistry—neuropsychology attempts to link behavioral data with regions of the brain, using mainly brain-damaged patients and brain-imaging techniques. Often, a notion of "weak modularity" (Kosslyn and Koenig, 1992) is used: even though networks compute input–output mappings, the same network may belong to several processing systems; and while there is a good measure of localization in the brain, it is also often the case that neurons participating in the same computation belong to different regions. As noted by Kosslyn and Koenig, weak modularity is conceptually close to Simon's concept of near-decomposability. The reader is referred to Churchland and Sejnowski (1992), Kosslyn and Koenig (1992), and Velichkovsky (chapter 16 in this volume) for further discussions of the use of modularity in the neuroscience literature.

Functional Meaning This class of meanings is highly prevalent in information-processing psychology, the leading approach to cognitive psychology. Here, modules refer to sets of processes or computations, and although some assumptions about biological implementation are typically made, these are not an essential part of the explanation. Perhaps the most influential proposal is Fodor's (1983), in which the mind is composed of information-processing devices that operate largely separately. Fodor proposed several attributes for such modules, including that they are hard-wired, rapid, encapsulated, and mostly innate. In his "society of mind" framework, Minsky (1985) proposed that cognition consists of a number of agents, which are all specialists in their domain and which act in parallel. Allport (1980) proposed that several production systems operate in parallel, each forming a module. Finally, in his theory of intelligence, Gardner (1983) proposed that the human mind contains multiple modules implementing aspects of intelligence, such as spatial intelligence, musical intelligence, and so on.

Even cognitive architectures such as ACT-R (Anderson, 1993) or Soar (Newell, 1990), often presented as nonmodular, have their share of functional modularity. For example, Anderson distinguishes three cognitive components: working memory, declarative memory, and procedural memory, and explicitly considers that perception and motor action are organized as modules.

Knowledge-Related Meaning The final class of meanings relates to the modular organization of knowledge, and shares some kinship with the way modularity is used in computer science and artificial intelligence. The question here is how knowledge is indexed, structured, organized, and retrieved. In this context, modularity "refers to the ability to add, modify, or delete individual data structures more or less independently of the remainder of the database, i.e., with clearly circumscribed effects on what the system 'knows'" (Barr and Feigenbaum, 1981, p. 149). While there is good evidence that modular and decomposable systems are easier for humans to understand (e.g., Simon, 1969), and that such properties are desirable in fields such as software engineering, it is an empirical question whether human knowledge demonstrates such features.

Several formalisms have been proposed in cognitive science to represent and implement knowledge. Modular representations include production systems, schemas, and discrimination nets. Nonmodular representations include distributed neural networks, holograms, and various mathematical representations based upon matrix algebra. This classification is rather rough, however. With the modular representations it can forcibly be argued that most of these representations do not satisfy the requirement of weak interaction among subsystems; for example, production rules are typically organized in problem spaces, and their interdependence can actually be considerable. With the nonmodular representations, several authors have suggested that, irrespective of the uniform organization of knowledge, modules exist as emergent properties occurring when the system develops or acquires information through learning.

Examples from Expertise Research

To illustrate the classification above, I will use the rapidly developing field of expertise psychology. This field is concerned with the question of how some individuals are able to excel in their domain of expertise, seemingly bending the capacity limits of memory and processing that plague most of us. The domains of expertise investigated include sciences, arts, games, sports, and professional occupations; in a word, almost the entire gamut of human activities.

The literature on *talent* has emphasized modularity in its biological meaning (i.e., the presence of cognitive faculties that are domain-specific, innate, and modular).[1] Recent developments in brain-imaging techniques have generated a renewed interest in this type of theorizing, where the human mind is seen as a collection of specialized modules operating almost independently—a tradition that goes back at least to Gall's phrenology. Evidence includes the pattern of data suggesting that ability in mathematics and music correlates with brain bilateralization (Obler and Fein, 1988). As just seen, a related claim is that these modules are innate, though the evidence for this claim has been strongly disputed (e.g., Ericsson and Faivre, 1988).

With the second meaning—functional modularity—one is interested more in whether some classes of mechanisms can operate independently from some other classes. There is now ample evidence that this is the case. For example, with chess expertise, Saariluoma (1992) has shown that interfering tasks impair memory and decision-making abilities when they are visuospatial, but not when they are verbal.

Finally, research into expertise has highlighted the role of knowledge modularity and has shown that experts "lump" information in ways different from novices. Perceptually, they are able to see the problem situation in terms of larger units; conceptually, they can relate it to a large amount of related and relevant information. Modular concepts, such as chunks or schemas, are typically used by psychologists to describe such an organization of perceptual and conceptual knowledge. A large part of this chapter will describe a theory explaining how experts represent and acquire knowledge, and will discuss the extent to which some aspects of this theory can be considered modular.

A Symbolic Computational Chunking Model: CHREST

CHREST (Chunk Hierarchy and REtrieval STructures) is a cognitive architecture belonging to the research tradition within cognitive science where understanding analytical, empirical data is supported by building synthetic models able to carry out the behavior studied. It is based on EPAM (Elementary Perceiver And Memorizer; Feigenbaum and Simon, 1984; Richman et al., 1995).[2] EPAM was developed to account for verbal learning phenomena and was later applied to other empirical phenomena where "chunking" (acquiring perceptual units of increasing size) plays an important role.

Brief Overview of the Model

CHREST mainly addresses high-level perception, learning, and memory, although various problem-solving mechanisms have been implemented recently. It comprises processes for acquiring low-level perceptual information, a short-term memory (STM), attentional mechanisms, a discrimination net for indexing items in long-term memory (LTM), and mechanisms for making associations in LTM, such as production rules or schemas.

Figure 15.1 contains a diagrammatic representation of the CHREST architecture. As can be seen, STM mediates the flow of information processing between the model's components. As an implemented theory, CHREST has little to say about low-level perception and feature extraction: implementations simply assume that

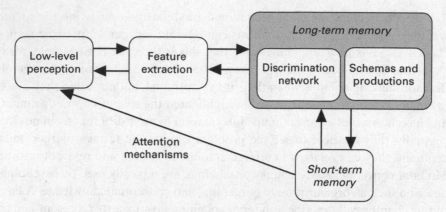

Figure 15.1
The main components of the EPAM/CHREST architecture.

they occur, supplying suitable features for discrimination. Processing in CHREST consists in the acquisition of a discrimination net based on high-level perceptual features picked up by attentional mechanisms and on the creation of links connecting nodes of this net.

Since it originates from EPAM, CHREST shows a clear influence of Simon's general view of human cognition (e.g., Simon, 1969). The central processing mechanisms operate serially; for example, the eye fixates on an object in the environment; features are extracted and processed in the discrimination net; based upon the result of the discrimination process, a further eye fixation is made; and the cycle continues. Moreover, STM operates as a queue (the first elements to enter are the first to leave). Visual STM has a limited capacity, which consists of four chunks (for empirical evidence, see Cowan, 2001; Gobet and Simon, 2000; Zhang and Simon, 1985). Additional influences from Simon include the idea that learning is sufficient to generate satisfactory, but not necessarily optimal, performance, and that processing is constrained by a number of restrictions. The limitations include various time parameters, such as the time to fixate a chunk in LTM or to store an element in STM, and various capacity parameters, such as the four-chunk limit of STM. Finally, the tree structure of CHREST's discrimination net is a remnant of Simon's interest in hierarchical organizations in human thinking and in nature generally.

Basic Structure of the Discrimination Net

The information experienced by the cognitive model is stored within a discrimination net. This net consists of *nodes*, which contain images (i.e., the internal repre-

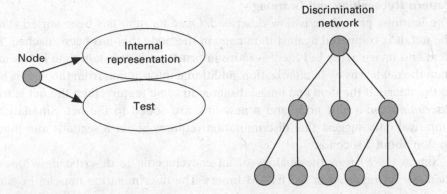

Figure 15.2
Left: In CHREST, a node consists of an *internal representation* and a *test*. Right: Example of a discrimination network (internal representations and tests are not shown).

sentations of the external objects). The nodes are interconnected by *links*, which contain *tests* by which a given item is sorted through the net. Figure 15.2 illustrates a node with its internal representation and its test (left) and shows the hierarchical structure of the discrimination net, with nodes connected by test links (right).

An item is sorted through a net as follows: starting from the root node of the net, the links from that node are considered to check if any of the tests apply to the next feature (or group of features) of the current item. If so, then that link is followed and the node thus reached becomes the new current node; the procedure is repeated from this current node until no further links can be followed.

Hierarchical Representation of Objects
All information in this cognitive model is stored as an ordered list of features. For example, the word "cat" may be stored as the ordered list of letters c-a-t. These features are initially extracted from the external object by the attentional mechanisms and the feature extraction module. However, these features may also be adjusted to incorporate what has been learned previously, leading to a hierarchical knowledge organization. For example, once words such as "cat" have been learned by the system, the sentence "The cat sat on the mat" would be represented as an ordered list of words and not individual letters.

The features within tests are simply taken from chunks within the discrimination net (in our example, individual letters or words). Each test applies simply to the next letter, or set of letters, in the stimulus. More complex tests are possible in which, for example, attributes such as color are tested.

Pattern Recognition and Learning

The learning procedure is now described. Once an item has been sorted through the net, it is compared against the image in the node that has been reached. If the item and image agree but there is more information in the item than in the image, then the net learns by familiarization: additional information from the item is added to the image. If the item and image disagree in some feature, then the net learns by *discrimination*: a new node and a new link are added to the net. Simulations of empirical data suggest that discrimination requires about 8 seconds and familiarization about 2 seconds.

Simon (1969) uses the analogy of an encyclopedia to describe how long-term memory is organized in the EPAM theory. The discrimination net can be seen as the "index" to the encyclopedia, and semantic memory, containing schemas, production rules, and other forms of knowledge, represents its "text." In comparison to the classical EPAM, which has focused upon the index of long-term memory, the research surrounding CHREST has paid more attention to its text. Augmenting semantic memory is seen as the creation of *lateral* links connecting nodes (see figure 15.3); in particular, these links are created when nodes are sufficiently similar ("similarity links") or when one node acts as the condition of another node ("production link"). All nodes used for creating new links must be in STM, a requirement that is consistent with the emphasis on processing limits present in both EPAM and CHREST. As we shall see later, mechanisms creating lateral links allow for the presence of a new type of knowledge module where elements are connected laterally, supplementing the hierarchical chunks created vertically by discrimination.

Domains of Application

As psychological theories, EPAM and CHREST have a number of strengths. They are simple and parsimonious, with few degrees of freedom (mainly subjects' strategies). They can make quantitative predictions—for example, about the number of errors committed or the time taken by a subject to carry out a task. Together, they are able to simulate in detail a wealth of empirical phenomena spanning various domains, such as verbal learning, context effects in letter perception, concept formation, expert behavior, acquisition of first language by children, and use of multiple representations in physics (see Gobet et al., 2001, for a review).

Modularity in CHREST

Not surprisingly, given Simon's involvement in both near-decomposability and CHREST, the research philosophy behind this cognitive architecture is consistent with Simon's pragmatic holism—human behavior can be studied at various levels

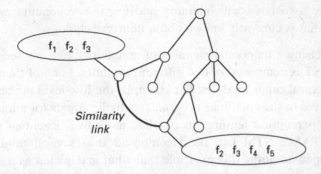

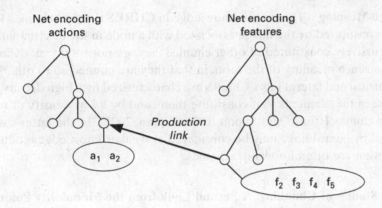

Figure 15.3
Examples of lateral links. Upper panel: a *similarity link*. Lower panel: a *production link*. Features are denoted by *f1*, *f2*, etc., and actions are denoted by *a1*, *a2*, etc.

of abstraction, which are all relatively insulated from the levels above and below. In the case of CHREST, the level of abstraction is how symbols and symbol structures are created and how they affect behavior, in the range of seconds. One goal of the following sections is to show, by pointing to an existing implementation, that symbolic, information-processing architectures can contain an important amount of modularity, contrary to a widely held misconception.

Biological Meaning   While biological implementation is not the main focus of CHREST, for the reason just mentioned, it should be realized that this architecture does not violate any biological constraints. As a matter of fact, Tabachnek-Schijf et al.'s (1997) analysis of their CaMeRa model shows how certain components of

CHREST, mainly those associated with memory and iconic components, can be mapped directly into what is currently known about neuropsychology.

Functional Meaning   Using a functional meaning of modularity, one can see from figure 15.1 that CHREST is composed of five different modules. Each of them can be decomposed into several other modules. For example, the low-level perception module (not implemented in the computer programs) actually stands for a number of modules extracting perceptual features in parallel; moreover, Richman et al. (1995) discuss how EPAM's STM can be decomposed into several modality-dependent components; in addition, it is plausible that what is depicted as a single discrimination net is actually a hierarchy of several nets processing information in parallel.

Knowledge Meaning   The unit of knowledge in CHREST is the "chunk." Chunks, which are constituted by the image associated with a node in the discrimination net, can be recursively constituted of other chunks; they are not totally modular, in the computer-science meaning of the term, in that they are connected to other chunks by both vertical and lateral links. Chunks are characterized by a high density of relations between the elements that constitute them, and by a low density of relations with other chunks (Chase and Simon, 1973; Cowan, 2001). In the latter case, clusters formed by lateral links may be considered as weaker knowledge modules, and map into what are often known as schemas.[3]

Empirical Studies of Chunking: A Second Look from the Modularity Point of View

The previous discussion has given a rather crisp definition of a chunk, and has shown to what extent this concept relates to that of modularity. However, the next step—and this is an important step—is to ground this theoretical analysis with empirical data. The main attempt to do so has been through research into chess expertise. In the following sections, I briefly describe the most important results.

Basic Phenomena in Chess Expertise

De Groot's (1965) monograph is often considered the seminal work on chess expertise, and on expertise in general, for that matter. He subjected chess players of different skill levels, including world champions, to a variety of memory and problem-solving experiments. The surprising result of his investigation of the thinking processes of chess players was that there were no large skill differences in the depth of search, the number of moves considered, or the heuristics of search employed. However, de Groot did find a difference in a short-term memory task where a chess position was presented for a few seconds: master-level chess players were able to recall almost the entire position correctly, while weaker players could recall only a few pieces. De Groot took this result as evidence that expertise does not depend upon any superior processing abilities but upon domain-specific knowledge.

Building upon de Groot's work, Chase and Simon (1973) studied the memory structures employed by chess players. They used two experimental paradigms: a *recall task* and a *copy task*. In the *recall task*, the same method as de Groot's (1965) was used. Subjects were allowed to inspect a chess position for 5 seconds. The position was then removed from view, and players had to reconstruct as much of the position as they could. In the *copy task*, subjects reconstructed a stimulus board position onto an empty board, while the stimulus board remained in view. The stimulus and the target boards could not be fixated simultaneously; hence, glances between the boards could be used to detect the use of memory chunks.

Chase and Simon replicated de Groot's (1965) finding that stronger players demonstrate greater recall. More important for our topic, the evidence from the copy and recall tasks was used to infer the memory structures used to mediate this superior performance. The key to Chase and Simon's approach in analyzing the reconstruction of positions was to simultaneously consider the latencies between placing two successive pieces and the pattern of semantic relations they shared. Chase and Simon's results and theoretical analyses have had a huge impact in cognitive psychology, and have been used to explain expert behavior in a variety of domains (see de Groot and Gobet, 1996, for a review).

Chase and Simon were interested not only in the role of knowledge in chess memory experiments, but also in how knowledge made it possible for masters to find good moves using a highly selective search. I will focus here on the memory aspects (for a discussion of the role of chunking in problem solving, see Chase and Simon, 1973; Gobet and Simon, 1998b).

Revisiting Experts' Chunks

Chase and Simon thus suggested an operational definition of chunks utilizing the latencies between successive piece placements and the relations (color, defense, attack, proximity, and type of piece) between them. It is worth looking at these analyses in detail, as they are illuminating about what constitute the building blocks of (chess) knowledge, and the extent to which such knowledge is modular. They also illustrate empirical techniques (and some of their limitations) that can be used to operationalize the concept of modular knowledge.

Figure 15.4
Examples of the kinds of position typically used in chess research on memory. On the left, a *game position* taken from a master game. On the right, a *random position* obtained by shuffling the piece locations of a game position.

Chase and Simon's (1973) seminal experiment has been recently replicated and extended by Gobet and Simon (1998a). The main difference between the two studies, apart from the number of subjects (three and twenty-six players, respectively), is that Gobet and Simon used a computer display to present the tasks instead of physical chessboards, as in the Chase and Simon experiment. In spite of these differences, there is an important overlap between the results of the two studies. I will focus upon Gobet and Simon's results, since they had a larger sample and better experimental controls.

In this experiment, subjects ranged from amateurs to professional grand masters, and were divided into three skill levels (masters, experts, and class A players). Two types of positions were used: game positions, taken from master games, and random positions, where the location of pieces had been shuffled (see figure 15.4 for an example of both types of positions). The results of the recall of game positions were in line with previous experiments: masters, experts and class A players correctly recalled 92 percent, 57 percent, and 32 percent of pieces, respectively. The corresponding results for random positions—19 percent, 14 percent, and 12 percent, respectively—were also consistent with previous results, showing a small but reliable skill effect even with meaningless positions (Gobet and Simon, 1998a).

Identifying Chunks

The logic behind the identification of chunks is somewhat long-winded, so it may be useful to spell it out from the beginning. There are three main steps: (1) to obtain a rough time threshold, which can be used to decide whether two pieces placed in succession belong to the same chunk; (2) to validate this threshold, using information about the semantic relations between successively placed pieces (the idea is that pieces belonging to the same chunk should both share several relations and be placed in rapid succession); and (3) to check that within-chunk placements show a different pattern of relations than between-chunk placements—the latter placements should be closer to what would be expected by chance than the former.

Step 1: Finding a Cutoff Value   The first step is to devise a method allowing one to use the latencies[4] between the piece placements in order to infer the presence of chunks. In order to do so, one needs to estimate a threshold below which a pair of pieces would be considered to belong to the same chunk, and above which they would be deemed to belong to two different chunks. Gobet and Simon followed Chase and Simon's approach, in which the copy task is employed to estimate this threshold. The basic idea is that pieces placed after a glance at the stimulus board (between-glance placements; BGP) belong to different chunks, while pieces placed without switching back to the stimulus position (within-glance placements; WGP) belong to the same chunk. Combining the latencies in placement with the presence or absence of a glance at the stimulus offers a way to approximate the time needed to retrieve chunks from LTM.

The results of the experiment were in good agreement with Chase and Simon's (1973), in spite of the differences in the experimental apparatus. The WGP and BGP distributions were quite different, and the 2-second cutoff segregated the two distributions well. Averaging across the three skill levels, 79.5 percent of the WGP latencies (versus only 1.1 percent of the BGP latencies) were less than 2 seconds, and 89.3 percent (versus 4.4 percent) were below 2.5 seconds. Finally, the median of the WGP latencies was 1.4 seconds. These times are close to Chase and Simon's, although a little slower.

Step 2: Correspondence Between Latencies and Relations   The next step is to demonstrate that the chunks defined by latencies can be validated by using the statistics of the meaningful relations between pairs of pieces that were placed on the board successively. The following relations are used: attack, defense, proximity, shared color, and shared type of piece. The hypothesis that chunks are modular implies that there would be many more relations between successive pieces within the same chunk than between successive pieces on opposite sides of a chunk boundary. In other words, an analysis of successive placements should show that relations between successive pieces are different depending on whether they are separated by short or long latencies. In addition, assuming that the same cognitive mechanisms underpin the latencies in the copy and recall tasks, the two tasks should show the same pattern of interaction between latencies and number of relations. That is, the relations for the within-glance placements in the copy tasks should correlate with those for rapid placements (≤2 sec) in the recall tasks, and the relations for between-glance placements in the former should correlate with those for slow placements (>2 sec), in the latter.

Figure 15.5 shows, averaged over all subjects (there was little difference between skill levels), the median latencies between the placement of two successive pieces as a function of combination of relations, for the game positions of the copy task. The data are partitioned according to the presence or absence of a glance at the stimulus position.

It can be seen that, within a chunk, small latencies correlate with large numbers of relations, while large latencies occur when there are few relations between successive pieces. No such relationship is apparent for successive pieces belonging to different chunks. The shortest latencies are found with the PCS and DPCS relations, which mainly occur with pawn formations. Moreover, in the within-glance condition, all but one of the latencies are below 2 seconds (the exception is the case with

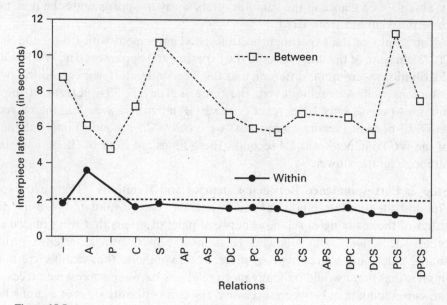

Figure 15.5
Median latencies between the placement of two successive pieces as a function of the combination of relations and the type of placement (within-glance and between-glance). The figure shows the results, averaged across skill levels, for the game positions in the copy task. A, Attack; D, Defense; P, Proximity; C, Color; S, Same type; –, no relation.

game positions where there is only a relation of attack; this case occurs rarely, in less than 0.5 percent of the observations). Statistical analysis confirms that there is a reliable negative linear correlation between number of relations and latencies when the two successive pieces belong to the same chunk, but not when they come from different chunks (Gobet and Simon, 1998a). As expected, the relationship between number of relations and interpiece latency also holds in the recall task.

Step 3: Observed and Expected Probabilities as a Function of Relations The third step consists in establishing whether the pattern of relation probabilities for within- and between-chunk placements differs from what could be expected by chance— that is, given the pair of pieces placed, what is the a priori probability of their relations occurring? As a means to estimate "chance," a priori probabilities for random and game positions were calculated by recording, for each position, all relations that exist between all possible pairs of pieces, then dividing the total number of occurrences of each relation by the total number of possible pairs. The a priori probabilities (for game and for random positions) were based on 100 positions and 26,801 pairs.

Figure 15.6 (upper panel) shows the correspondence between the number of chess relations and the deviations from a priori probabilities, computed by subtracting the a priori probabilities from the observed frequencies of a given condition. In the figure, the results of the recall task have been pooled across skill levels according to whether they are within-chunk or between-chunk. Based on the notion of modularity, it should be expected that the within-chunk deviations from a priori probabilities would be highly correlated with the number of relations, while this would not be the case for the between-chunk deviations. This is exactly what was found. From the figure, it is clear that, for within-chunk conditions, the placements having few relations are below chance, while the placements having several relations are above chance. There is no such clear relation for the between-chunks placements.

The "Glue" Between Chunk Elements

What, then, is the "glue" that keeps chunks together? Or, put differently, which of the five relations largely account for the differences in latencies? For the within-chunk data (pooled over skill levels, tasks, and types of positions), a stepwise regression removes two of the relations (Defense and Attack) from the equation, because they do not contribute significantly to the overall variance. Computed with the remaining relations, the multiple-regression equation is

Latency = 1.75 − 0.27 Same_Type − 0.29 Color − 0.18 Proximity

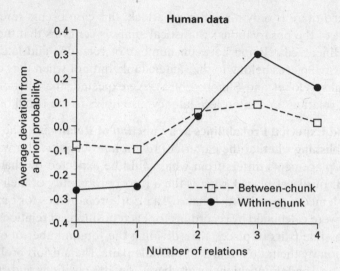

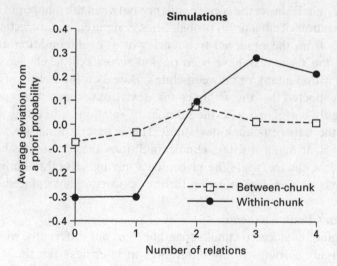

Figure 15.6
Correspondence between chess relation probabilities and the number of relations shared by two pieces, successively placed, in the recall task (results for game and random positions are combined). (Upper panel) Human data (data pooled over skill levels); (lower panel) computer simulations.

This equation accounts for about 65 percent of the variance. By contrast, for the between-chunk data, the stepwise regression removes all relations but Same Type as nonsignificant; furthermore, the regression equation with Same Type as predictor is not statistically significant. These results suggest that the glue keeping pieces so strongly together within a chunk does not apply to pieces belonging to different chunks. Thus, the relations of Same Type, Color, and Proximity are crucial in predicting the latency when successive placements belong to the same chunk, but not when they belong to two different chunks.

The surprisingly weak contribution of the Attack and Defense relations cannot be explained only as an artifact in the multiple regression (for example, that relations of Defense can occur only when a relation of Color is also present); used as a single predictor, Defense accounts only for about 10 percent of the variance, compared with about 25 percent for Color and for Kind. The results are similar when computed separately for each skill level, with the qualification that masters have a comparatively larger regression coefficient, in absolute value, for Color and a smaller one for Proximity. However, the results do not necessarily lead to the conclusion that chess chunks are constructed only by basic Gestalt principles of organization; this is because the relations of Proximity, Color, and Kind are important in the semantics of the chess game.

Hierarchical Structure of Chunks

A final question is whether chunks are organized hierarchically. In this experiment, subjects replaced chunks much larger than in Chase and Simon's (1973) study. For example, with masters in the recall task, the median largest chunk was about seventeen pieces, compared with five pieces in the former study. The small size found by Chase and Simon seems to be explained by the limited number of pieces that the hand can grasp, a limit that is obviously not present with the computer display.

Informal analysis of the large chunks suggests that they are recursively composed of smaller chunks. However, the weak signal-to-noise ratio makes it difficult to pick up hierarchical levels using interpiece latencies.

Summary of the Experiment

The preceding sections have presented techniques identifying modular elements of knowledge in chess players. The operational definition of chunks proposed by Chase and Simon utilized both the latencies between successive piece placements and the semantic relations between them. It was found that the two criteria correlated highly: pieces placed successively with a time interval of less than 2 seconds (or within glances in the copy task) tended to share several relations. Chunks can then be operationalized by either of these criteria. However, while useful for studying statistical regularities, these techniques are not adequate for identifying individual chunks with certainty, because chunks can overlap with or be a subset of other chunks. Other methods, reviewed in Gobet and Simon (1998a), support the concept of chunk arrived at by the analysis presented in this chapter.

Computer Simulations

The overall picture, already present in Chase and Simon's analysis, is that chunks are acquired over years of practice and study within a domain. In chess, chunks allow experts to rapidly recognize known (parts of) positions, and to access information about potential moves or other useful information. CHREST, the cognitive architecture presented earlier, was developed to capture these features. In this section, we are interested in whether CHREST can capture the main features of the empirical data presented in the previous section, in particular the composition of chunks.

Main Processes
During the learning phase, the program is presented with a large database of chess positions taken from masters' games. Fixating squares with its simulated eye movements, it acquires chunks using the discrimination and familiarization mechanisms described earlier. Three nets were selected which, placed in a recall task with a 5-second presentation time, obtained a recall percentage similar to that of class A players, experts, and masters. These nets had 1,000 nodes, 10,000 nodes, and 100,000 nodes, respectively.

For the simulations of the performance phase, the program was tested with 100 game positions and 100 random positions. Learning was turned off during the 5-second presentation of the position. Two sets of processes are crucial: eye-movement mechanisms, and mechanisms managing information encoding and storage in STM. During the presentation of a position, CHREST moves its simulated eye, using instructions from the eye-movement module. Each eye fixation defines a visual field (all squares within two squares from the square fixated); the pieces within the visual field are treated as a single pattern and sorted through the discrimination net. If a chunk is found in the discrimination net, a pointer to it is placed in STM.

During the reconstruction of a position, the information stored in STM is used to place pieces. In the cases when a piece belongs to several chunks, it is replaced only once. When conflicts occur (e.g., several pieces are proposed for the same square),

CHREST resolves them using the frequency with which each placement is proposed. The program may therefore make several different proposals about the location of a piece or about the contents of a square—humans also change their minds. Finally, some heuristics are used; for example, the program knows that there can be only one white king in a position (see Gobet and Simon, 2000, for additional details).

Results

CHREST can reproduce a number of features of the behavior of human players as a function of their skill level, such as their eye movements, the percentage correct, the size of chunks, the number of chunks, the number and types of errors, and the differential recall of game and random positions (de Groot and Gobet, 1996; Gobet and Simon, 2000). Here, I focus upon the pattern of semantic relations of within-chunk and between-chunk placements.

Figure 15.6 (lower panel) shows the average deviation from the a priori probabilities as a function of the number of relations, for both between-chunk and within-chunk placements. As is apparent, CHREST replicates the human data (upper panel) well, showing that its learning mechanisms acquire chunks that have the same relational properties as humans'. Interestingly, much of the data can be accounted for by the fact that most chunks are constructed using information within the visual field.

Conclusion

In this chapter, I have first shown how the concept of modularity is used in different guises in psychology. I have then applied several of its meanings to the CHREST architecture, which has made it possible to provide a crisp, computational definition of knowledge modules (chunks). In order to ground chunks with empirical data, I have discussed Chase and Simon's operationalization method, and have sought evidence for these structures from research into chess expertise. Results confirmed that relations and latencies between pieces offer converging evidence for the psychological reality of chunks.

These results, as well as others, indicate that the recursive structure of the discrimination net used by CHREST captures a key feature of human cognition, and that its redundancy (small variations of the same information can be stored in various nodes) ensures that CHREST is not brittle or oversensitive to details. Since CHREST, a symbolic architecture, essentially learns in a bottom-up fashion, it follows that it is incorrect to equate symbolic processing with top-down processing, as is often done.

What is the relationship between the three types of modularity that have shaped this chapter? In particular, how do functional and knowledge modularity interact within CHREST (or within other symbolic systems)? This is a question on which little research has been carried out. In CHREST, the functional property that most clearly invites modularity at the knowledge level is the limited capacity of short-term memory. Given this limited capacity, complex objects can be represented only as a hierarchy built up from smaller objects. Whether this limited capacity has the same effect on other symbolic architectures, such as ACT-R (Anderson, 1993), or on nonsymbolic architectures, such as those based on neural networks, is a fascinating question which has barely begun to be addressed.

If trying to link two levels of modularity, such as functional modularity and knowledge modularity, opens up Pandora's box, more and harder questions are bound to pop up when additional levels are considered, starting with biological modularity. Given the complexity of these questions, it is unlikely that the human (modular) mind alone will be sufficient to provide a satisfactory answer. Artificial intelligence systems, most certainly modular and perhaps based upon the EPAM/CHREST architecture, will be needed to help this quest.

Acknowledgments

I am grateful to Herb Simon for his involvement in many aspects of this research, and to Peter Lane and Julian Pine for comments on this chapter.

Notes

1. Some theorists, such as Spearman (1927), have proposed general innate determinants of talent. The current trend clearly favors modular determinants; hence, I will focus on these.

2. Most of what I will have to say about CHREST also applies to EPAM. However, to simplify the presentation, I will focus on the former.

3. Even more than modularity, the concept of schema has a number of meanings in psychology. Some of these meanings, particularly those related to computational models, are discussed in Lane et al. (2000).

4. In the analyses, the latencies were corrected for by taking into account the time to move the mouse. See Gobet and Simon (1998a) for details.

References

Allport DA (1980) Patterns and actions: Cognitive mechanisms are content-specific. In: Cognitive Psychology: New Directions (Claxton, GL, ed), 26–64. London: Routledge.

Anderson JR (1993) Rules of the Mind. Hillsdale, NJ: Erlbaum.

Barr A, Feigenbaum EA (eds) (1981) The Handbook of Artificial Intelligence vol 1. New York: Addison-Wesley.

Chase WG, Simon HA (1973) Perception in chess. Cog Psych 4: 55–81.

Churchland PS, Sejnowski TJ (1992) The Computational Brain. Cambridge, MA: MIT Press.

Cowan N (2001) The magical number 4 in short-term memory: A reconsideration of mental storage capacity. Behav Brain Sci 24: 87–114.

De Groot AD (1965) Thought and Choice in Chess. The Hague: Mouton.

De Groot AD, Gobet F (1996) Perception and Memory in Chess. Assen: Van Gorcum.

Ericsson KA, Faivre IA (1988) What's exceptional about exceptional abilities? In: The Exceptional Brain: Neuropsychology of Talent and Special Abilities (Obler LK, Fein D, eds), 436–473. New York: Guilford.

Feigenbaum EA, Simon HA (1984) EPAM-like models of recognition and learning. Cog Sci 8: 305–336.

Fodor JA (1983) The Modularity of Mind. Cambridge, MA: MIT Press.

Gardner H (1983) Frames of Mind: The Theory of Multiple Intelligences. New York: Basic Books.

Gobet F, Lane PCR, Croker S, Cheng PCH, Jones G, Oliver I, Pine J (2001). Chunking mechanisms in human learning. Trends Cog Sci 5: 236–243.

Gobet F, Simon HA (1998a) Expert chess memory: Revisiting the chunking hypothesis. Memory 6: 225–255.

Gobet F, Simon HA (1998b) Pattern recognition makes search possible: Comments on Holding (1992). Psych Res 61: 204–208.

Gobet F, Simon HA (2000) Five seconds or sixty? Presentation time in expert memory. Cog Sci 24: 651–682.

Kosslyn SM, Koenig O (1992) Wet Mind. New York: Free Press.

Lane PCR, Gobet F, Cheng PCH (2000) Learning-based constraints on schemata. In: Proceedings of the Twenty-second Annual Meeting of the Cognitive Science Society (Gleitman LR, Joshi AK, eds), 776–781. Mahwah, NJ: Erlbaum.

Lashley KS (1950) In search of the engram. Symp Soc Exper Biol 4: 454–482.

Minsky M (1985) The Society of Mind. New York: Simon and Schuster.

Newell A (1990) Unified Theories of Cognition. Cambridge, MA: Harvard University Press.

Obler LK, Fein D (eds) (1988) The Exceptional Brain: Neuropsychology of Talent and Special Abilities. New York: Guilford.

Richman HB, Staszewski JJ, Simon HA (1995) Simulation of expert memory with EPAM IV. Psych Rev 102: 305–330.

Saariluoma P (1992) Visuospatial and articulatory interference in chess players' information intake. Appl Cog Psych 6: 77–89.

Simon HA (1969) The Sciences of the Artificial. Cambridge, MA: MIT Press.

Spearman C (1927) The Abilities of Man. London: Macmillan.

Tabachneck-Schijf HJM, Leonardo AM, Simon HA (1997) CaMeRa: A computational model of multiple representations. Cog Sci 21: 305–350.

Zhang G, Simon HA (1985) STM capacity for Chinese words and idioms: Chunking and acoustical loop hypothesis. Mem Cog 13: 193–201.

16 Modularity of Cognitive Organization: Why It Is So Appealing and Why It Is Wrong

Boris M. Velichkovsky

All animals are equal but some animals are more equal than others.
—George Orwell

"Module" and "modularity" are frequent words in the vocabulary of contemporary psychology and cognitive neuroscience. This is a symptom of the currently dominant scientific paradigm, which is based on the search for double dissociations and such underlying mind/brain mechanisms that are presumably independent of each other. The modularity-of-mind hypothesis was formulated in the 1980s on a wave of skepticism about the possibility of a single design for different cognitive phenomena. This "grand design" of the 1960s and 1970s was a technologically influenced distinction of the parts of a conventional computer—input and output interfaces, limited central processing unit (known as short-term or working memory), and passive information store with virtually unlimited capacity (long-term memory). But should doubts about the computer metaphor architecture be kept apart from the denial of any "grand design"?

In answering this question, we will concentrate, first of all, on the modularity-of-mind paradigm, which can be interpreted as a conception of random (or flat) mosaics of partially autonomous mechanisms. There is increasing evidence, however, that in many cases the relationship between alleged "cognitive modules" is not as symmetrical as it should be in a perfect parallel architecture. Rather, elements of gradient-like or hierarchical organization of processing are clearly visible, both in psychological data and in results of neurophysiological studies (Craik et al., 1999; Gabrieli et al., 1996; Velichkovsky, Klemm et al., 1996). Our analysis of the ongoing research in several domains, such as memory, perception, and control of eye movements, demonstrates that the idea of a vertical modularity or stratification of cognitive organization is more suitable to explain actual controversies.

We will describe outlines of such a design based on the concept of levels of processing. These emerging multilevel models of human cognition will be discussed from the point of view of the evolutionary roots of our cognitive abilities (for a corresponding analysis of affective processes, see Panksepp and Panksepp, 2000) as well as of basic dilemmas of activity organization in a changing biological and social environment. The purpose of this chapter is to make explicit this general approach aimed at a reconstruction of hierarchical evolutionary mechanisms of mind and brain (1) with respect to some of its theoretical contexts and (2) with respect to new empirical data.

What Are "Modules" in Psychology and Cognitive Science?

There are three and a half different understandings of modules in contemporary cognitive research. The first is relatively unspecific and simply means "functional parts." This is the interpretation contained in David Marr's (1976, 1982) principle of modular design. Marr proposed that

Any large computation should be split up and implemented as a collection of small sub-parts that are nearly independent of one another. . . . If a process is not designed in this way, a small change in one place will have consequences in many other places. This means that the process as a whole becomes extremely difficult to debug or to improve, whether by a human designer or in the course of natural evolution, because a small change to improve one part has to be accomplished by many simultaneous compensating changes elsewhere. (Marr, 1976, p. 485)

Such relatively separate parts are perfectly conceivable in the computer metaphor architecture. For instance, one could treat short-term and long-term memory stores of early cognitive psychology as different "modules." No one—not even followers of connectionism—will contest this interpretation of modularity in principle. The only question may be how stable these parts are and whether they may be transformed by different conditions and subject's strategies. At least in some cases, these parts could be modified by some conditions, such as workload and fatigue (Leonova, 1998), and, of course, by developmental changes (Karmiloff-Smith, 1992). It is this broad interpretation of "modules" which is related to Herbert Simon's idea of near decomposability (in the foreword to this volume and elsewhere). It is shared by the vast majority of experimentally working authors. There are also different attempts (see, e.g., Sternberg, 1998, as the latest one) to define general criteria for something being a separate part of a process or a mechanism. We do not think, however, that such a loose understanding is of any immediate relevance for this volume's interdisciplinary concerns.

The second interpretation of modularity is more specific. It amounts to the claim that there is no single multipurpose mechanism of cognition, but several—perhaps many—mechanisms, which are domain-specific and, in a sense, parallel. The theoretical background of this understanding is not clearly delineated; one simply expects differences in performance and possibly in mechanisms in different areas of inquiry. Increasingly more data of the sort were collected with the progress of cognitive research. For instance, memory in nonverbal domains of naturalistic pictures, sounds, and odors seems to be rather different from the memory for words or alphanumeric sequences. Any such evidence of domain-specificity has (negative) consequences for general models like those advocated by the first generation of cognitive scientists, Newell and Simon (1972) among them; but this does not imply some definite new theoretical view of underlying architecture.

However, there is also a third understanding—the one that became extremely popular since the 1990s. One should thank Jerry Fodor (1983) for formulating this view of the modularity of mind. It includes both above interpretations but goes farther in describing a set of criteria of cognitive modules. These criteria are domain specificity (sic!), informational encapsulation, high speed, mandatory character, shallow representation as product, specific pattern of disturbances, fixed neuroanatomical localization, and a similar pattern of growth in different species (which implies cognitive modules' inborn character). Now, according to Fodor, all these features are those of input systems. He also postulates the existence of central systems that use shallow perceptual representations delivered by cognitive modules for unlimited and unrestricted associations. These central components of cognitive architecture are, after Fodor, nonmodular and isotropic. What we previously called the "half" interpretation is the attempt of some of the radical followers of Fodor —for instance, from evolutionary psychology (Cosmides et al., 1997; Tooby and Cosmides, 2000; see also Panksepp and Panksepp, 2000)—to extend the modularity idea to the whole set of cognitive processes, including those which Fodor would count as nonmodular "central systems."[1]

Note that the last, Fodorian interpretation includes the previous ones. In the following, we will concentrate on an analysis of domain specificity, which seems to be better testable empirically than the highly disputed claims about the inborn characters of underlying mechanisms (e.g., Elman et al., 1996). In other words, we will basically ignore some of the data relevant to the modularity discussion; for example, countless demonstrations of inherent plasticity of the brain's functional mechanisms (see Deacon, 1996; Sterr et al., 1998). Our strategy will be very simple—any flaw discovered in the domain-specificity arguments would also mean criticism of the strong interpretation of modularity. Let us now sketch the outlines of an alternative theory of cognitive architecture based on evolutionary data and akin to some general notions of hierarchical organization formulated by Herbert Simon in the foreword to this volume.

Horizontal Versus Vertical Modularity: The Multilevel Model

In the history of psychology there have been several attempts to speak about a hierarchy of psychological mechanisms—for instance, by Karl Bühler (1918), who postulated three stages of a child's mental development—instinct, learning, and intelligence—and by Jens Rasmussen (1986), who discussed, in the context of human factors studies, three different groups of regulatory mechanisms: skill-based, rule-based, and knowledge-based. In all these cases no method was proposed to differentiate contributions or effects of the postulated structural units. Craik and Lockhart (1972) first coined the notion of levels of processing (LOP) within the cognitive community. In their paper they expressed doubts about the usual (within the computer-metaphor paradigm) consideration of memory as a structure of interconnected stores or boxes. These authors developed rather simple method to demonstrate LOP effects empirically. Certainly this could be one of the reasons why their paper became the second most often cited publication in experimental psychology (Roediger and Gallo, 2001).

In a typical LOP experiment, subjects are at the beginning unaware of memory context of the study, and simply have to process stimulus material, usually lists of words but sometimes also nonverbal information, under a "shallow" or a "deep" encoding instruction. Shallow encoding is related to processing of perceptual, form-related features of the material (e.g., notifying of all letters of a word graphically deviating from the line, like "p" and "h"). Deep encoding is a semantic categorization of the word. After some temporal delay, subjects are confronted with a sudden memory test such as recognition, or free or cued recall. What has been found in several hundred such experiments is better memory performance after deep, as compared to a shallow, encoding (the LOP effect). Theoretically, memory is interpreted here as a by-product of a subject's activity, which can evolve on at least two levels. Unfortunately, for a long time this was a purely functional interpretation, without any reliance on the possible brain mechanisms of encoding effects. This bring us closer to another long tradition—that of evolutionary consideration of different brain structures, which was started by John Hughlings-Jackson and continued by Alexander Luria and Paul MacLean (see Velichkovsky, 1990, 1994).

In particular, the founder of modern biomechanics and one of the early protagonists of ecological approach in modern psychology, Nikolai Bernstein, published a book whose title translates as "On the Construction of Movements" (1947; reprinted 1990) describing in detail four levels of neurophysiological mechanisms, from level A to level D, involved in coordination of human movements.[2] While Bernstein's emphasis was on the sensorimotor processes and not on cognition, he mentioned the possibility of "one or two higher levels of symbolic coordinations" selectively connected with associative areas of cortex and with the frontal lobes. To show the sophistication of his multilevel approach, we provide a translation of the table of contents of this remarkable book in appendix 1 and some of the introductory pages, relevant for the present discussions, in appendix 2.

This case of parallel development is more than a historical curiosity—it shows that there may be reasons for a reoccurrence of similar ideas in a different scientific context. It may be useful to spell out the emerging architecture once again, in order to make it clear which segments of the hierarchical brain mechanisms are associated with the typical experimental manipulations. As many as six global levels of organization can be differentiated if one takes into account both Bernstein's work on the construction of movements and the recent progress in studies of cognition and the higher forms of consciousness (Wheeler et al., 1997; Velichkovsky, 1990). The first group (levels from A to D) is built up by the primarily sensorimotor mechanisms clearly identified by Bernstein (1947/1990). The second group of mechanisms (E and F) consists of two levels anticipated by Bernstein as higher levels of symbolic coordinations.

Level A, *paleokinetic regulations.*   Bernstein called this also the "rubro-spinal" level, having in mind the lowest structures of the spinal cord and brain stem regulating the tonus, paleovestibular reflexes, and basic defensive responses. The awareness of functioning is reduced to Head's (1920) prothopathic sensitivity, so diffuse and lacking in local signs that even the term "sensation" seems to be too intellectual in this case.

Level B, *synergies.*   Due to involvement of new neurological structures—the "thalamo-pallidar system," according to Bernstein—the regulation of an organism's movements as a whole is now possible; it becomes a "locomotory machine." The specializations of this level are rhythmic and cyclic patterns of motion underlying all forms of locomotion. Possibilities of awareness are limited to proprioceptive and tangoreceptoric sensations.

Level C, *spatial field.*   A new spiral of evolution adds exteroception with the striatum and primary stimulotopically organized areas of the cortex as the control instances. This opens outer space and makes possible one-time extemporaneous goal-directed motions in the near environment. The corresponding subjective experience is that of a stable, voluminous surrounding filled with localized but only globally sketched objects.

Level D, *object actions.*   The next round of evolution leads to the building of secondary areas of neocortex that permit detailed form perception and object-adjusted manipulations. Individualized objects affording some actions but not others are the focus of attention. Formation and tuning of higher-order sensorimotor and perceptual skills are supported by a huge procedural-type memory. Phenomenal experience is the perceptive image.

Level E, *conceptual structures*. Supramodal associative cortices provide the highest integration of various modalities supporting the ability to identify objects and events as members of generic classes. Development of language and human culture foster this ability and virtually lead to formation of powerful declarative-procedural mechanisms of symbolic representation of the world. Common consciousness is the mode of awareness at this level.

Level F, *metacognitive coordinations*. In advanced stages, changes in conceptual structures result not only from accretion of experience but also from experimentation with ontological parameters of knowledge. Necessary support for this "personal view of the world" is provided by those parts of the neocortex that show excessive growth in anthropogenesis, notably by the frontal and right prefrontal regions. This level is behind personal and interpersonal reference, reflective consciousness, and productive imagination.

There can be an abundance of domain-specific effects, even apparently modular components, in such a type of architecture. Furthermore, most of these levels are simultaneously providing their specific competences and resources to the task solution. What makes a difference to the modular theories is that these contributions are not all equivalent by virtue of their being embedded in a kind of *gradient* of larger-order evolutionary mechanisms, so that *systematic asymmetries* in relations of local, domain-specific effects should be expected.

Levels "Below" Perceptual (Form-Oriented) Processing

The concept of levels presupposes some consistency of effects across different modalities, domains, and modes of processing. In a time dominated by the search for presumably isolated mechanisms, this multilevel paradigm plays a clear integrative function. According to Fodor (1983), cognitive modules are fast and localized mostly at the input side of cognitive machinery. Let us consider first of all experimental results that are related to the functioning of one of the oldest and fastest biological "input systems." An exciting chapter of evolutionary biology is the analysis of behavioral effects of a sudden change of situation—for instance, when certain butterflies display their colorful wing patterns to *psychologically immobilize* and escape from a predator (Schlenoff, 1985). In a series of experiments we (Pannasch et al., 2001; Unema and Velichkovsky, 2000) investigated some influences of such changes on eye movements, perception, and memory in human beings.

The starting point for us was the *distractor effect*, a phasic "freezing" of visual fixation (i.e., a delay of the next saccade) in the case of a sudden visual event

(Levy-Shoen, 1969; Findlay and Walker, 1999). It is currently explained in a strictly modular way as a oculomotor reflex involving structures of the superior colliculus (Reingold and Stampe, 2000). This is the lowest-level circuit (corresponding to level A) for eye movement responses. The remarkably short latency of only 100 ms, the fastest behavioral responses of the human organism, leaves no alternative to the proposed subcortical origin of this effect. However, we found that the distractor effect can also be produced—during exploration of visual scenes and in simulated driving—by acoustic events in which the response time obtained is the same as or even faster than that obtained with visual cues. In addition, we demonstrated a new *relevance effect*—the rise in duration of the subsequent fixation when a change is relevant to the activity of the subject—for instance, when subject should brake in response to some of them. This second effect could be observed at longer latencies of 300 to 400 ms (figure 16.1).

Two lessons can be learned from these data. First, observation of behavior (via high-resolution eye tracking) seems to demonstrate information-processing stages in approximately the same way as event-related potentials reveal them. The fast oculomotor responses to changes in stimulation, manifested in distractor effect, are similar to the early N100 component of ERPs. Although this may be a numerical coincidence, our relevance effect corresponds to another known aspect of ERPs: that semantic and metacognitive variables usually influence the relatively late components P300, P3b, and N400 (Kutas and King, 1996; Simons and Perlstein, 1997). The second lesson is that though these effects clearly belong to the realm of "input systems" (see Fodor, 1983), they are not domain-specific and ipso facto not modular; rather, they are intermodal and hierarchically organized. This similarity across domains, and particularly the presence of the vertical dimension, make the phenomenon interesting for the stratification approach. In fact, our recent results testify that relevance leads to a better reproduction of information, even if critical events are produced in a close temporal vicinity to other changes, saccades or blinks (Velichkovsky et al., 2002).

There is growing evidence that visual fixations, depending on their durations, can be controlled by different hierarchically related mechanisms. Some fixations have extremely short durations of about 100 ms or less (Velichkovsky et al., 1997). They can be produced by the same fast biological mechanism that is responsible for the distractor effect and for other, mostly inhibitory, manifestations of orienting and startle reactions (Sokolov, 1963). A rather different, partially reciprocal type of behavior can be observed in the case of the longest fixations, 500 to 1000 ms. First of all, they are underrepresented during changes and at the beginning of presentation of new visual information, as well as during any additional cognitive load (for

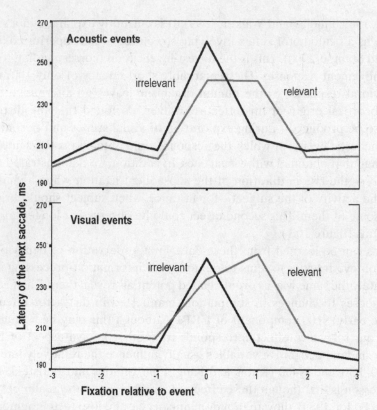

Figure 16.1
Distractor and relevance effects manifested in duration of visual fixations for two modalities of events.
(After Unema and Velichkovsky, 2000.)

example, in the phone communication that is simultaneous with visual processing). Second, the proportion of these longer fixations steadily grows up to the end of every period of presentation that can be treated as a sign of an expectation of change. All in all, there can be up to four functionally specific groups of fixations, according to their duration, in a seemingly unified distribution (Deubel, 1999; Velichkovsky, 1999).

The investigation of free visual exploration of dynamic scenes and pictures, which is not restricted to abrupt repetitive trials, is a challenge for experimental psychology. A differentiation of several levels in picture processing may rely on one of the first distinctions in the field: that between ambient and focal visual processing, introduced by Trevarthen (1968). This insightful distinction has been largely forgotten due to later proposals stressing cortical localization of spatial vision in humans. The situation has begun to change again more recently, with increasing acceptances of the role of subcortical mechanisms, particularly the striatum. The distinction between ambient and focal processing is consistent with our general levels model: it divides level C from level D and higher symbolic mechanisms (Bridgeman 1999; Velichkovsky, 1982). Finally, the same distinction seems to be closest to the current connectionist and evolutionary modeling of perception (Rueckl et al., 1989; see also chapter 14 in this volume).

In order to separate ambient and focal visual processing, it is useful to consider the relationships between duration of fixation and amplitude of the next saccade (Velichkovsky et al., 2000). Some of fixations *are followed* by saccades that are at least twice as large as the radius of the fovea (>4°). This means that these saccades cannot be oriented by any detailed or "focal" visual representation of objects. Interestingly, corresponding fixations also have a relatively short duration, from 120 to 250 ms. If the spatial distributions of these and longer fixations over the surface of a picture are computed separately and applied as filters to process the picture itself (see Velichkovsky, Pomplun et al., 1996), one gets the representations shown in Figure 16.2b and c. (The task of the subjects in this particular case consisted in evaluation of social relations of the actors presented in this scene). These representations can be called *ambient* and *focal views* of the picture, respectively.

It is important for the LOP approach that the surface covered by the ambient (or "shallow") fixations not only is much larger than in the focal counterparts, but also that it remains relatively constant with different encoding instructions. On the contrary, spatial distribution of the focal (or "deep") fixations seems to be rather sensitive to processing instructions (i.e., it is clearly task-dependent). This is why the method of "attentional landscapes filtering" seems to provide us with a reconstruction of subjective perspectives which different hierarchical levels have on the same situation *at the same time*.

Thus, data on eye movements provide strong evidence on some additional, earlier forms of processing, preceding that of form-oriented (level D) or "perceptual" encoding from the classical LOP studies. Its function can be the detection of changes in stimulation and initiating of startle and orienting responses. The extraordinary speed and a distinct peak in the otherwise relatively smooth distribution of fixation durations (Velichkovsky, 1982, 1999) speak for a deep subcortical localization of corresponding mechanisms, somewhere at the top of Bernstein's level A (e.g., Fischer and Weber, 1993; Reingold and Stampe, 2000). Other early forms of processing, revealed by the same eye-tracking experiments, can be associated with level C of our global model (i.e., with a global spatial, or "ambient," localization of sensory events).

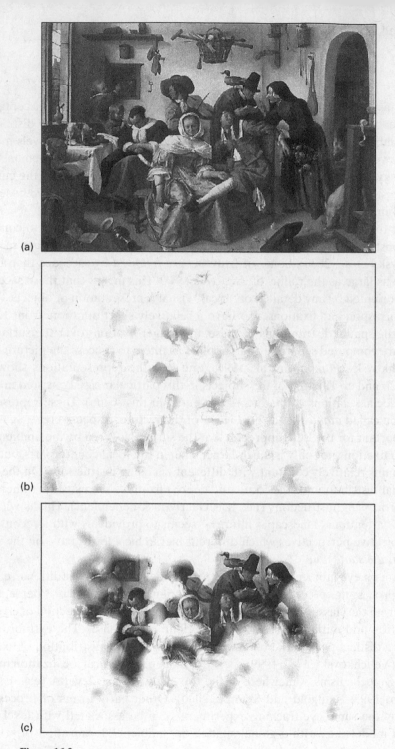

(a)

(b)

(c)

Figure 16.2
The Twisted World, by Jan Steen (1663): (a) copy of the original; (b) ambient view; (c) focal view (see text for details).

There Are Forms of Processing "Above" Semantic Categorization

In the multilevel conceptualization of Craik and Lockhart's (1972) initial ideas, human memory is in service and is a part of global, hierarchically organized activities that can be related (though very approximately) to different forms of processing, from sensory and sensorimotor to semantic and metacognitive (see Velichkovsky, 1999). The usual LOP manipulations—shallow versus deep—can be identified with processes involving mainly levels D and E.

An objection to the LOP approach was related to the initially neglected role of interactions between encoding and memory tests. These considerations led to concepts such as transfer-appropriate processing and encoding specificity (Roediger and Gallo, 2001). According to this variant of domain-specific/modular thinking, there is no inherent continuum of higher versus lower levels, but several qualitatively different forms of processing that are parallel. With respect to memory, the prediction was that even "shallow" encoding can demonstrate a better retention than "deep" encoding if the specific memory test initiates retrieval processes similar to (e.g., in terms of overlap between component mental operations) the processes which have been involved at encoding stage.

Indeed, there have been several demonstrations of a trend in this direction (e.g., Morris et al., 1977), but even with an appropriate "shallow" test the amount of memory performance seems to be lower than after an appropriate "deep" test such as free recall or recognition (Fisher and Craik, 1977). In other words, empirically some types of encoding ("shallow") never yield excellent recognition or free recall with any retrieval cues, however "specific" or "appropriate." We shall soon return to this particular problem when discussing experiments specially designed for the comparison of interactions of encoding and retrieval conditions.

Empirical data deviating from the expectations of the LOP approach in its initial form have been found particularly often with the broad introduction of implicit or indirect tests of memory. In this type of task, subjects are unaware that their memory is tested, so that only a performance change (priming) in the next confrontation with the task is of importance. As a rule, these tests demonstrate few if any influences of encoding manipulations on the corresponding (perceptual) priming measures (Roediger et al., 1992). Even when post-hoc meta-analytic studies (Challis and Brodbeck, 1992) were able to show a tendency to stronger (perceptual) priming with semantic encoding, this effect was weak and less systematic than in direct memory measures. The most parsimonious explanation of the residual influence of manipulation of levels is that of explicit awareness of some subjects in some experiments of the proper goal of implicit tests (Stadler and Roediger, 1997). However, more recent data cast doubt on this simple option (Gardiner et al., 2001); thus the differential influence of the LOP manipulations on explicit and implicit tests will be one of the questions discussed in this chapter.

Self-referential encoding is another challenge; it can have even larger memory effect than semantic encoding (e.g., Bower and Gilligan, 1979; Miall, 1986). The growing evidence for metacognitive processing lends itself to an integration with the concept of levels, particularly with respect to this controversial issue (see Lockhart and Craik, 1990; Velichkovsky, 1994). Is it a genuinely new level, or only an elaboration of the old? More for theoretical than empirical reasons, we included it as a separate level F in the general hierarchical model. The question could be empirically answered by contrasting neurophysiological mechanisms behind semantic and self-referential encoding or, as this distinction was called in Russian activity psychology (Leont'ev, 1978), behind representations of *meaning* and *personal sense*. Unfortunately, as already mentioned, until recently LOP effects were interpreted with a functional bias (i.e., not as effects involving different neurophysiological structures).

In this respect, memory processes obviously have different "functional anatomy." The *leading* level (see appendix 2) of all the direct (explicit) tests—whether they are called "perceptual" or "conceptual" tests—usually is the same level F. It is not surprising, therefore, that one mostly finds a right prefrontal activation (the "retrieval volition" component; e.g., Gardiner et al., 2001) with this group of memory tasks. They are, however, different in their *background* coordinations (appendix 2), so that, for instance, recognition and free recall—both conceptual explicit tests—differ (1) in their use of level E facilities and (2) in the involvement of levels D and C with only recognition and not free recall. As to indirect (implicit) memory tests, their levels composition is defined by the content of the nonmemory "carrier" tasks—the leading levels of these tasks are D or E (if subjects do not discover that their memory is tested—discovery would immediately involve level F's coordination and make the test explicit). Thus, the list of levels involved in the implicit memory tasks usually is shorter and the projection of their mechanisms on the vertical dimension of mental processing is *narrower* than in the case of explicit tasks.

Before the discussion of experimental results, one methodical drawback of the most recent studies of memory should be mentioned. This is the 2 × 2 experimental design. The problem with such experiments is that their results are suitable only for rather rudimentary theoretical considerations: dichotomous distinctions, the search for double dissociations or for "in principle" interactions of encoding and retrieval tasks.

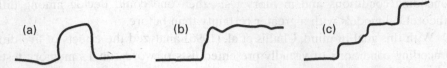

(a)   (b)   (c)

Figure 16.3
Three hypothetical functions of memory performance depending on encoding conditions (see text for details).

Figure 16.3 shows three hypothetical functions of performance in a memory task involving dependence on encoding conditions, arranged according to the hypothetical "vertical" dimension. Though extremely simplified (for a detailed analysis—see Challis et al., 1996, table 1), these cases cannot be differentiated by the typical variation of two encoding conditions. Unfortunately, all three cases are meaningful and have different theoretical implications. The nonmonotonic function (a) is problematic for the LOP approach because it demonstrates a strong domain specificity, or even modularity of the respective encoding-retrieval coupling. The next case, that of the "one-step" or "flat" LOP function (b) combines predictions of the basic approaches: one the one hand, there is a difference in efficiency of "shallow" and "deep" encoding, and on the other hand, when one of the encoding conditions improves memory performance, this contribution seems to be preserved in all the higher levels of processing—an interesting example of the "modular voice" within a hierarchy. The last function (c) is what all true "believers in the hierarchical order" would expect. However, this case is ambiguous. One possibility (the replacement model) is that the gradient-like function is produced by consecutive involvement of new, more powerful mechanisms that demonstrate their performance without interacting with lower-level structures. Another alternative is that of *vertical integration* (the integration model): memory performance at some "deep" level of processing integrates contributions from previous levels.

Blaxton (1989) was the first to investigate a large group of memory tests under the same experimental conditions. In this study, she found that there are no essential differences between explicit and implicit tests if one other dimension is taken into account: the emphasis on perceptual versus conceptual (semantic) processing. One aspect of her study, the difference in behavior of perceptual implicit tests and conceptual implicit tests, has been confirmed in several later experiments (e.g., Tulving and Schacter, 1990). Though these results seem to be relevant to the present discussion, Blaxton did not manipulate encoding conditions. If it would be possible in the same experiment to vary—systematically and in a broad range—both encoding conditions and memory tests, then one could decide among different theoretical models with a greater certainty than before.

With this goal in mind, Challis et al. (1996) analyzed the effects of five different encoding conditions of visually presented lists of words on 13 memory tests (see also Velichkovsky, 1999). The memory tests were selected according to major theoretical points of view, such as perceptual versus conceptual and implicit versus explicit. With the large matrix of encoding and retrieval conditions, several predictions are possible. From the strong domain-specific or modular view, one would expect several more or less equivalent clusters of interaction between encoding and retrieval in memory performance. That is, performance would be relatively high when encoding and retrieval processing "matched," and relatively low elsewhere. From the LOP view, one would expect the clusters of interaction to be asymmetric through the creation of something like a gradient from weak to strong and perhaps even stronger memory effects.

Our data, shown in figure 16.4, were quite compelling in relation to these predictions. To see the trends clearly, one should, first of all, reorder incidental encoding conditions in the sequence of perceptual processing (counting of letters deviating in form), phonological processing (counting of syllables), semantic processing (categorizing as living thing), and metacognitive processing (evaluation of personal sense). Second, explicit and implicit memory tests have to be considered separately. It is in the first group of four explicit memory tasks (recognition, free recall, semantically and graphemically cued recall) where resulting interactions demonstrate a systematic, gradient-like growth—the perfect LOP effect—across at least most of the higher-order encoding conditions.

Thus, when memory tests are explicit, performance functions are of type (c) in figure 16.3, but in implicit tests they resemble case (b). Indeed, priming functions in this last group of tests, which were word-fragment completion and general knowledge tests, look different: they are much flatter, or there seems to be no effect at all. Some systematicity can, however, be discovered with respect to the point in the row of encoding conditions where variation in encoding starts to influence memory performance. In a general knowledge test, this starting point is the semantic encoding. In the word-fragment completion test, priming is present in all encoding conditions. One can argue, in the spirit of a mixture of hierarchical and modular ideas, that the crucial influence is already present at the stage of perceptual encoding (our level D), and it is included in all further, higher-level encodings as well.

Empirically, the more traditional case of intentional encoding finds its place somewhere between semantic and metacognitive conditions. It is quite amazing that in all but one case, incidental metacognitive processing leads to better memory

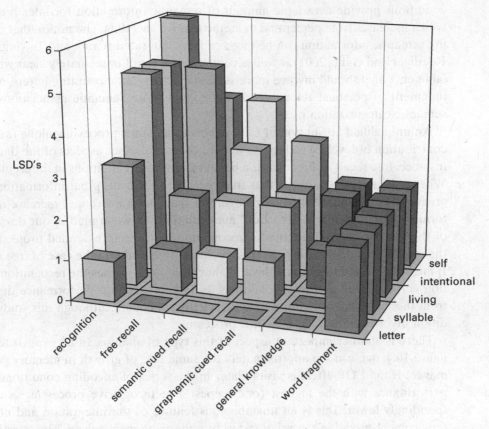

Figure 16.4
The number of statistically significant deviations from baseline (least significance differences) on six memory tasks dependending on five encoding conditions (after Challis et al., 1996).

performance than even intentional encoding. The only deviation is found in the free-recall task.[3] This small, albeit significant, deviation from the overall gradient-like picture can be attributed to the fact that intentional instructions invoke organizational strategies of particular benefit for free recall (Brown, 1979). Intentional encoding is, however, plastic and corresponds to the actually expected test. To emphasize this small effect as the case for the transfer-appropriate principle would mean simply not seeing the forest behind *one tree*.

This pattern of results is compatible with the hierarchical view of underlying mechanisms and also supports the broad idea of multiple levels of processing. A dichotomous interpretation of our data would be that incidental encoding conditions provide a variable amount of semantic information for later retrieval, which increases from perceptual to metacognitive encoding. The notion that encoding semantic information can be more or less elaborated (Craik and Tulving, 1975; Roediger and Gallo, 2001) is well accepted, but it is not immediately clear why syllable encoding should involve more semantic activity than counting letters, or why judgment of personal relevance should recruit more semantic associations than semantic categorization itself.

An unqualified attribution of LOP effects to semantic processing alone (i.e., to a contribution of level E) is dubious in light of several other aspects of the data. For instance, free recall and recognition benefit from study conditions in different ways. Whereas it might be argued that free recall utilizes conceptual information to a greater degree, empirically it is recognition that shows a stronger increase in performance as a function of the LOP manipulation. As we argued in our discussion of the general hierarchical model, recognition has a more extended projection to the multilevel ladder: from C to F—in contrast to E and F in the case of free recall. It might be the contribution of level D that additionally raises the recognition function across all the encoding conditions. In fact, recognition performance deviates from the baseline even after perceptual (letter) encoding, although this study condition involves little if any semantic processing.

There is another important aspect of this type of analysis. In all explicit tests in figure 16.4, one can see approximately the same rate of growth in memory performance: If the LOP effect is visible later in the series of encoding conditions, then performance with the highest (or "deepest") metacognitive processing is correspondingly lower. This is an unambiguous feature of the integration and not the replacement model in the relationship of underlying mechanisms. The question is, of course, why there are no similar signs of integration in implicit tasks. Perhaps implicit retrieval tasks address very narrowly tuned mechanisms—for instance, only level D or E, so that there is simply no basis for such an integration. If the patterns of flat or one-steplike LOP functions found with implicit tests are connected with activation of relatively narrow segments of the hierarchical cognitive ladder, then it might be possible to produce similar results *even with explicit tasks* when their retrieval criteria are highly restrictive, narrowing mechanisms involved to a single level of processing.

One of our experiments (Challis et al., 1996, exp. · 3) provides support for this interpretation of data from explicit and implicit memory tests. After studying lists of words under the same five encoding conditions, subjects were presented with the test material, which consisted of the new words, 50 percent of which were graphemically similar and 50 percent semantically similar to those previously shown. In three parallel groups of subjects, the tasks were to find words similar to previously presented words graphemically, phonologically, or semantically. Although these tasks were explicit, the results presented in figure 16.5 demonstrate the whole set of flat and one-step LOP functions that in all previous studies were typical of implicit tests. Again, it is the encoding condition where memory performance starts to deviate from the baseline, which is important in showing the relation to underlying hierarchical architecture. This starting point shifts for graphemic, phonological, and semantic similarity tests to correspondingly higher levels of encoding.

In general, our results demonstrate that in the core area of cognitive research (psychological investigation of memory), the stratification approach can produce data that are surprisingly consistent—indeed, to a degree that is rare in psychological research. This consistency is a strong argument against reducing multilevel memory effects to idiosyncratic domain-specific phenomena, such as transfer-appropriate processing.

Parts Are Embedded in the Gradients of Phylogenetic Growth

In response to the natural questions about the neurophysiological basis of LOP effects, there have been a number of attempts to find their independent measures or correlates in psychophysiological data, such as event-related potentials (Naumann, 1985; Sanquist et al., 1980) and cardiovascular reactions (Vincent et al., 1993). Particularly compelling are recent studies of memory encoding and retrieval using brain imaging.

In one of the first such neuroimaging studies, Blaxton and her colleagues (Blaxton et al., 1996) conducted a PET-scan analysis of regional cerebral blood flow in four memory tests. Memory tasks for the study were chosen in such a way that they represented two theoretically important dimensions: perceptual versus conceptual processing and implicit versus explicit testing. Larger topographical changes were found between perceptual and conceptual tasks. Memory effects for perceptual fragment-completion tests (both implicit and explicit) were localized in posterior regions including occipital cortex with some slight right-side asymmetry. In contrast, the analysis of conceptual tests of semantic cued recall and word association revealed metabolic changes in the medial and superior temporal cortex as well as in the left frontal cortex. Of course, these neuroanatomical differences have been found with respect to localization of retrieval processes, so it can be interesting to compare them with results of encoding studies.

Gabrieli et al. (1996) used functional magnetic resonance imaging (fMRI) for a comparison of perceptual and semantic encodings of visually presented words. They

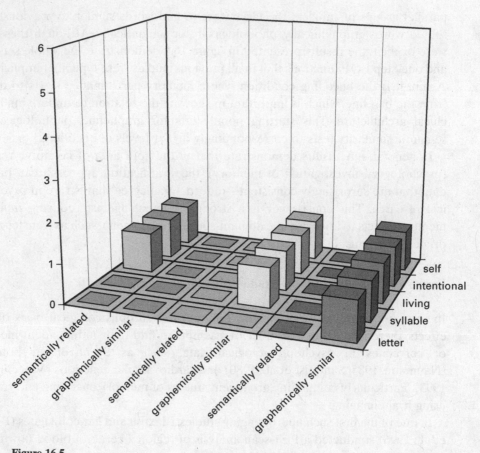

Figure 16.5
The number of significant deviations from baseline (least significance differences) in three new explicit memory tasks in dependence on five encoding conditions. (After Challis et al., 1996.)

discovered a greater activation of left inferior prefrontal cortex for semantic encoding. This makes it possible that the level I have called "conceptual structures" (or level E) should have connections with the left frontal regions and perhaps with the temporal lobes of the cortex.[4]

One of our investigations of encoding has been based on a psychophysiological method known as evoked coherences analysis of EEG—a cheaper and a faster alternative to PET scan (Velichkovsky, Klemm, et al., 1996). In a series of LOP experiments, we investigated reproduction of visually and acoustically presented words in terms of dependence on three encoding conditions: perceptive, semantic, and metacognitive (self-referential). Although the database of EEG analysis is completely different from that of PET scan or fMRI, the loci of global incoherence[5] in perceptual and semantic orienting tasks were found in approximately the same regions where Blaxton et al. (1996) discovered significant change in metabolism for the corresponding memory tests. In the perceptual (form-oriented) encoding of visually presented words, the major incoherences are localized in the occipital and right occipito-temporal area. In semantic encoding, they expand to the more anteriorly located region, particularly bilateral temporal and left frontal areas (see similar data from another investigation of neuroanatomical correlates of LOP effects by Kapur et al., 1994). In the third condition–self-referential encoding—even more anteriorly located regions within frontal and right prefrontal lobes are involved.

The trend from the left anterior localization in semantic encoding to a predominantly *right prefrontal* activation (especially Brodmann's areas 9, 10, and 45) in self-referential encoding has been confirmed in a recent PET study comparing loci of self-referential encoding with those of other-referential and phonological processing (Craik et al., 1999).

This overall picture contradicts some attempts of attributing encoding exclusively to the left and (episodic) retrieval to the right frontal lobes (Tulving, 1998; Tulving et al., 1994)—an exciting idea from the modularity of mind perspective, perhaps, but also one for which we can hardly see any evolutionary justification. At the same time, the described neuroanatomical changes are not at all astonishing from the point of view of neuropsychology (Goldberg, 1991; Luria, 1966) and neurophysiology (Christoff and Gabrieli, 2000; Mishkin et al., 1999). Theoretically important, in our opinion, is not the neuroanatomical change *per se*, but its direction. The posterior-anterior gradient corresponds to the main direction of evolutionary growth of the cortex (e.g., Deacon, 1996) as it is visualized in figure 16.6.

In other words, our data suggest that a purely functional interpretation of LOP effects does not go far enough and should be revisited, perhaps, along with a simple

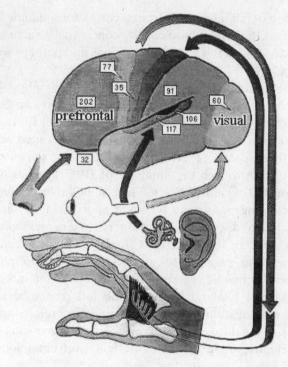

Figure 16.6
Relative growth (in percents) of different areas of human cortex in anthropogenesis. (After Deacon, 1996.)

principle: The "deeper" (or "higher") a particular "level of processing" is, the more massive is an involvement of *phylogenetically recent* brain mechanisms in the task's solution. This correspondence of genetic, functional, and structural parameters is of course only a heuristic rule—our knowledge of brain functional evolution is too fragmentary. However, the rule may work in several other cases. Let us show this with one additional example.

Until now we have discussed the posterior–anterior gradient. There are other lines of brain evolutionary development, such as one reflecting differences between subcortical and cortical mechanisms (for such analysis of motor control, see Bernstein, 1947/1990, 1996, and appendix 1 in this volume). One of the peculiarities of human cortex is a relatively strongly developed right prefrontal lobe (Holloway and De la Coste-Lareymondie, 1982). The role of these structures in autobiographical and episodic remembering is well established (Cimino et al., 1991; Gardiner et al., 2001; Tulving et al., 1994). However, the role cannot be limited to higher forms or aspects of memory. The same regions seem to be involved in the pragmatics of speech communication and understanding of fresh metaphors, irony, and humor, as well as in self-awareness and in the aspects of reflective social behavior known as manifestation of the "theory of mind" (Bihle et al., 1986; Shammi and Stuss, 1999; Velichkovsky, 1994; Wheeler et al., 1997). This list goes well beyond the span of memory phenomena approaching the "metacognitive coordinations," or level F in the above classification.

Conclusions

The history of science abounds with examples of mutual misunderstanding among scientists belonging to different schools of thought or following different scientific paradigms. Psychology with its abundance of complex and ambiguous phenomena has become a kind of huge projective test where everybody can find confirmation for his/her views. For centuries and decades proponents of different theoretical schemata—associationism and behaviorism, gestalt theory and connectionism, functionalism and cognitive psychology—were all (moderately) successful in describing diverse aspects of sensory, affective, willful, and cognitive processes. So, obviously, is the modularity-of-mind conception as the modern version of nineteenth-century phrenology.

In this chapter, we presented experimental data from investigations of human eye movements, perception, and memory demonstrating that domain specificity (modularity) is only one aspect of human cognitive architecture. These effects are embedded into gradient-like or hierarchical mechanisms, which can be preliminarily identified in relation to their evolutionary origins. In particular, we demonstrated that there are mechanisms below perceptual and above semantic processing. In other words, even the mild version of modularity of mind—the one stressing differences between perceptual input systems and semantic central systems (e.g., Fodor, 1983)—seems to be incomplete qua model of human cognition, especially with respect to its ignorance of this vertical dimension of mental functioning. Our strong impression therefore is that the psychological and neurophysiological analysis aimed at developmental stratification of the evolving mind-and-brain mechanisms is a viable alternative to the currently popular search for double dissociations and presumably isolated modules of processing. A framework for the stratification endeavor, based on the ideas of Bernstein's motor control theory (Bernstein, 1947/1990), is presented and supported by empirical data.

Acknowledgments

I wish to thank Fergus Craik and Jaak Panskepp for discussion of the paper, and my colleagues Beata Bauer, Sascha Dornhoefer, Sebastian Pannasch, Pieter Unema, and Hans-Juergen Volke for participation in research.

Notes

1. This understanding of modularity is still rather different from those used in biology (chapter 2 in this volume) or, for example, in graphic art and mathematics (chapter 12 this volume), since psychological modules can be heterogeneous. The common features are autonomy and parallelism.

2. The theory is virtually unknown in the West. Bernstein's main oeuvre was suppressed in the Soviet Union (because of his criticism of the official Pavlovian doctrine) and reprinted for the first time in 1990. A popular version of his theory was translated into English only a few years ago (Bernstein, 1996).

3. This was actually the *only* nonmonotonic deviation from flat, one-step, gradient-like memory effects in the array of sixty-five comparisons of encoding and retrieval in this study (Challis et al., 1996).

4 Gardiner et al. (2001) report right prefrontal and parietal activation in a combined EEG and MEG study of a word-stem completion task. This suggests the involvement of two levels: F as the leading level and D as the major background structure. The selection of level D for this type of task may be connected with its rich assortment of pattern-recognition procedures.

5. In this method of EEG analysis, global incoherence is the nearest analogue of the metabolic activation from brain imaging studies. An incoherence of the EEG between some region and the rest of the cortex, separately evaluated for different bands of the spectrum, can testify to its involvement in a particularly intensive processing.

Appendix 1

Nikolai A. Bernstein, *On the Construction of Movements*

Moscow: Medgiz, 1947, pp. v + 255 (in Russian)
(Translation by B. M. Velichkovsky)
Table of Contents
Foreword
Part I. Movements
Chapter 1. On the Origin of Motor Function
Evolutionary significance of motor function. Enrichment of coordinational resources. Development of the structure of the nervous system. Origin and development of the levels of movement construction. Coordinational contingents of movements.

Chapter 2. On the Construction of Movements
Kinematic chains of the organism and degrees of freedom of mobility. Difficulties of movement control in a system with more than one degree of freedom. Main task of coordination. Importance of elasticity of skeletal muscles and peripheral cycle of interaction. Examples of complicated relationships between muscle tension and movement. The principle of sensory correction. The reflex circle. Internal, reactive, and external forces. Definition of the coordination of movements. Levels of movement construction. Leading and background levels. Outline of the levels of construction.

Part II. Levels of Movement Construction

Chapter 3. Subcortical Levels of Construction: The Rubro-spinal Level of Paleokinetic Regulation (A)
Paleokinetic and neokinetic systems. Properties of the nervous process in both systems. Synapses of the neokinetic system. Alternating shifts of the characteristics. Paleoregulation of the neokinetic process. Substrata of nervous processes.

Afferentation. Characteristics of nervous processes. Functions of rubro-spinal level A. Subordination. Muscular tonus. Alpha waves and paleokinetic regulation. Self-sufficient movements and background components of level A. Dysfunctions.

Chapter 4. Subcortical Levels of Construction: Level of Synergies and Engrams, or thalamo-pallidum level (B)
Phylogenesis of level B. Substrata. Leading afferentation. Coordinating qualities. Self-sufficient movements. Background role. Typical dysfunctions.

Chapter 5. Cortical Levels of Construction: Pyramido-striatum level of the spatial field (C)
Ambiguity of this level. Afferentation. The spatial field. Character of movements of level C. Spatial determination of movements. Variability, switchability, extemporality. Substrata. Self-sufficient movements. Background role. Dysfunctions.

Chapter 6. Cortical Levels of Construction: Parieto-premotor Level of Action (D). Highest Cortical Levels
Specifically human properties of level D. The group of apraxias. Substrata. Afferentation. Meaningful structure of actions. The space of level D. Evolution of interaction with objects. Organization of motor acts of level D. Motor specification of an action. Highest automatisms. The role of premotor systems. Sensory and kinetic apraxias. Deautomatization. Classification of motor acts of level D. Levels above the level of action (group E). Coordinational properties of group E.

Part III. Development and Destruction

Chapter 7. Origin and Development of the Levels of Construction
The biogenetic law and its limitation. Embryogenesis of the motor centers of the brain. Phylogenesis of the main nuclei of the brain. Schema of the development of vertebrate motion. Ontogenesis of human motions in the first half-year of life. Maturation of the striate system. Ontogenesis of grasping movements. Development of locomotion. Maturation of level D. Development of movements in childhood. Puberty.

Chapter 8. Development of Motor Skills
The conditioned-reflex theory of motor skill development and its mistakes. Definition of motor skill. Two periods in development of a skill. Finding of the leading level. Motor specification. Pickup of sensory corrections. The phase of automatization. Backgrounds-as-such and automatisms. Transfer of practice by skill. Lowering of thresholds of signal receptors. The phase of involvement of corrections. Standardization. Three stages in development of skills with synergetic backgrounds. Dynamically stable movements. Discontinuity and "all humanity" of dynamically stable forms. The phase of stabilization. Factors impeding automatization. Increase of switchability. Transfer by organ and by trick: generalization of a skill. Preliminary corrections. The structure of writing skill. The development of writing skill. Reautomatization and breaking in.

Chapter 9. Features of the Levels Structure Under Pathological and Under Normal Conditions
Required features of a coordinating structure. Phenomena determining the complexity of pathological syndromes. The influence of location of a lesion on the reflex circle. Hypodynamias and ephrenations. Hypodynamic and hyperdynamic syndromes according to levels. Perseverations. Perseverations in the normal state. The grouping of normal features after two periods of skill development. Main questions concerning the levels structure of normal movement. Features of precision and variability. Manifestation of levels in the accuracy index. Levels characteristic in the factors of deautomatization.

Appendix 2

(Translation by M. Mirsky and B. M. Velichkovsky)

... Thus, coordination of movements consists, in its precise definition, in overcoming the excessive degrees of freedom of the moving organ, i.e., in turning the latter into a controllable system. The solution to this problem is based on the principle of sensory corrections, which are provided jointly by a variety of afferent systems and executed within the basic structural formula of a reflex circle. We define the construction of a given movement as a composition of those afferent ensembles that participate in coordinating this movement, in carrying out the corrections required, and in providing the adequate recoding for the efferent impulses, as well as the totality of systemic interactions between these components.

... Sensory corrections are always executed by the already *whole syntheses* whose complexity increases progressively from the bottom to the top [of the evolutionary hierarchy—M.M. and B.V.]. The syntheses are built from a great variety of sensory signals, which have undergone a deep integration processing. These syntheses are exactly what we call *levels of construction* of movements. *Each motor task finds for itself, depending upon its content, or meaningful structure, a certain level (in other words, a certain sensory synthesis), which, by the features and the composition of afferentation forming it, as well as by the principle of their synthetic integration, is the most adequate for the solution of this task.* This level is defined as the leading one for a given movement in regard to carrying out the most significant, decisive sensory corrections and to performing the recoding needed for it.

The best way to clarify the concept of the various leading, or key, levels of construction is to compare examples of movements similar in their external shaping but drastically different in their level composition.

A person is capable of executing a circular arm and/or hand movement in a number of situations that are extraordinarily different: A. Often, when an outstanding virtuoso pianist performs a so-called "vibrato," i.e., repeats the same note or octave at a frequency of 6 or 8 cycles per second, the trajectories of points on a moving wrist or forearm are forming small circles (or ovals). B. It is possible to draw a circle by moving one's arm during a gymnastic exercise or a choreographic performance. C. One can trace, with a pencil, a circle drawn in advance or etched on a piece of paper (C1) or a copy a circle (C2) placed in front of him/her. D. A circular movement can be executed in the process of making a stitch with a needle, or of untying a knot. E. To prove a geometric theorem, one can draw a circle on a blackboard as one of the components of the drawing used for showing the proof.

Every one of these movements is a circle or closely similar, but nevertheless in all the above examples, the circles' origins, their central nervous "roots," and the levels of their construction are substantially different, as will be shown below. In all the cases mentioned above, we encounter both differences in the mechanics of movement, of its external, spatial-dynamic pattern, and—even more important—deep differences in the mechanisms of coordination determining these movements. It is completely evident, first of all, that each of these circular movements is related to a *different afferentation*.

Type A circles are formed involuntarily, brought about by a *proprioceptive reflex*, which is more or less unconscious. A circle of the gymnastic or dancing type (B) also is executed mainly on the basis of proprioceptive corrections, but it is in considerable part consciously realized and already reveals the prevalence of the joint space over the muscle-force components of proprio-afferentation. Drawing of a circle which is traced (C1) or copied (C2) is led by visual control, more direct and primitive in the former case, and carried out by a complexly synthesized afferent "spatial field" system in the latter. In the case of D, the leading afferent system is the representation of *an object*, and its apperception is connected with knowledge of its shape and functional significance. This knowledge results in *an action* or series of actions aimed at a purposeful manipulation with the object. Finally, in the case E, a circle drawn on a blackboard by the math lecturer, the central moment is the semiconventional or symbolic representation of its relations to other elements of a mathematical drawing, rather than the reproduction of its geometrical shape.

... Distortion of the correct circular shape will not disturb the original idea of the math lecturer, nor will it initiate any corrective impulses in his/her motor control centers ... All these movements (from A through E) are circles in regard to their muscle-joint schemes, but their *construction* as performed by the central nervous system is executed at different levels for each version.

... There is no movement (perhaps with extremely rare exceptions) that is served by one and only one leading level of construction in all its coordinatory details. Indeed, at the beginning stages of forming of a new individual motor skill, almost all corrections are executed, in a surrogate mode, by the initiating leading level—but soon the situation changes. Every single technical part and detail of the complex movement sooner or later finds such a level, *among the lower levels*, whose afferentation is the most adequate for this detail because of the features of sensory corrections they provide. Thus, gradually, by means of a sequence of switches and step-by-step changes, a complex, multilevel construction is being formed, headed by the *leading level* adequate for the *meaningful structure* of the motor act and providing only those very basic corrections that are decisive with respect to the sense of a movement. Under its guidance, a number of *background levels* participate in the further execution of a movement, subserving such background or technical details of the movement as tone, innervation and denervation, reciprocal inhibition, and complex synergies. *It is the process of switching the technical components of controlling the movement to the lower, background levels that is usually called automatization.*

In any movement, regardless of the absolute "height" of its level, *only the leading levels*, as well as only those corrections that are executed at this level, are *consciously available*. For example, if the current motor act is tying a knot, which is performed at level D, its technical components from the level of the spatial field (C) as a rule do not reach the threshold of consciousness. However, if the next movement is a stretch or a smile, which occur at level B, this movement is consciously realized despite being of a level lower in the hierarchy than C. It does not mean, of course, that the degree of conscious apprehension should be the same for each leading level. To the contrary, *both the extent of conscious apprehension and the degree of volitional control increase with moving through the levels from the bottom up.*

Switching of a technical component from the leading level to one of the lower, background levels, in accordance with the above, results in the component's leaving the field of consciousness, and it is exactly this process that deserves the name "automatization." The advantages of automatization are completely evident, for it results in freeing the consciousness from secondary technical material, enabling it to concentrate on the significant and crucial facets of the movement, which are, as a rule, the potent source of a variety of "unpredictables" requiring quick and adequate switches. The process of a temporary or total decay of automatization, opposite to the one described, is called *deautomatization* ... (Bernstein, 1947/1990, pp. 40–43).

... It may be reasonable for our further analysis to introduce two notions: the notion of *meaningful structure* of an action and that of its *motor composition*. Meaningful structure of a motor act is defined by the content of the task raised and, in its turn, it defines the sensory or sensory-cognitive synthesis that is adequate to the task and can uphold the solution. In other words, the sensory-cognitive synthesis attuned with the task is the leading level of construction. Furthermore, motor composition of an action is the result of a confrontation (as it were, a substitution of parameters in some general equation) between requirements of the task and the situation, in the first line, the motor options being in possession of the organism for its solution. Motor composition includes many components, such as the listing of sequential elements of a chain (if we deal with a chain action), definition of motor procedures, supporting these elements, and collecting the simultaneous background coordinations necessary for a fluent realization of a complex movement. Motor composition is also defined by the situation—by the biomechanical design of levers and kinematic chains of the body, by resources of innervation, by the inventory of sensory corrections, and finally, by the tools that may be applied for the needed action. Thus, motor composition is a function not only of the task's intention but also of the task-solver. Let us illustrate this by the same task of a fast relocation in space. A human will solve it by sprinting (or, for instance, by riding a bicycle); a horse, by galloping; a bird, by flight; etc ..." (Bernstein, 1947/1990, pp. 112–113).

References

Bernstein NA (1947) O postrojenii dvizhenij [On the construction of movements]. Moscow: Medgiz. Reprinted in N.A. Bernstein, Physiologija aktivnosti [Physiology of activity]. Moscow: Nauka, 1990.

Bernstein NA (1967) The Coordination and Regulation of Movements. Oxford: Pergamon.

Bernstein NA (1996) Dexterity and Its Development. Mahwah, NJ: Erlbaum.

Bihle AM, Brownell HH, Powelson JA, Gardner H (1986) Comprehension of humorous and non-humorous materials by left and right brain damaged patients. Brain Cognit 5: 399–411.

Blaxton TA (1989) Investigating dissociations among memory measures. J Exp Psychol Learn 15: 657–668.

Blaxton TA, Bookheimer SY, Zefiro ThA, Figlozzi CM, Gaillard WD, Theodore WH (1996) Functional mapping of human memory using PET: Comparisons of conceptual and perceptual tests. Can J Exp Psych 50: 42–56.

Bower H, Gilligan SG (1979) Remembering information related to one's self. J Res Personal 13: 420–432.

Bridgeman B (1999) Vertical modularity in the visual system. In: Stratification in Cognition and Consciousness (Challis BH, Velichkovsky BM, eds), 19–41. Amsterdam and Philadelphia: John Benjamins.

Brown AL (1979) Theories of memory and the problems of development: Activity, growth, and knowledge. In: Levels of Processing in Human Memory (Cermak LA, Craik FIM, eds), 225–228. Hillsdale, NJ: Erlbaum.

Bühler K (1918) Die geistige Entwicklung des Kindes [The Mental Development of the Child]. Leipzig: Quelle und Mayer.

Challis BH, Brodbeck DR (1992) Level of processing affects priming in word fragment completion. J Exp Psychol Learn 18: 595–607.

Challis BH, Velichkovsky BM, Craik FIM (1996) Levels-of-processing effects on a variety of memory tasks: New findings and theoretical implications. Conscious Cognit 5: 142–164.

Christoff K, Gabrieli JDE (2000) The frontopolar cortex and human cognition: Evidence for a rostrocodal hierarchical organization within the human prefrontal cortex. Psychobiology 28: 168–186.

Cimino CR, Verfaillie M, Bowers D, Heilmann KM (1991) Autobiographical memory: Influence of right hemisphere damage on emotionality and specificity. Brain Cognit 15: 106–118.

Cosmides L, Tooby J, Turner JH, Velichkovsky BM (1997) Biology and psychology. In: Human by Nature: Between Biology and Social Science (Weingart P, Mitchell S, Richerson PJ, Maasen S, eds), 52–64. Mahwah, NJ: Erlbaum.

Craik FIM, Lockhart R (1972) Levels of processing: A framework for memory research. J Verb Learn Verb Behav 11: 671–684.

Craik FIM, Moroz TM, Moscovitch M, Stuss DT, Vinokur G, Tulving E, Kapur S (1999) In search of the self: A positron emission tomography study. Psych Sci 10: 26–34.

Craik FIM, Tulving E (1975) Depth of processing and the retention of words in episodic memory. J Exp Psych Gen 104: 268–294.

Deacon TW (1996) Prefrontal cortex and symbolic learning: Why a brain capable of language evolved only once. In: Communicating Meaning: The Evolution and Development of Language (Velichkovsky BM, Rumbaugh DM, eds), 103–138. Mahwah, NJ: Erlbaum.

Deubel H (1999) Separate mechanisms for the adaptive control of reactive, volitional, and memory-guided saccadic eye movements. In: Attention and Performance XVII (Gopher D, Koriat A, eds), 697–721. Cambridge, MA: MIT Press.

Elman JL, Bates EA, Johnson MH, Karmiloff-Smith A, Parisi D, Plunkett K (1996) Rethinking Innateness: A Connectionist Perspective on Development. Cambridge, MA: MIT Press.

Findlay JM, Walker R (1999) A model of saccadic generation based on parallel processing and competitive inhibition. Behav Brain Sci 22: 661–674.

Fischer B, Weber H (1993) Express saccades and visual attention. Behav Brain Sci 16: 553–610.

Fisher RP, Craik FIM (1977) The interaction between encoding and retrieval operations in cued recall. J Exp Psych Hum 3: 153–171.

Fodor JA (1983) The Modularity of Mind. Cambridge, MA: MIT Press.

Gabrieli JDE, Desmond JE, Demb JB, Wagner AD, Stone MV, Vaidya CJ, Glover GH (1996) Functional magnetic resonance imaging of semantic memory processes in the frontal lobes. Psych Sci 7: 278–283.

Gardiner JM, Richardson-Klavehn A, Ramponi C, Brooks BM (2001) Involuntary level-of-processing effects in perceptual and conceptual priming. In: Perspectives on Human Memory and Cognitive Aging: Essays in Honor of Fergus Craik (Naveh-Benjamin M, Moscovitch M, Roediger HL III, eds), 71–82. Philadelphia: Psychology Press.

Goldberg E (1991) Higher cortical functions in humans: The gradiental approach. In: Contemporary Neuropsychology and the Legacy of Luria (Goldberg E, ed), 229–276. Hillsdale, NJ: Erlbaum.

Head H (1920) Studies in Neurology. Oxford: Oxford University Press.

Holloway RL, de la Coste-Lareymondie M (1982) Brain endocast asymmetry in pongids and hominids. Amer J Phys Anthropol 58: 108–116.

Kapur S, Craik FIM, Tulving E, Wilson AA, Houle S, Brown GM (1994) Neuroanatomical correlates of encoding in episodic memory: Levels of processing effect. Proc Natl Acad Sci USA 91: 2008–2011.

Karmiloff-Smith A (1992) Beyond Modularity. Cambridge, MA: MIT Press.

Kennard C (1989) Hierarchical aspects of eye movement disorders. In: Hierarchies in Neurology: A Reappraisal of a Jacksonian Concept (Kennard C, Swash M, eds), 151–158. London: Springer.

Kutas M, King JW (1996) The potentials for basic sentence processing: Differentiating integrative processes. In: Attention and Performance XVI (Toshio I, McClelland JL, eds), 501–546. Cambridge, MA: MIT Press.

Leonova AB (1998) Basic issues in applied stress research. In: Advances in Psychological Sciences, vol. 1: Social, Personal and Cultural Aspects (Adair JG, Belanger D, Dion K, eds), 302–331. London and Philadelphia: Psychology Press.

Leont'ev AN (1978) Activity, Consciousness, and Personality. Englewood Cliffs, NJ: Prentice-Hall.

Levy-Shoen A (1969) Détermination et latence de la réponse oculomotrice à deux stimulus. Ann Psychol 69: 373–392.

Lockhart R, Craik FIM (1990) Levels of processing: A retrospective commentary on a framework for memory research. Can J Psych 44: 87–112.

Luria AR (1966) Higher Cortical Functions in Man. London: Tavistock.

Marr D (1976) Early processing of visual information. Phil Trans Roy Soc London B275: 483–524.

Marr D (1982) Vision. San Francisco: Freeman.

Miall DS (1986) Emotion and the Self: The context of remembering. Br J Psychol 77: 389–397.

Mishkin M, Suzuki WA, Gardian DG, Vargha-Khadem F (1999) Hierarchical organization of cognitive memory. Phil Trans Roy Soc London B352: 1461–1467, 1997.

Morris CD, Bransford JD, Franks JJ (1977) Levels of processing versus transfer appropriate processing. J Verb Learn Verb Behav 16: 87–112.

Naumann E (1985) Ereigniskorrelierte Potentiale und Gedächtnis [Event-related potentials and memory]. Frankfurt am Main: Lang.

Newell A, Simon HA (1972) Human Problem Solving. Englewood Cliffs, NJ: Prentice-Hall.

Panksepp J, Panksepp JB (2000) The seven sins of evolutionary psychology. Evol Cognit 6: 108–131.

Pannasch S, Dornhoefer SM, Unema PJA, Velichkovsky BM (2001) The omnipresent prolongation of visual fixations: Saccades are inhibited by changes in situation and in subject's activity. Vis Res 41: 3345–3351.

Pylyshyn ZW (1984) Computation and Cognition. Cambridge, MA: MIT Press.

Rasmussen J (1986) Information Processing and Human–Machine Interaction. Amsterdam and New York: North-Holland.

Reingold E, Stampe D (2000) Saccadic inhibition and gaze-contingent research paradigms. In: Reading as a Perceptual Process (Kennedy A, Raddach R, Heller H, Pynte J, eds), 1–23. Amsterdam: Elsevier.

Roediger HL III, Gallo D (2001) Levels of processing: Some unanswered questions. In: Perspectives on Human Memory and Cognitive Aging: Essays in Honor of Fergus Craik (Naveh-Benjamin M, Moscovitch M, Roediger HL III, eds), 28–47. Philadelphia: Psychology Press.

Roediger HL III, Weldon MS, Stadler ML, Riegler GL (1992) Direct comparison of two implicit memory tests. J Exp Psychol Learn 18: 1251–1269.

Rueckl JG, Cave KR, Kosslyn SM (1989) Why are "what" and "where" processed by separate cortical visual systems? A computational investigation. J Cog Neurosci 1: 171–186.

Sanquist TF, Rohrbaugh JW, Syndulko K, Lindsley DB (1980) Electrocortical signs of levels of processing: Perceptual analysis and recognition memory. Psychophysiology 17: 568–576.

Schlenoff DH (1985) The startle responses of blue jays to *Catocala* (Lepidoptera: Noctuidae) prey models. Anim Behav 33: 1057–1067.

Shammi P, Stuss DT (1999) Humor appreciation: A role of the right frontal lobe. Brain 122: 657–666.

Simons RF, Perlstein WM (1997) A tale of two reflexes: An ERP analysis of prepulse inhibition and orienting. In: Attention and Orienting: Sensory and Motivational Processes (Lang P, Simons RF, Balaban MT, eds), 229–255. Mahwah, NJ: Erlbaum.

Sokolov EN (1963) Perception and the Conditioned Reflex. Oxford: Pergamon Press.

Stadler MA, Roediger HL III (1997) The question of awareness in research on implicit learning. In: Handbook of Implicit Learning (Stadler MA, Frensch PA, eds), 105–132. London: Sage.

Sternberg S (1998) Discovering mental processing stages: The method of additive factors. In: An Invitation to Cognitive Science: Methods, Models, and Conceptual Issues (Scarborough D, Sternberg S, eds), 703–863. Cambridge, MA: MIT Press.

Sterr A, Mueller MM, Elbert T, Rockstroh B, Pantev C, Taub E (1998) Perceptual correlates of changes in cortical representation of fingers in blind multifinger Braille readers. J Neurosci 18: 4417–4423.

Tooby J, Cosmides L (2000). Toward mapping the evolved functional organization of mind and brain. In: The New Cognitive Neurosciences, 2nd ed. (Gazzaniga MS, ed). Cambridge, MA: MIT Press.

Trevarthen C (1968) Two visual systems in primates. Psych Forsch 31: 321–337.

Tulving E (1998) Brain/mind correlates of human memory. In: Advances in Psychological Sciences, vol. 2: Biological and Cognitive Aspects (Sabourin M, Craik FIM, Robert M, eds), 441–460. Hove, UK: Psychology Press.

Tulving E, Kapur S, Craik FIM, Moscovitch M, Houle S (1994) Hemispheric encoding/retrieval asymmetry in episodic memory. Proc Natl Acad Sci USA 91: 2016–2020.

Tulving E, Schacter DL (1990) Priming and human memory systems. Science 247: 301–305.

Turvey MT, Shaw RE, Mace W (1978) Issues in the theory of action. In: Attention and Performance VII (Requin J, ed), 557–595. Hillsdale, NJ: Erlbaum.

Unema P, Velichkovsky BM (2000) Processing stages are revealed by dynamics of visual fixations: Distractor versus relevance effects. Abstr Psychonom Soc 5: 113.

Velichkovsky BM (1982) Visual cognition and its spatial-temporal context. In: Cognitive Research in Psychology (Klix F, Hoffmann J, van der Meer E, eds), 55–64. Amsterdam: North-Holland.

Velichkovsky BM (1990) The vertical dimension of mental functioning. Psych Res 52: 282–289.

Velichkovsky BM (1994) The levels endeavour in psychology and cognitive science. In: International Perspectives in Psychological Science: Leading Themes (Bertelson P, Eelen P, d'Ydewalle G, eds), 143–158. Hove, UK: Erlbaum.

Velichkovsky BM (1995) Communicating attention: Gaze position transfer in cooperative problem solving. Pragmat Cognit 3: 199–222.

Velichkovsky BM (1999) From levels of processing to stratification of cognition. In: Stratification in Cognition and Consciousness (Challis BH, Velichkovsky BM, eds), 203–226. Amsterdam and Philadelphia: John Benjamins.

Velichkovsky BM, Dornhöfer SM, Kopf M, Helmert J, Joos M (2002) Change detection and occlusion modes in road-traffic scenarios. Transport Res F 5: 99–109.

Velichkovsky BM, Dornhoefer S, Pannasch S, Unema P (2000) Visual fixations and level of attentional processing. In: Eye Tracking Research and Applications (Duhowski A, ed), 79–85. Palm Beach Gardens, FL: ACM Press.

Velichkovsky BM, Klemm T, Dettmar P, Volke H-J (1996) Evoked coherence of EEG II: Communication of brain areas and depth of processing. Zeit EEG-EMG 27: 111–119.

Velichkovsky BM, Pomplun M, Rieser H (1996) Attention and communication. In: Visual Attention and Cognition (Zangemeister WH, Stiel S, Freksa C, eds), 135–154. Amsterdam and New York: Elsevier.

Velichkovsky BM, Sprenger A, Unema P (1997) Towards gaze-mediated interaction: Collecting solutions of the "Midas touch problem." In: Human–Computer Interaction (Howard S, Hammond J, Lindgaard G, eds), 509–516. London: Chapman asnd Hall.

Vincent A, Craik FIM, Furedy JJ (1993) Sensitivity of heart rate and T-wave amplitude to effort and processing level in a memory task. J Psychophysiol 7: 202–208.

Wheeler MA, Stuss DT, Tulving E (1997) Toward a theory of episodic memory: The frontal lobes and autonoetic consciousness. Psych Bull 121: 331–354.

17 Decomposability and Modularity of Economic Interactions

Luigi Marengo, Corrado Pasquali, and Marco Valente

Introduction

Talking about modularity and decentralization in economics is a surprisingly difficult task; it involves going into the very heart of the nature of the institutions that govern economic life. As reported in Hurwicz (1971), the discussion dates back at least to Plato's defense of central planning in the *Republic* and Aristotle's warning against the dangers and disadvantages of collective ownership. Far from presuming to directly enter a discussion with these big names of Western thought, we present a model and some results dealing with how well different institutional settings, characterized by different degrees of decentralization, perform as backgrounds for distributed problem solving.

The main focus of our work is on solving problems whose solution derives from coordinating a large number of interacting and interdependent entities which together contribute to forming a solution to the problem. The key issue and difficulty addressed here is the opacity of single entities' functional relations and the partial understanding of their context-dependent individual contributions in forming a solution to the problem at hand. Our model accounts for the relationships between problem complexity, task decentralization, and problem-solving efficiency.

The issue of interdependencies and how they shape search processes in a space of solutions is also faced in Kauffman's NK model of selection dynamics in biological domains with heterogeneous interdependent traits (Kauffman, 1993). Kauffman's approach to the exploration of a fitness landscape, however, does not necessarily fit well with the realm of social evolution. The main reason for this inadequacy is that social actors might well engage in adaptive walks based on a far richer and virtually huge class of algorithms other than single-bit mutations, the only possibility considered by Kauffman.

In particular, following Simon (1983), our focus is on those problem-solving strategies which decompose a large problem into a set of smaller subproblems that can be treated independently by promoting what we have come to call a *division of problem-solving labor*. Searches in the problem space based on one-bit mutations amount to fully decomposing the problem into its smaller components, while coarser decompositions correspond to mutating more bits together. In this chapter we study some general properties of decompositions and the effect of adopting mutational algorithms of any size on the exploration of a fitness landscape.

Imagine, for instance, an N-dimensional problem. Adopting a point-mutation algorithm corresponds to exploring the space of configurations by flipping elements' values one at a time. A point-mutation algorithm corresponds, in our view, to a maximally decentralized search strategy in which each of the N bits forming the problem will be given a value independently from all the other $N - 1$ bits. In this sense the whole problem is (trivially) decomposed into the set of the smallest subproblems. On the other hand, the same problem could be left totally undecomposed and a search algorithm might be adopted which explores all the N dimensions of the problem. This strategy corresponds to mutating up to all the N components of the problem at a time. Between the finest and the coarsest lie all other possible decompositions of the problem (i.e., all the possible algorithms of any cardinality), each corresponding to a different division of labor.

As we shall see in greater detail, the division of problem-solving labor determines to a large extent which solutions will be generated and eventually selected. This leaves the possibility open that the optimal solution will never be generated and a fortiori selected, since it might well be that it will never be possible to reach it starting from a given decomposition of the problem. It thus turns out that while decomposing a problem is necessary in order to reduce the dimension of the search space, it also shapes and constrains a search process to a specific subspace of possible solutions, thus making it possible for optimal solutions not to be generated and for systems to be locked into suboptimal solutions.

We believe that the main significance of our work is that it casts further doubts on any "optimality through selection" argument from a specific point of view. Evolutionary arguments in economic theory have often taken a rather Panglossian form, according to which the sole existence of, say, an organizational form or of a technological design can be reliably taken as proof of its optimality. The view according to which market forces are always able to select away suboptimal types is a widely accepted idea in economic theory. It is also noteworthy that most of the historically grounded attempts to show the limits that selection encounters in the economic realm are grounded on a claim concerning some sort of weakness of selective pressures.

A famous example is Paul David's work on the persistence of the highly inefficient QWERTY keyboard on typewriters and computers. In his seminal work, David (1985) showed how a technological standard adopted for its efficiency in a given set of constraints—those imposed by mechanical typewriters, with the need to reduce the frequency of lever jams—was not displaced by standards which had become far superior when those constraints had totally disappeared: in electric typewriters and, later on, computers, where the keyboard layout could be freely designed in order to make the most frequently struck keys more easily accessible. However, the case discussed by David deals with a situation in which the "optimal" configuration exists and is actually available as a possible choice, but selective pressures are not strong enough to favor it over the others.

Quite to the contrary, our focus is on those cases in which the optimal alternative is not available at all as a consequence of a particular way of exploring a landscape and of a specific accessibility relation between solutions that holds as a consequence of the adopted search algorithms. In our approach, these considerations are highlighted by the fact that exploring a landscape according to different decomposition patterns implies that the geometry of the fitness landscape changes, and a landscape that is very rugged when explored with a one-bit algorithm might become smooth when explored with an algorithm of greater cardinality.

Many authors in the field of theoretical biology emphasize how many evolutionary asymmetries—such as patterns and processes of phenotypic evolution, punctuated evolution, developmental constraints, homology, and irreversibility—do not fit naturally with evolutionary theories as implemented in the neo-Darwinian paradigm because it is based purely on (heritable) diversity generation and selection.

An approach that we regard as particularly consonant with our own is that proposed by Stadler et al. (2000). In this work, the authors emphasize the quest for a theory of accessibility between structures in terms of a precise theory of genotype–phenotype maps and claim that the aforementioned asymmetries are in some way rooted in the structure of the map itself and are nothing but "statements about the accessibility topology of the phenotype space" (Stadler et al., 2000, p. 4). This way of looking at things in a sense relates to the idea that accessing or constructing a structure (say, a phenotype, a technological design, or a solution to a combinatorial problem) by means of a set of variational mechanisms is logically and empirically prior to any selective pressure and its outcomes.

We thus take the perspective according to which asymmetries in selection and evolution are not dependent solely on selective pressures plus the structure of fitness landscapes, but rather on accessibility relations between objects as defined by and grounded on specific mutational operators that allow the exploration of a solution space by transforming objects into new objects. Once again, what can be reached from what is highly dependent on some notions of neighborhood and distance is prior to and independent from any bare fitness consideration.

The problem of accessibility in current economic theory is approached (and radically solved) in a very clear way whose main effect is to make the problem disappear. Basically, for both individual consumers and producers, solution spaces (i.e., the space of consumption and production bundles) are imagined to be uniformly, symmetrically, and everywhere accessible, so that a continuous path can always be imagined to exist connecting each and every point in those spaces. Expressed in these terms, our main focus is on seriously considering the tangled (though neglected) interaction between economic interaction as such and organizational structures and constraints adopted to complement ambiguous market signals.

A Bird's-Eye View on Economic Theory

Lionel Robbins (1932) defined economic theory as "the science which studies human behavior as a relationship between given ends and scarce means which have alternative uses." Focusing on "scarcity" and "alternative uses" will give us the possibility of articulating a brief overview of the realm and character of economic theory and its approach to the problem of modularity.

That resources are not unlimitedly available to us and that limits to their utilization force us to allocate them among alternative uses are facts that surround each of us in the experience of everyday life. These two facts taken together imply social forms of competition for resources in order to give oneself the possibility of using them.

To the end that competition will be mediated and efficiently organized, a system is needed that governs competition for resources and helps in efficiently allocating them to different possible utilizations. The "system" can be, for instance, an authoritarian one (such as in a military state or in a dictatorship) or a decentralized one (such as a market).

Adam Smith emphasized that some form of societal or interpersonal organization is needed in order to exploit the advantages of cooperation and social interaction. In particular, he observed that individuals are different, and have different "talents"; and, at the same time, their skills and capacities to pursue their ends increases with specialization.

To this end, trade and division of labor are necessary so that a butcher is not forced to live solely on meat, and can exchange meat for bread with a baker who, in turn, will not have to live solely on bread.

Economic theory has among its aims the study of the properties of the outcomes resulting from different societal organizations. In order to tell a better system from a worse one, economists rely on the notion of Pareto efficiency: whatever we mean by "better" or "efficient," we surely mean that a system or a particular allocation is better than another one if no individual perceives it as a worse one and at least one individual perceives it as a better one. That is, there is no allocation in which all persons would prefer to find themselves.

Leaving aside any reference to a (fairly big) set of assumptions that economic theory relies on, the very heart of the theory is the proof that efficiency so defined can be reached thanks to a specific societal system: the price system. The working of the system can be roughly described as follows. Goods are transferred, and an income is derived from selling them at given prices. Income is in turn utilized to buy other goods at given prices.

The general idea is that in a setting in which all persons try to maximize their own welfare and utility, it naturally happens, given certain rather specific hypotheses, that the whole body of a society comes to a point at which there is a remarkable degree of coherence among a number of different and decentralized decisions to sell and buy different commodities. "Coherence" here means a state in which markets "clear" and supply equals demand for every commodity. To be slightly more precise: with a few (though very strong) assumptions concerning individual rationality, it is possible to show the existence of an equilibrium. That is, for a given economy it is possible to find a vector of prices and an allocation to each individual such that the excess demand function of the economy is zero for every good.

Pressures and information coming from markets are what turn selfish behaviors into socially desirable outcomes. Both pressures and information are represented and implemented by the price system, whereby prices reflect the relative scarcity of commodities. Thus, what happens, according to this picture of the economic world, is that economic interaction takes place through individual agents reacting (i.e., adjusting demand and supply) to quantitative information coming from the market.

The faith in the capacity for markets to reach a coherent state in which different, and possibly conflicting, actions are compensated for is crucially based on strong notions of individual and collective rationality and on a very clear distinction between individual agents and institutional contexts. The latter (i.e., the rules and the "places" in which individual interaction lives) are assumed to be given before any interaction takes place and to be either totally transparent to individuals or irrelevant at the level of the theory. In other words, given that a class of hypotheses is fulfilled on individual rationality, on collective rationality, and on how the latter is grounded on the former, "equilibrium analysis" postulates that every dynamic trajectory of an economy leads to an equilibrium state.

As a matter of fact, an ex-ante definition of interaction structures and institutional contexts amounts to grounding markets' compensation possibilities on issues and factors that are independent from economic interaction. Not only that: the sharp separation between economic agents and the contexts in which they interact, and the fact that these are given before and independently from their interactions entails that agents are imagined as perfectly "adapted" to their environment and that interaction structures are perfectly suited (or irrelevant) to any task at hand. The only relevant thing is the transparency of institutional contexts and settings, and the fact that prices accurately reflect all the relevant information. This consideration will be the first cornerstone of our argument and the starting point of our analysis.

The "Granularity" of the Economic World

Even from our rather sketchy description of economic life as depicted by orthodoxy, it is fairly evident how a strong and most peculiar notion of modularity underlies the whole thing. The main tenet of this approach is representing economic agents as autonomous and anonymous individuals that make decisions independently from one another and that interact only through the price system. Thus, all the relevant information is encapsulated within individual economic agents and coordination is achieved within markets by the use of prices (i.e., by pure selection).

One of the most fundamental questions asked by economic theory (and by welfare economics in particular) is about the extent to which perfect competition can lead to an optimal allocation of resources. Among the assumptions made by the theory in order to derive such results is that indivisibilities must never show up in consumption or production. The key problem here, and the main relevance of this point for our discussion, is that a decentralized coordination mechanism based on prices is no longer available when indivisibilities appear.

Classical results of welfare economics on the possibility of decentralization rely heavily on an assumption of perfect decomposability of the underlying allocation problem. The proof of one of the most important results of welfare economics, the Second Welfare Theorem, depends critically on a separation argument, and the same proof no longer holds in the presence of externalities (interdependencies which lead to social interactions not mediated through the market, i.e., in the presence of one of the most fundamental instances of nonseparability).

In the picture of economic life described so far, every actor is an "island" for any given set of prices. In particular, never is his utility affected by choices made by other actors apart from those choices—decisions to buy or sell marketable goods—which have a direct reflection on prices. Externalities, on the other hand, are those situations in which actions taken by others might well affect the outcome of our actions but do not affect prices, and thus cannot be coordinated through the action of markets. An ideal "orthodox" market ceases to work as perfectly as prescribed by the theory as soon as externalities are present. It is not by chance that those situations related to their presence are referred to as "market failures."

This idea of a sort of perfect modularization of an economy is pushed to its extreme by the so-called Coase theorem (Coase, 1937). Coase's theorem is a circular argument which states that if every single activity that affects agents' welfare can be exchanged and allocated in a perfectly competitive market, then nonseparability ceases to be a problem. For instance, consider an externality, say a factory which produces plastic buckets and in doing so, pollutes the environment and negatively affects the welfare of people living nearby. The problem exists because while the socially optimal number of buckets to be produced can be determined in the perfectly competitive market for buckets, the socially optimal amount of pollution cannot be determined because a market for that does not exist.

But if that is the problem, the solution—at least in principle—can be very easy: create a market for pollution rights, and then they will be allocated in a socially optimal way, like every other good.[1] In order to have such a market, we need to allow buckets and pollution to be allocated independently from each other, by establishing negotiable property rights on pollution rights and allowing them to be traded in a competitive market separate from the one in which buckets are traded. In the language of modularity, we could say that the problem of externalities arises because we are working with modules that are too large. Thus, the solution is to disassemble them and let market selection operate on finer units.

When interpreted in terms of degrees of correlation or, to use the biological terminology, degrees of epistasis, the Coase theorem asserts that, under a set of rather weak assumptions, in any situation the degree of interdependence can always be made minimal. The relevance of property rights in our discussion lies in the fact that with the modularization of property rights, we modularize social interaction, which becomes mediated through the interface of voluntary exchange.

In particular, the theorem can be read as saying that every tangled situation can be transformed into one whose degree of granularity is the atomistic one prescribed by the theory, that is, a degree in which every atomic entity is properly encapsulated and bears no correlation with other entities. It is solely this degree that allows competitive markets to work perfectly as decentralized (atomistically)—as modular mechanisms.

In a sense Coase's argument could pose an apparently odd but perhaps not so absurd question to biology: If selection is such a powerful device of coordination, adaptation, and optimization, why do we observe so little of it? In particular, why is selection applied to rather large ensembles of modules and not to each module separately in order to exploit its power to optimize each module? In other words, why do we observe multifunctional, complex living creatures and not much simpler entities specialized in single tasks and coordinated through selection?

Thus we are left with two analogous questions: in biology one could ask why multifunctional complex organisms exist, just as in economics we can ask why firms exist. Is it just because, as Coase would say, there is a cost associated with the use of the price system, i.e., with the imperfections of the (market) selection mechanism (that Coase calls transaction costs)? And if transaction costs are the costs of a bad modularization, what could possibly go wrong with a totally atomistic modularization?

Adopting a Panglossian attitude (which is fairly common in economics), we might imagine that Coase asserted that in the absence of transaction costs, rights (modules) cut too thin or too coarse will quickly reassemble or disassemble into optimal bundles. The starting point of this kind of analysis would be to imagine a completely atomistic, modular, and market-based way of production, consumption, and economic interaction (i.e., the finest possible decomposition of economic activities). Williamson (1975) well epitomizes this view with the position that "in the beginning there were markets." A possible explanation for nonseparable organizations is what we might call the *nonmonotonicity of marginal productivity*. An example is Alchian and Demsetz's (1972) discussion of team work, which illustrates how some instances of coordination problems require kinds of transmission of information well beyond what can be sent through the interface of the price system.

We might thus argue that organizations arise as a nonmodular response to the fact of, and the need for, non-price-mediated interactions among different modules, whatever they may be. An organization is always a demodularization and a repartitioning that limits the right of alienation from at least some rights of decision.

In addition, the view of economic systems as ideally or in principle organized around a minimal level of granularity is clearly confuted by the straightforward observation of the existence and importance of economic entities, such as business firms, whose grain is much coarser than the one prescribed by a theory which praises the virtues of decentralization.

Actually, economic life for the most part occurs in forms and structures that go largely beyond the atomistic limits envisaged by orthodox economic theory. As recognized by many economists (e.g., the Foreword in this volume), a large part of economic life occurs in organizations composed of a large number of entities that are neither regulated nor coordinated by the price system. Within organizations, other coordination mechanisms also rule activities, such as hierarchies, power, or authority.

Surprisingly enough, economic theory has little to say about organizations as such. Rather, it assimilates them to the action structure of individual agents. Thus economic textbooks devote very little room to an analysis of how, for instance, different organizational structures might differ in performance, efficiency, or speed of adaptation. The most general questions with respect to the organizations/markets opposition are what determines the boundaries between markets and organizations? and What is the real difference between what happens in a market and what happens in an organization?

In this sense, Coase's main concern can be thought to be about the way negotiation can repair externalities and interdependencies. His argument is broader and can be reformulated thus: if nothing obstructs efficient bargaining, then people can negotiate until they reach Pareto efficiency. So, after all, it might well seem that the good functioning of a competitive market is a matter of finding the right "granularity," the level that ensures individual interaction will be so effective that optimal outcomes are reached. After all, the ultimate meaning of Coase's argument is that coordination is always achievable via market mechanisms, provided that the "granularity" of the system is fine enough to encompass all atomic entities and a proper market exists (or is created) for all of them.

"Granularity" will be a big issue in our discussion. With this term, we refer, on the one hand, to the level of analysis adopted by economic theory (i.e., studying consumers' and firms' behavior as individuals, and postulating that each and every phenomenon occurring at the aggregate level can be traced back to individual actions and behaviors). On the other hand, we straightforwardly refer to economic reality and we ask why it has settled at the actual level of aggregation: Why are there firms producing goods rather than people buying raw materials and themselves building the objects they need? Why is there a multitude of firms rather than a single huge one? How are new markets created? In other words, why and how has economic life settled at the present level and degree of organization?

With respect to these points, we will argue that the two main forces that drive the whole process are integration and disintegration. These two historical forces we regard to be the ultimate reasons for economic reality having settled at the present level of aggregation. We will further maintain that integration and disintegration processes are the two main economic forces that contribute to the creation of new economic entities. With respect to this point, it is worth pointing out that even from a historical perspective (painting it with an extremely broad brush, of course), the evolutionary path undertaken by capitalist economies has been one in which newborn technologies, productive processes, commodities, and whole industries and markets enter the scene in highly integrated settings (see Langlois and Robertson, 1989; Klepper, 1997) for a detailed discussion). It is only at successive stages that a finer-grained organization of, say, production takes place, thus creating new markets, new specialized functions, and further division of labor. Also, Adam Smith referred to this theme when stressing that one of the main causes of the development of productive capacities is specialization deriving from finer-grained divisions of labor.

Causes and mechanisms by which new entities are created and enter the economic scene are not at the center of economic theory as we know it. This deficiency was stressed by Joseph Schumpeter, who pointed out that the real phenomenon that should surprise and interest economists is not how a firm is run, but how it was created in the first place. Economics thus seems to be affected by the same "object problem" (Fontana and Buss, 1994) that plagues biology as understood by the Modern Synthesis. This is not surprising when we realize that economic theory was born with a strong faith in the possibility of applying the main tools and abstractions of dynamic system theory to the domain of social interaction. This has driven this discipline to take the existence of its objects and the nature of their dynamic couplings as given and immutable from the very start, thus committing itself to purely quantitative analyses and theories, and neglecting constructive ones. Transaction-costs economics and principal–agent theories have partly tried to fill this gap. But they still—or so we believe, are missing a number of fundamental points.

Transaction-costs economics has developed an explanation of vertical integration in strictly Coasian terms (actually, just expanding upon Coase's ideas). According to this explanation, "in the beginning there were markets": there was full, atomistic modularity; and integration phenomena, viewed as processes of assembling modules, took place whenever the working of the price system was bound to face comparatively higher costs than those associated with, say, bureaucratic governance. Firms exist because of greater allocative efficiency. This kind of explanation leaves unexplained a class of very relevant phenomena and seems to be contradicted by a fairly large class of evident, though neglected, facts.

First of all, it conflicts with the historical development of technologies and industries which have mostly developed according to a path going from initial states of high vertical integration to a progressive disintegration. Also, the very process of division of labor may create opportunities for new markets to exist.

Second, it should be pointed out that both logical arguments and empirical evidence exist to support the view that markets, far from being an original state of nature, require certain sets of conditions to be met in order to emerge, some of which are determined by explicit forms of organizational planning.

Third, in many transaction-costs-based approaches it is usually implicit that efficiency can be considered as an explanation for existence. That is, proving the relative greater efficiency of an organizational form is in some way considered to be an explanation of its emergence. Actually, it might well be true that selection forces are strong enough to select fitter structures; but these have to exist to be selected: in Fontana and Buss's terms (1994), they must arrive before being selected. Thus, selective forces can and do account for a population's convergence to a specific form, but not for the existence and emergence of such a form.

In addition, we know that when the entities subject to selection have internal structure and components that present a strong degree of interdependence, their selection landscape will be correspondingly rugged and uncorrelated, thus possibly making selection forces unable to drive such entities to global optima. It then follows that selective forces may not be strong enough to select suboptimal structures out (hence providing an at least partial explanation for the persistent diversity of organizational forms).

We thus ask, Can optimal organizational structures and division of labor patterns emerge out of decentralized interactions? We show this to be possible only under some very special conditions, and at the same time we prove that decentralization usually has an associated cost in terms of suboptimality.

Our second fundamental question will be about the "necessary" character of the actual grain of the economic world. What would we see if we ran the tape of economic history twice? Could other evolutionary paths have been taken by the development of economic history? Or, rather, are there are some fundamental properties that we would be bound to see at every run of the evolutionary tape? Is it true that evolution leads to increasing modular structures in the economic realm?

The Task Ahead

In the rest of this chapter we analyze some of the issues raised above by means of a very sketchy and abstract model. We analyze the properties of adaptive walks on N-dimensional fitness landscapes by entities which decompose the N dimensions of the problem into modules and adapt such modules independently of each other. Thus intermodular coordination is performed by selection mechanisms which we assume to be "perfect" in the economic sense (i.e., not subject to any friction, inertia, or any other source of transaction cost).

In the next section we develop a methodology which determines, for any given landscape, its smaller "perfect" decomposition, the set of the smallest modules which can be optimized by autonomous adaptation with selection. Then, in the following section, we extend this methodology to near-decompositions (i.e., decompositions which, in a Simonian sense, isolate into separate modules only the most fitness-relevant interdependencies. We show that in general, such decompositions determine a loss of optimality, but sharply increases the speed of adaptation. The section after that uses such methodologies to build landscapes whose decompositions and near-decompositions are known, and simulate competition among entities which search the landscape with algorithms based on different decompositions. This enables us to test the evolutionary properties of different decompositions. In the final section we draw some conclusions.

Decomposition and Coordination

Problems and Decompositions

We assume that solving a given problem requires the coordination of N atomic elements, which we generically call components, each of which can assume some number of alternative states. For simplicity, we assume that each element can assume only two states, labeled 0 and 1. Note that all the properties presented below for the two-states case can be very easily extended to the case of any finite number of states.

More precisely, we characterize a problem by the following three elements:

1. The set of *components*: $\aleph = \{x_1, x_2, \ldots, x_N\}$, with $x_i \in \{0,1\}$

2. The set of *configurations*: $X = \{x^1, x^2, \ldots, x^{2^N}\}$, where a configuration (i.e., a possible solution to the problem) is a string $x^i = x_1^i x_2^i \ldots x_N^i$

3. An *ordering* over the set of possible configurations: we write $x^i \geq x^j$ (or $x^i > x^j$) whenever x^i is weakly (or strictly) preferred to x^j.

In order to avoid some technical complications, we assume for the time being that there exists only one configuration which is strictly preferred to all the other configurations (i.e., a unique global optimum). This simplifying assumption will be dropped in the next-to-last section.

A *problem* is defined by the pain $(X, \geq)$, and solving it amounts to finding the $x^i \in X$ which is maximal according to $\geq$.

Since the size of the set of configurations is exponential in the number of components, whenever the latter is large, the state space of the search problem becomes much too vast to be extensively searched by agents with bounded computational capabilities. One way of reducing its size it to decompose[2] it into subspaces:

Let $\Im = \{1,2,3,\ldots,N\}$ be the set of indexes, and let a block $d_i \subseteq \Im$ be a nonempty subset of it, and let $|d_i|$ be the size of block d_i, i.e., its cardinality.[3]

We define a *decomposition scheme* (or simply decomposition) of the space $\aleph$ as a set of blocks:

$$D = \{d_1, d_2, \ldots, d_k\} \text{ such that } \bigvee_{i=1}^{k} d_i = \Im.$$

Note that a decomposition does not necessarily have to be a partition (blocks may have nonempty intersections).

Given a configuration x^i and a block d_k, we call *block configuration* $x^j(d_k)$ the substring of length $|d_k|$ containing the components of configuration x^j belonging to block d_k:

$$x^j(d_k) = x^j_{k_1} x^j_{k_2} \dots x^j_{k_{|d_k|}} \quad \text{for all } k_i \in d_k.$$

We also use the notation $x^j(d_{-k})$ to indicate the substring of length $N - |d_k|$ containing the components of configuration x^j not belonging to block d_k:

$$x^j(d_{-k}) = x^j_{k_1} x^j_{k_2} \dots x^j_{k_{N-|d_k|}} \quad \text{for all } k_i \notin d_k.$$

Two block configurations can be united into a larger block configuration by means of the $\wedge$ operator:

$$x(d_i) \wedge y(d_j) = z(d_i \cup d_j) \quad \text{where} \quad z_h = x_h \text{ if } h \in d_i \text{ and } z_h = y_h \text{ otherwise.}$$

We can therefore write $x^j = x^j(d_k) \wedge x^j(d_{-k}) = x^j(d_{-k}) \wedge x^j(d_k)$ for any d_k.

We define the *size of a decomposition scheme* as the size of its largest defining block:

$$|D| = \max\{|d_1|, |d_2|, \dots, |d_k|\}.$$

A decomposition scheme and its size are important indicators of the complexity of the algorithm which is being employed to solve a problem:

• Problems that can be successfully solved by adopting the finest-grained decomposition according to the scheme $D = \{\{1\},\{2\},\{3\}, \dots, \{N\}\}$ have minimum complexity, while a problem which cannot be decomposed has maximum complexity and it can only be searched extensively.

• A problem that can be decomposed according to the scheme $D = \{\{1\},\{2\},\{3\}, \dots, \{N\}\}$ can be solved optimally in linear time, while a problem which cannot be decomposed can be solved optimally only in exponential time.

• On the other hand, a problem that has not been decomposed can always be solved optimally while, as it will be shown below, a problem which has been decomposed according to the scheme $D = \{\{1\},\{2\},\{3\}, \dots, \{N\}\}$—or, for that matter, according to any scheme whose size is smaller than N—can be solved optimally only under some special conditions, which, as we will show, become generally (though with important exceptions) more restrictive as the complexity of the problem increases and the size of the decomposition scheme decreases.

Thus there is a trade-off between complexity and optimality for which we will provide a precise measure.

Selection and Coordination Mechanisms

We suppose that coordination among blocks in a decomposition scheme takes place through marketlike selection mechanisms (i.e., there are markets which select at no cost and without any friction over alternative block configurations).

More precisely, assume that the current configuration is x^j and take block d_k with its current block configuration $x^j(d_k)$. Consider now a new configuration $x^h(d_k)$ for the same block. If

$$x^h(d_k) \wedge x^j(d_{-k}) > x^j(d_k) \wedge x^j(d_{-k}),$$

then $x^h(d_k)$ is selected and the new configuration $x^h(d_k) \wedge x^j(d_{-k})$ is kept in the place of x^j; otherwise $x^h(d_k) \wedge x^j(d_{-k})$ is discarded and x^j is kept.

It might help to think in terms of a given structure of division of labor (the decomposition scheme), with firms or workers specialized in the various segments of the production process (a single block) and competing in a market which selects those firms or workers whose characteristics make the highest contribution to the overall production process.

We can now analyze the properties of decomposition schemes in terms of their capacities to generate and select better configurations.

Selection and Search Paths

A decomposition scheme is a sort of template which determines how new configurations are generated, and can therefore be tested by market selection. In large search spaces in which only a very small subset of all possible configurations can be tested, the procedure employed to generate such new configurations plays a key role in defining the set of attainable final configurations.

We will assume that boundedly rational agents can only search locally in directions which are given by the decomposition scheme: new configurations are generated, and tested in the neighborhood of the given one, where neighbors are new configurations obtained by changing some (possibly all) components within a given block.

Given a decomposition scheme $D = \{d_1, d_2, \ldots, d_k\}$, we define the following. A configuration $x^i = x_1^i x_2^i \ldots x_N^i$ is a *preferred neighbor*—or a neighbor, for short—of configuration $x^j = x_1^j x_2^j \ldots x_N^j$ for a block $d_h \in D$ if

1. $x^i \geq x^j$
2. $x_v^i = x_v^j \; \forall v \notin d_h$
3. $x^i \neq x^j$.

Conditions 2 and 3 require that the two configurations differ only by components which belong to block d_h.

According to the definition, a neighbor can be reached from a given configuration through the operation of a single market-selection mechanism.

We call $H_i(x,d_i)$ the *set of neighbors* of a configuration x for block d_i. The *set of best neighbors* $B_i(x,d_i) \subseteq H_i(x,d_i)$ of a configuration x for block d_i is the set of the most preferred configurations in the set of neighbors:

$$B_i(x,d_i) = \{y \in H_i(x,d_i) \text{ such that } y \geq z \quad \forall z \in H_i(x,d_i)\}$$

By extension from single blocks to entire decomposition schemes, we can give the following definition of neighbors for a decomposition scheme:

$$H(x,D) = \bigcup_{i=1}^{k} H_i(x,d_i) \text{ is the set of neighbors of configuration } x \text{ for decomposition}$$
scheme D.

We say that a configuration x is a *local optimum* for the decomposition scheme D if there does not exist a configuration $y \in H(x,D)$ such that $y > x$.

A *search path*—or a path, for short—$P(x^\delta,D)$ from a configuration x^δ and for a decomposition scheme D is a sequence, starting from x^δ, of neighbors:

$$P(x^\delta,D) = x^\delta, x^{\delta+1}, x^{\delta+2}, \ldots \quad \text{with} \quad x^{\delta+i+1} \in H(x^{\delta+i},D)$$

A configuration y is reachable from another configuration x and for decomposition D if there exists a path $P(x,D)$ such that $y \in P(x,D)$.

Suppose configuration x^j is a local optimum for decomposition D. We then call *basin of attraction* $\Psi(x^j,D)$ of x^j for decomposition D the set of all configurations from which x^j is reachable:

$$\Psi(x^j,D) = \{y, \text{ such that } \exists P(y,D) \text{ with } x^j \in P(y,D)\}$$

A *best neighbor path* $\Phi(x^\delta,D)$ from a configuration x^δ and for a decomposition scheme D is a sequence, starting from x^δ, of best neighbors:

$$\Phi(x^\delta,D) = x^\delta, x^{\delta+1}, x^{\delta+2}, \ldots \quad \text{with} \quad x^{\delta+i+1} \in B_h(x^{\delta+i},d_h) \text{ and } d_h \in D$$

The following proposition states that reachability of local optima can be analyzed by referring only to best-neighbor paths. This greatly reduces the set of paths we have to test in order to check for reachability.

PROPOSITION 1.   Let x^0 be the global optimum. If x^0 is reachable from x^δ for decomposition D, then there exists a best-neighbor-path from x^δ to x^0.

Proof. By hypothesis, x^δ belongs to the basin of attraction $\Psi(x^0, D)$ of x^0. Let us order all the configurations in $\Psi(x^0, D)$ by descending rank:

$$\Psi(x^0, D) = \{x^0, x^{0+1}, x^{0+2}, \ldots\}, \text{ with } x^h \geq x^{h+1}.$$

Now proceed by induction on x^δ. If $x^\delta = x^{0+1}$ then, by definition, x^0 must be a best neighbor of x^δ for a block in D (in fact, by hypothesis, x^0 does not have any strictly preferred neighbor). If $x^\delta = x^{0+2}$, then either x^0 is a best neighbor of x^δ or it is not. In the latter case x^{0+1} must necessarily be a best neighbor of x^δ. And so on.

Now let x^0 be the global optimum[4] and let $Z \subseteq X$ be a subset of the set of configurations with $x^0 \in Z$. We say that the problem $(X, \geq)$ is locally decomposable in Z by the scheme D if $Z \subseteq \Psi(x^0, D)$. If $Z = X$, we say that the problem is globally decomposable[5] by the scheme D.

Thus, according to the previous proposition, if the problem is locally decomposable in Z, there must exist a best-neighbor path for decomposition D leading to the global optimum from every configuration in Z. If it is globally decomposable, such a path must exist from every starting configuration.

Among all the decomposition schemes of a given problem, we are especially interested in those for which the global optimum becomes reachable from any starting configuration. One such decomposition always exists, and is the degenerate decomposition $D = \{\{1,2,3, \ldots, N\}\}$, for which there exists only one local optimum, and it coincides with the global one. But obviously we are interested in—if they exist—smaller decompositions, particularly in those of minimum size. The latter decompositions represent the maximum extent to which problem-solving can be subdivided into independent subproblems coordinated by marketlike selection, with the property that such selection processes can eventually lead to optimality from any starting condition. Finer decompositions will not, in general (unless the starting configuration is "by luck" within the basin of attraction of the global optimum), allow decentralized selection processes to optimize.

The following proposition shows that there are problems which are globally decomposable only by the degenerate decomposition $D = \{\{1,2,3, \ldots, N\}\}$.

PROPOSITION 2.   There exist problems which are globally decomposable only by the degenerate decomposition $D = \{\{1,2,3, \ldots, N\}\}$.

Proof.   We prove it by providing an example. Consider a problem whose globally optimum configuration is the string $x^0 = x_1^0 x_2^0 \ldots x_N^0$ and whose second-best configuration is $x^j = x_1^j x_2^j \ldots x_N^j$, where $x_h^j = |1 - x_h^0| \; \forall \; h = 1, 2, \ldots N$. It is obvious that the global optimum can be reached from the second-best only by mutating all components together; any other mutation gives an inferior configuration.

The following proposition establishes a rather obvious but important property of decomposition schemes: as we move into the basin of attraction of a local optimum for a decomposition D which is not the finest one, then finer decomposition schemes can be introduced which allow to reach the same local optimum.

PROPOSITION 3. Let $\Psi(x^\alpha, D) = \{x^\alpha, x^{\alpha+1}, \ldots, x^\delta\}$ be the ordered basin of attraction of a local optimum x^α, and define $\Psi^i(x^\alpha, D) = \Psi(x^\alpha, D) \backslash \{x^{\alpha+i}, x^{\alpha+i+1}, \ldots, x^\delta\}$ for $0 < i \le \delta$. Then if $D \ne \{\{1\}, \{2\}, \{3\}, \ldots, \{N\}\}$, there exists an i such that for $\Psi^i(x^\alpha, D)$ a decomposition $D' \ne D$ can be found with $|D'| < |D|$.

Proof. If $i = 1$, x^α is trivially reachable from x^α for all decompositions, including the finest one, $D = \{\{1\}, \{2\}, \{3\}, \ldots, \{N\}\}$.

Minimum-size decomposition schemes can be found recursively with the following procedure. Let us rearrange all the configurations in X by descending rank $X = \{x^0, x^1, \ldots, x^{2^N-1}\}$, where $x^i \ge x^{i+1}$.

The algorithm can be described informally[6] as follows:

1. Start with the finest decomposition, $D^0 = \{\{1\}, \{2\}, \{3\}, \ldots, \{N\}\}$.

2. Check whether there is a best-neighbor path leading to x^1 from x^i, for $i = 2, 3, \ldots 2^N$. If yes, stop.

3. If no, build a new decomposition D^1 by union of the smallest blocks for which condition 2 was violated and go back to 2.

Let us finally provide an example for illustration:

Example. Table 17.1 contains a hypothetical ranking (where 1 is the most preferred) of configurations for $N = 3$.

If search proceeds according to the decomposition scheme $D = \{\{1\}, \{2\}, \{3\}\}$, there exist two local optima: 100 (which is also the global optimum) and 010. The basins of attraction of the two local optima are, respectively:

$$\Psi(100) = \{100, 110, 000, 111, 101\}$$

$$\Psi(010) = \{010, 110, 011, 001, 000, 111, 101\}.$$

Note that the worst local optimum has a larger basin of attraction[7] because it covers all possible configurations except the global optimum itself. Thus, only a search which starts at the global optimum will (trivially) stop at the global optimum itself with certainty, while for four initial configurations, a search might end up in either local optima (depending on the sequence of mutations), and for the

Table 17.1
A problem decomposable into blocks {1,2} and {3}

Configurations	Ranking
100	1
010	2
110	3
011	4
001	5
000	6
111	7
101	8

remaining three initial configurations, a search will end up at the worst local optimum with certainty.

Using the notion of dominance (see Page, 1996), it is possible to find out that the only dominant block configuration is actually the globally optimum string itself, corresponding to the degenerate decomposition scheme of size 3, $D = \{\{1,2,3\}\}$. Thus, apparently no decentralized search structure always allows location of the global optimum from every starting configuration.

Actually, this is not true: the decomposition scheme $D = \{\{1,2\},\{3\}\}$ also allows decentralized selection to reach the global optimum. For instance, if we start from configuration 111, we can first locate 011 (using block {1,2}), then 010 (using block {3}), and finally 100 (again with block {1,2}); alternatively, we can locate 110 (using block {3}) and 100 (with block {1,2}). It can easily be verified that the same blocks actually "work" for all other starting configurations. The algorithm just presented will find this decomposition.

Near-Decomposability

So far, when building a decomposition scheme for a problem, we have looked for perfect decomposability, in the sense that we require that all blocks can be optimized in a totally independently from all others. In this way it is guaranteed that we can decompose the problem into perfectly isolated components (in the sense that each of them can be solved independently). This is a very stringent requirement: even when interdependencies are rather weak, but diffused across all components, we tend to observe problems for which no perfect decomposition exists.

For instance, in Kauffman's NK landscapes (see Kauffman, 1993), for such small values of K as 1 or 2—that is, for highly correlated landscapes—the above-described algorithm finds only decomposition schemes of size N or just below N.

We can soften the requirement of perfect decomposability into one of near-decomposability: we do not want the problem to be decomposed into completely separated subproblems (i.e., subproblems which contain all interdependencies); we want subproblems to contain only the most "relevant" interdependencies, though less relevant ones can persist across subproblems. In this way, optimizing each subproblem independently will not necessarily lead to the global optimum, but to one of the best solutions.[8] In other words, we construct "near-decompositions" which give a precise measure of the trade-off between decentralization and optimality: higher degrees of decentralization and market coordination, and therefore higher speed of adaptation, can be obtained at expense of the optimality of the solutions which can be reached.

Let $X_\mu = \{x^0, x^1, \ldots, x^{\mu-1}\}$ with $0 \le \mu \le 2^N - 1$ be the set of the best μ configurations. We say that X_μ is reachable from a configuration x and for a decomposition D if there exists at least one $y \in X_\mu$ such that y is reachable from x.

We call basin of attraction $\Psi(X_\mu, D)$ of X_μ for decomposition D the set of all configurations from which X_μ is reachable. If $\Psi(X_\mu, D) = X$, we say that D is a μ-decomposition for the problem.

μ-decompositions of minimum size can be found algorithmically with a straightforward generalization of the above algorithm, which computes minimum size decomposition schemes for optimal decompositions.

The following proposition gives the most important property of minimum μ-decompositions.

PROPOSITION 4. If D_μ is a minimum μ-decomposition size, then $|D_\mu|$ is monotonically weakly decreasing in μ.

Proof. If $\mu = 2^N - 1$, then X_μ includes all configurations and it is trivially reachable for any decomposition, including the finest, $D_\mu = \{\{1\}, \{2\}, \{3\}, \ldots, \{N\}\}$ with $|D_\mu| = 1$. If $\mu = 1$ X_μ includes only the global optimum, the minimum decomposition size is $1 \le |D_\mu| \le N$. We still have to show that it cannot be $|D_{\mu+1}| > |D_\mu|$: if this were the case, X_μ could not be reached from $X_{\mu+1}$ for decomposition D_μ, which contradicts the assumption that X_μ is reachable from any configuration in X for decomposition D_μ.

The latter proposition shows that higher degrees of decomposition and decentralization can be attained by giving up optimality and allows provision of a precise measure for this trade-off. In order to provide an example, we generated random

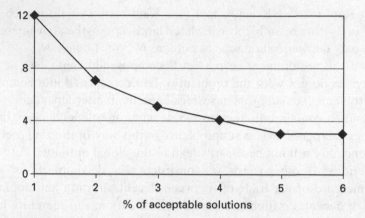

Figure 17.1
Near-decomposability.

problems of size $N = 12$, all of them characterized by $|D| = 12$ (i.e., they are not decomposable). Figure 17.1 shows the sizes of the minimum decomposition schemes as we vary the number μ of acceptable configurations (average on 100 random landscapes).

Speed and Accuracy of Search: Some Consequences for Organizational Structures

The trade-off between decomposability, reduction of complexity, and speed of search on one hand, and optimality on the other, as outlined in the previous section, enables us to discuss some interesting evolutionary properties of various organizational structures competing in a given problem environment. The properties and algorithms analyzed in the preceding section allow us to build problems which can be decomposed with any decomposition scheme decided by the modeler. We can thus run simulations in which various organizational structures compete in finding solutions to a problem whose characteristics are entirely controlled by the experimenter.

In this section we briefly discuss how such simulations have been built and the main results they have produced.[9] First of all, in order to reduce the space of possible decompositions, we have supposed that only decompositions which are partitions of the problem into subproblems of the same size are possible and that only organizational structures which fulfill this constraint are viable. For instance, if

$N = 12$ (as in most simulations), only the following six decompositions, named for their size, are possible:

$D1 = \{\{1\}, \{2\}, \{3\}, \{4\}, \{5\}, \{6\}, \{7\}, \{8\}, \{9\}, \{10\}, \{11\}, \{12\}\}$

$D2 = \{\{1,2\}, \{3,4\}, \{5,6\}, \{7,8\}, \{9,10\}, \{11,12\}\}$

$D3 = \{\{1,2,3\}, \{4,5,6\}, \{7,8,9\}, \{10,11,12\}\}$

$D4 = \{\{1,2,3,4\}, \{5,6,7,8\}, \{9,10,11,12\}\}$

$D6 = \{\{1,2,3,4,5,6\}, \{7,8,9,10,11,12\}\}$

$D12 = \{\{1,2,3,4,5,6,7,8,9,10,11,12\}\}$

Problems characterized by one of these decomposition schemes can be created and populations of agents, each characterized by one of these decomposition schemes, compete in a simple selection environment to find better solutions. Such competition works as follows:

1. A problem is randomly generated whose minimum size decomposition scheme is one of the six possible.

2. A population of agents is created: each agent is characterized by one of the six possible decomposition schemes and is located in a randomly chosen configuration (normally we have used populations of 180 agents, 30 for each possible decomposition scheme).

3. Each agent picks a randomly chosen block, and by mutating at least one bit (and up to all bits) within such a block, generates a new configuration. If the latter is preferred to the previous one, the agent moves to it; otherwise, it stays put.

4. At given time intervals, all the agents are ranked: the ones located on the worst configurations are deleted from the populations and replaced by copies of the agents located on the best configurations. Such copies inherit the decomposition scheme of the parent, but are positioned on a different, randomly chosen, configuration.

Thus we have a selection environment in which decompositions compete and are reproduced from an initial population in which one-sixth of the decompositions are the "right" ones and the others are wrong.

The main results can be summarized as follows. First, it is not necessarily the "right" decomposition which invades the population: as the size of the right decomposition becomes big enough, agents characterized by decompositions which are finer than the right one tend to prevail. In fact, only agents with the right decomposition can find the global optimum wherever they start from, but their search process is very slow and can be invaded by agent which cannot reach the global optimum but does reach good local optima relatively fast. Indeed, potential optimizers can die out before they reach even good solutions. This result is even stronger in problems that we could define "modular" (i.e., characterized by blocks with strong interdependencies within blocks and much weaker—but nonzero— interdependencies between blocks. In these problems, higher levels of decomposition can be achieved at lower costs in terms of suboptimality.

In general, simulations show two persistent features, which are present in all but the most simple (highly decomposable) problems: a persistent suboptimality and a persistent diversity of agents, both in terms of the configurations achieved and in terms of the decomposition schemes which define them. This is the outcome of the multiplicity of local optima which characterizes environments with high degrees of interdependencies (see Levinthal, 1997). Simulations therefore tend to support the view that heterogeneity and suboptimality of organizational structures can indeed be a persistent feature of organizational evolution.

We have run other simulations in which we have, at given intervals, changed the current problem for one having exactly the same structure in terms of decomposability, but with different, randomly generated, orders. This can be taken as a condition of uncertainty: for instance, consumers have changing preferences over a stable set of characteristics. Interestingly, it turns out that even with totally decomposable problems, as the change of the order becomes more frequent, the population is entirely invaded by agents characterized by coarser and coarser decompositions, and at the limit by agents which do not decompose at all. It therefore seems that growing uncertainty has similar consequences of growing interdependency.

Finally, two more points are worth considering. First, higher degrees of decomposition allow the selection process to work effectively with less underlying variety. Blocks of size k can be optimized by selecting upon 2^k types of individuals: as k grows, the variety requirement becomes stronger, and thus less plausible.

Second, decentralized market coordination mechanisms can indeed exploit the advantages of parallelism and increase the speed of adaptation, but such parallelism can prevent important reductions of complexity in systems composed of nested or overlapping subsystems.[10] Consider as an example the extreme case of a problem which can be decomposed by nested blocks as in $D = \{\{1\},\{1,2\},\{1,2,3\},\ldots,\{1,2,3,\ldots,N\}\}$. The size of such a decomposition scheme is N, and thus markets working in parallel will face a problem of maximum complexity. However the problem would have minimum complexity if it were solved according to the following sequence: first block $\{1\}$ can be optimized; then block $\{1,2\}$ can exploit the optimal configuration of block {1} and optimize only the second component; and so on. In order to exploit such a reduction of complexity, a precise, unique sequence must be followed: it seems quite unlikely that it could spontaneously emerge starting from isolated markets working in parallel.

We already mentioned that there exists a trade-off between speed of increments and optimality, in that highly decomposed strategies increment quickly but frequently get stuck in local optima. However, there is another interesting finding: in the very early stages, when simulated agents find themselves in very poor areas of the landscape, they are *slower* than nondecomposed strategies. This is due to the fact that their moves out of "wells" have a limited range compared to nondecomposed strategies. This result suggests an interesting interpretation of the historical development of real-world market organizations. Typically, when a product has just been invented, a vertically integrated firm is the most common producer. Later a process of vertical disintegration introduces market interactions between producers who limit their activities to portions of the whole production process.

Conclusions

In this chapter we claimed that economic theory has at its very core a strong, albeit implicit, idea of modularity. Perhaps there is nothing closer to an ideal modular system than the picture of the economic system which has been developed since the early theorizing of Adam Smith and long before the birth of such ideas as object-oriented programming, modular software development, agent-based systems, and the like. This implicit theory of modularity, which finds its higher expression in the theory of perfectly competitive markets, is grounded on two basic concepts: division of labor and market selection. The former is the process by which modules are created; the latter is the process by which they are coordinated.

Economic theory has much to say about the latter mechanism (the theory of markets, partial and general equilibrium theory, and on up to more "heterodox" and evolutionary theories of market selection[11]); but it has much less to say about the former, which has fallen well below the top position it held on the research agenda attributed to Adam Smith. Economic theory has even less to say about the connections and interdependencies between the two mechanisms: the existence and "granularity" of modules has always been assumed to be exogenously given, and irrelevant for the outcome of the coordination process.

In this chapter we argued that it assumption is unwarranted: the "morphogenetic" process of construction of modules determines the effectiveness of the market coordination mechanism. Only when the economic system is fully decomposable does it achieve optimal coordination, irrespective of the underlying modules into which labor has been divided. In all the other cases, particularly in nearly decomposable systems (which, following Simon, we argue to be the almost universal case), there is a trade-off between the scope of market selection (i.e., the depth of the division of labor into autonomous units) and its efficiency. Economic organizations and institutions are solutions of this dilemma, and their genesis and evolution can be fully understood in terms of this delicate balance.

Finally, there is also a deeper sense in which economic organizations and institutions enter the scene, and it has to do with the "representation problem" in evolutionary computation. Adaptation and evolution can work effectively only if variations are not too frequently very disruptive and not too rarely favorable. In computer science it is well understood that this depends crucially on how the task environment is represented and coded. By acting on the representation, we can make simple adaptive mechanisms such as mutation and recombination either very effective or totally ineffective.

In biology a similar problem concerns the "evolvability" of organisms: if, as biologists claim, evolution is the result of the interplay between mutation and selection, then we must ask where these mutations come from, and particularly whether an evolvable genome is itself a product of evolution. A representation problem also arises in biology, because the crucial factor affecting "evolvability" is the genotype–phenotype map, which determines how variations at the genotypic level affect the phenotypic characteristics, and thus the fitness of organisms. Since mutations in the biological world are believed to be totally random, this randomness must be able to produce some improvement. Again the likelihood of producing ordered and stable structures out of totally random events depends crucially on the representation: modularity and near-decomposability can be a way to solve the representation problem (see Wagner and Altenberg, 1996).

The interesting feature of the evolution of socioeconomic systems is that the representation and encoding of the landscape is not (unless to a minimum extent) exogenously given by some "law of nature," but is a social construct embedded in organizations and institutions, which in a sense are the loci of socially distributed representations of the socioeconomic problems we face. We believe that it could be very fruitful to pursue a line of research in organizational economics which starts from the acknowledgment that the way in which economic activities are organized, particularly their distribution between markets and hierarchies, not only determines the static efficiency of the economic system but also—and more importantly and with inevitable trade-offs—shapes the possible dynamic "morphological" paths which the system can follow, its space of possibilities.[12]

Acknowledgments

We thank Esben Andersen, Mauro Caminati, Giovanni Dosi, Massimo Egidi, Koen Frenken, Yuri Kaniovski, Thorbjørn Knudsen, Daniel Levinthal, Paolo Legrenzi, Scott Page, and Herbert Simon for useful discussions, remarks, and suggestions. This research is part of the project NORMEC (SERD-2000-00316), funded by the European Commission, Research Directorate, 5th framework program. The authors are solely responsible for the opinions expressed in this chapter, which are not necessarily those of the European Union or the European Commission.

Notes

1. As Coase himself stressed, in many cases this solution holds only in principle. In reality there might exist many reasons why the creation of such a market is difficult, costly, or downright impossible. These reasons are referred to as *transactions costs*.

2. A decomposition can be considered as a special case of search heuristic. Search heuristics are ways of reducing the number of configurations to be considered in a search process.

3. We intend to use intrablock as a proxy for hierarchy or centralized organization, and interblock as a proxy for market or decentralized interaction.

4. We remind the reader of the assumption of uniqueness of the global optimum.

5. A special case of global decomposability, called dominance, is presented in Page (1996) and is generalized here. In our terminology, a block configuration $x^h(d_k)$ is dominant if $x^h(d_k) \wedge x^j(d_{-k}) > x^j(d_k) \wedge x^j(d_{-k})$ for every configuration $x^j(d_{-k})$ (i.e., if it is always preferred to all the other configurations of that block, irrespective of the configuration of the rest of the string).

6. The complete algorithm is quite lengthy to describe in exhaustive and precise terms. Its Pascal implementation is available from the authors upon request.

7. Kauffman (1993) provided some general properties of one-bit-mutation search algorithms (equivalent to our bitwise decomposition schemes) on string fitness functions with varying degrees of interdependencies among components. In particular, he found that as the span of interdependencies increases, the number of local optima increases as well, whereas the size of the basin of attraction of the global optimum shrinks.

8. This procedure allows one also to deal with the case of multiple global optima. Thus we can now drop the assumption of a unique global optimum.

9. Due to limitations of space, we can give only a succinct summary of all the simulations here. Programs and detailed results can be obtained directly from the authors upon request.

10. A similar property is measured by Page (1996) as the *ascent size* of a problem.

11. See, e.g., Nelson and Winter (1982).

12. See Marengo (2003) for a tentative formal investigation of this issue.

References

Alchian A, Demsetz H (1972) Production, information costs and economic organization, Amer Econ Rev 62: 772–795.

Coase RM (1937) The nature of the firm. Economica 4: 386–405.

David PA (1985) Clio and economics of QWERTY. Amer Econ Rev 75: 332–337.

Fontana W, Buss LW (1994) "The arrival of the fittest": Toward a theory of biological organization. Bull Math Biol 56: 1–64.

Frenken K, Marengo L, Valente M (1999) Interdependencies, nearly-decomposability and adaptation. In: Computational Techniques for Modelling Learning in Economics (Brenner, T, ed), 145–165. Dordrecht: Kluwer.

Hurwicz L (1971) Centralization and decentralization in economic processes. In: Comparison of Economic Systems (Eckstein, A, ed), 79–102. Berkeley and Los Angeles: University of California Press.

Kauffman SA (1993) The Origins of Order. Oxford: Oxford University Press.

Klepper S (1997) Industry life cycles. Ind Corp Change 6: 145–181.

Langlois R, Robertson P (1989) Explaining vertical integration: Lessons from the American automobile industry. J Econ Hist 49: 361–375.

Levinthal D (1997) Adaptation on rugged landscapes. Manage Sci 43: 934–950.

Marengo L (2003) Problem complexity and problem representation: Some implications for the theory of economic institutions. Cognitive Economics (Rizzello, S, ed), 371–388. Dordrecht: Routledge.

Nelson R, Winter S (1982) An Evolutionary Theory of Economic Change. Cambridge, MA: Harvard University Press.

Page SE (1996) Two measures of difficulty. Econ Theory 8: 321–346.

Robbins L (1932) An Essay on the Nature and Significance of Economic Science. London: Macmillan.

Simon HA (1969) The Sciences of the Artificial. Cambridge, MA: MIT Press.

Simon HA (1983) Reason in Human Affairs. Stanford, CA: Stanford University Press.

Simon HA (1991) Organizations and markets. J Econ Perspect 5: 25–44.

Smith A (1976) An Inquiry into the Nature and Causes of the Wealth of Nations. Oxford: Clarendon, Press.

Stadler B, Stadler P, Wagner G, Fontana W (2000) The topology of the possible: Formal spaces underlying patterns of evolutionary change. Santa Fe, NM: Santa Fe Institute Working Paper #00–12–070.

Wagner GP, Altenberg L (1996) Perspective: Complex adaptations and the evolution of evolvability. Evolution 50(3): 967–976.

Williamson O (1975) Markets and Hierarchies: Analysis and Antitrust Implications. New York: Free Press.

18 The Natural Logic of Communicative Possibilities: Modularity and Presupposition

D. Kimbrough Oller

Logical Distinctions and Modularity

The richness of human language and the way it is utilized both in social discourse and in nonverbal mental operations hint at the possibility that much of human cognitive structure is rooted in the same structures that make language possible. Language expresses thought and seems to support thought; in addition, the structures and structural limitations of both appear to have much in common. Still, it is important not to confuse the idea that language and thought are closely intertwined in humans (an idea that seems unassailable) with the untenable notion that thought is not possible without language, nor with the equally unacceptable idea that thought is nothing but language. The two are quite distinct from a logical standpoint, and this fact can be illustrated in many concrete ways. For example, we can perceive, recognize, and imagine objects for which we have no names. Further, thought can be expressed in extremely complex ways even in animals that do not possess "language" (Griffin, 1992; Hauser, 1996).

It is convenient to raise the logical distinction between thought and language in the context of the discussion of modularity, because the distinction points in the direction I wish to take this chapter. Logical distinctions between functional mental systems (such as language and thought) must be addressed in any theory about the nature of the mind. At a high order of abstraction, such distinctions specify modules of function, which, while they may be related in important ways, still must be treated with notable separation. As we approach language in particular, we find additional ways that logical distinctions and logical possibilities must guide the specification of modularity in its broadest strokes. The fundamental contentions of this chapter, and of my recent theoretical endeavors, are based upon the notion that the natural logic of potential communicative systems must be accounted for in order to understand language structure (Oller, 2000). The natural logic as I envision it includes a series of distinctions that constitute modular elements of communicative capability. The implications of the approach are far-reaching. The infrastructural natural logic, in my view, offers the following opportunities:

• Understanding, at a very general level, how language is structured in the mind in terms of modules of function

• Relating the mature structure of language to its primitive beginnings in the human infant

• Comparing richly the nature of human vocal communication with that of nonhumans

• Providing the basis for fruitful speculations about the evolution of language in our species

• Laying the groundwork for a theory of possible communicative evolution in any species, human, nonhuman, or—yes—extraterrestrial.

An immodest set of goals perhaps, but then, I am not alone. For example, when Richard Montague, the progenitor of one of the several rich formal systems of linguistic description of the last few decades (see, e.g., Montague 1970), was informed that linguistic rationalists of the Chomskyan tradition wished to construct a theory of all human languages (but only human languages), he wondered aloud why they should wish "to disqualify themselves from fieldwork on other inhabited planets" (Partee, 1979, 95).

The model I propose emphasizes the notion that there is a limited range of logical possibilities for structures that can occur in natural systems, including language, regardless of whether they are in humans or on earth. These logical possibilities specify necessary modules of function, but it may be important to note here that the notion of modularity that I invoke is not isomorphic with the usage found in much linguistic literature of the Chomskyan tradition. For one thing, in my view, many systematic, modular characteristics of human language (and we will consider examples below) do not necessarily have to be innate, even if they cannot be learned strictly on the basis of pattern analysis of language data presented to the learner (and it has often been argued that they could not be learned on such a basis; Chomsky, 1966, 1968, 1986).

The modules and their systematic relations could instead, in many cases, represent optimized solutions to the task at hand: that of forming a rich and efficient system of communication (and perhaps a rich and efficient system of thought, to the extent that thought and language have common structural features). In this view, the human mind might be thought to adapt itself or to self-organize (and traits might be naturally selected in evolution) to reflect modules of function that are required for rich communication. Assuming such adaptations are guided by the logic of possibilities, the development of (even domain-specific) language modules, at least in some cases, might not have to be prespecified in the genotype.

This idea does not imply that there are *no* genetically specified linguistic characteristics in humans, but it does imply that some critical features of language may emerge without genetic specificity for language when a relatively intelligent organism (a small child, for example) actively seeks a solution to the problem of

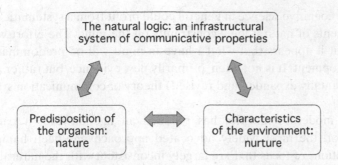

Figure 18.1
A general theoretical account of structure and origins in language and other rich communication systems.

rich communication. The idea implies that we cannot assume innateness of observed language modules in humans merely because they do not appear to be learnable on the basis of pattern analysis of linguistic information presented to the potential learner. There *is* another possibility: that the modularity and complexity of language results from an interaction among three influences: the predispositions of the organism (nature), the input to the learner (nurture), *and* the natural logic of possibilities (see figure 18.1). It is the last of the three elements that has been largely ignored in prior models, even in approaches to language that emphasize self-organization and dynamical systems.

The Natural Logic of Communication Systems and Universal Grammar

The natural logic approach requires development of an "infrastructural" model including hidden units, properties, and processes that are not directly observable but that account deeply for the function of systems. The model is intended to resemble other infrastructural scientific models in that its hidden components provide the primary theoretical apparatus. In the science of language it will be necessary to develop infrastructural modeling for all three elements of figure 18.1 (nature, nurture, and natural logic). However, in the present chapter I shall focus on infrastructural modeling only for the natural logic component depicted in the figure.

In my opinion, what has been missing in the science of language is a thoroughly infrastructural model. In previous theoretical approaches the natural logic of possibilities has been left out altogether or it has been accorded only ad hoc status to account for special problems of explanation. The systematic development of the natural logic is, in my view, the most pressing issue in linguistic science, and I am inclined to add that cognitive science in general could profit from a systematic effort to specify the elements of natural logic constraints on the mind. The effort will be partly empirical, but it appears that what I have in mind will be predominantly a philosophical development. It is not, then, primarily new evidence, but rather a new (or at least fundamentally expanded and revised) theory of communication systems that is needed.

While much of modern linguistics has pursued an agenda with a partially infrastructural flavor, the most widely advocated approach has been hampered by important limitations of focus that are largely inconsistent with the natural logic framework. The widely recognized approach which seeks to characterize the universal grammar (UG) postulated to underlie similarities among human languages has been identified since the 1950s heavily with Chomsky and his followers (for review see Haley and Lunsford, 1994; Newmeyer, 1996). The insularity of this framework has a variety of facets. On the one hand the approach has not only deemphasized, but has actively avoided, issues of evolution and development of communication systems in other creatures.[1]

In general it has provided no comment on how posited linguistic structures might have emerged in evolution. The model has posited innateness of linguistic structures, but has offered no model for how the innate structures might have come to exist in terms of progression from more primitive forms of communication. Further, proponents of the model have tended to insist that their empirical evidence has ruled out the possibility that any solution other than genetic innateness could possibly account for many of the characteristics of human grammar. As an example from a recent review by a proponent of the position, Smith (1999) notes that children seem quickly to acquire certain well-defined features of grammar even though the language environment (the conjunct of utterances produced in children's presence by grownups) seems to provide no teaching for those features and inconsistent evidence upon which learning might be based. He claims it is reasonable to believe these features are innate because, he says, research has "removed the plausible alternative explanation that we are taught these things. If correct, this demonstrates that we must ascribe a large part of the knowledge we end up with to the initial state, to UG rather than to the effect of the linguistic input we are directly exposed to. In brief, it is innate" (p. 42).

The evidence marshaled to support this view is, I believe, overinterpreted, and we shall see examples of alternative interpretations suggested by an infrastructural, natural logic approach below.

A final aspect of the insularity in the radical innateness approach is somewhat more subtle, but it is crucial to the understanding of the general limitations of the framework. While proponents of the model often claim to be interested in "explaining" language acquisition, they tend to do so exclusively by reference to models composed solely of mature units. These mature units can be grammatical categories, levels of representation, principles of grammatical theory, or parameters of universal grammar, to name a few of the possibilities. In a variety of recent works detailing proposed formal characteristics of grammars within the model (and the proposals have been modified numerous times since the 1960s; for a perspective on the vicissitudes of the perspective see Newmeyer, 1996), the only sense in which children have been portrayed as differing grammatically from adults is in *whether or not* certain presumable universal grammar parameters have yet been set (have yet assumed language-specific values) in the children.

The emphasis on the words "whether or not" is important, because the model presumes that children do not acquire precursors to grammatical knowledge. The parameters of grammatical function must be fully preprogrammed for adultlike function since, according to the reasoning, they cannot be learned. The input to the learner in terms of sentences produced in the environment is deemed too unstructured by itself to be used in the learning of a functional grammar. This is the "poverty of the stimulus" argument, encompassing the assertion that human grammars are unlearnable.

The pervasive property of "poverty of stimulus" is striking even in the case of simple lexical items. Their semantic properties are highly articulated and intricate and known in detail that vastly transcends relevant experience, and is largely independent of variations of experience and of specific neural structures over a broad range (Chomsky, 1993a, 24).

Thus, it is concluded that the parameters of grammar must be present in the infant mind, and that the only function of input (learning) must be to set switches specifying values that are innately available as options for each parameter.

There is no room in this model for a baby to have a prelanguage capability with properties that might be incomplete or immature in the same way that, for example, a bud possesses characteristics that are immature with respect to the characteristics of a mature flower. A botanist would not be inclined to describe the bud as possessing a mature stamen, mature petals, and mature sepals with on-off switches to determine their length, color, and so on. Instead, the bud would be seen as possessing precursors to all of these mature floral structures. The structures of the flower are emerging (and emergent) in the bud, and the bud thus possesses a type of organization and a set of structures that cannot be equated, one to one, with those of a mature flower.

Yet in the linguistic perspective I am critiquing here, the child is seen as possessing mature grammatical capabilities (or the unset parameters that specify them) innately, and within the model these can be tweaked during language development only in a sort of switch-setting that allows the child to choose the particular mature characteristics of the language being learned. Learning is reduced to acquiring lexical items and setting switches for mature parameters because the mature grammar is presumed already to exist in the child's head before any learning begins. This sort of approach is patently preformationist in character, resembling a theory of developmental anatomy from more than two centuries ago (Bonnet, 1762). In this antiquated and uninsightful anatomical theory, a preformed homunculus was posited to exist in the fertilized ovum; the homunculus was believed to contain within it the mature structures of anatomy—fingers, toes, heart, liver, and so on. In this way Bonnet sought to "explain" the emergence of mature anatomical features, but clearly all he actually did was to beg the question of how the structures might have arisen.

The approach I advocate, in contrast, is designed in such a way as to provide a basis for characterizing the infant on the infant's terms, just as we are inclined nowadays to characterize a zygote on a zygote's terms, and the bud of a plant on the bud's terms. In this view, the infant does not necessarily possess adult communicative or linguistic characteristics, but instead commands only primitive modules of function that have to be differentiated as development proceeds in order for them to attain mature status. These primitive communicative systems are specifiable (at least in part) within the infrastructural, natural logic approach.

The natural logic model provides a basis for characterizing the range of possible solutions to the problem of forming a "rich communication system." It specifies a hierarchy of properties and principles that must be commanded, and thus offers the opportunity to characterize richly the nature of primitive as well as more elaborate systems in terms of placement within the hierarchy; primitive communication systems are not required to possess fully formed or "mature" characteristics of very elaborate systems such as human language; they can instead be characterized in terms of precursor capabilities that are describable in the context of the natural logic. I provide a brief sketch of certain fragments of the hierarchy below, but for more elaborate characterization of the model see Oller (2000).

It should be added that since the implementation of a rich communication system offers a range of opportunities consistent with the properties and principles, the number of possible communication systems will surely prove to be indefinitely large. A broad range of opportunities appears to be available for both primitive communication systems (as in human infants or in nonhuman primates, for example) and more complex ones (as in preschool children or mature humans). However, to say that there is a range of possible communication systems does not negate the value of the logic in constraining the types of possibilities that exist, and in specifying sequences of potential evolution and/or development at a general level.

An advantage of the infrastructural approach is that it details the properties of systems that possess power intermediate between those of, for example, nonhuman primates and modern mature humans. The natural logic approach affords comparative perspective across the infrastructural components of a wide array of possible language and languagelike systems.

It is ironic that in the context of this infrastructural notion, I should be inclined to characterize Chomsky's approach as explanatorily superficial, because Chomsky has maintained from the very beginning that the goal of his linguistic enterprise is to characterize the deepest principles of which human grammar is composed (Chomsky, 1957, 1965). And credit is due: important achievements have been made in formulation of increasingly general principles and parameters of function that characterize *mature* human languages. In the "minimalist program" that is the recent focus of much of the UG endeavor, linguists are seeking principles "external" to the grammar that may determine its structure (Chomsky, 1993b). And yet, even in this context, the claim that the essential human gift for grammar must be innate persists.

Roots of the Infrastructural, Natural Logic Model

While the infrastructural approach is novel in some regards within the context of linguistics and psycholinguistics, it does have notable theoretical antecedents, some of them going back many years (see Bates and MacWhinney [1982] for a review of functionalist approaches to language, approaches that share much with this effort). The crux of the proposal is, thus, far from new. For an example of an antecedent that comes from outside the realm of language, consider Koffka's account of the forms of certain natural physical structures: for example, honeycomb cells and echinoid plates are both hexagonal, not because there is a direct genetic endowment in bees and other creatures for hexagon creation, but because there is a limited range of geometric forms that can fill up a space (Koffka, 1928). The bee may be assumed to have a predisposition to create a hive with cells, but the regular shape of cells does not itself have to be genetically specified or even to be specified as a predisposition, because the hexagonal shape is predetermined logically as long as the bee pursues the building of cells that fill up spaces. Considerations of natural logic thus account in significant ways for the form that beehives assume, and these considerations are entirely independent of the bee's physical makeup or preprogramming, as well as being independent of the "learning" environment of the bee.

In modern paleontology, a concept has developed that also has much in common with the sort of natural logic I propose for language. The morphology of evolving creatures is not entirely unpredictable, and the lines of possible evolution for physical structures appear to be restricted by logical possibilities based on considerations of, for example, physics and economy of organization. Consequently, as one examines the physical structures that have occurred in nature, the pattern of occurrence is not by any means random, but appears instead to be limited by logical possibilities that are themselves a primary object of development in the modern theory of morphology (see chapters 9, 10, and 11 in this volume). The theory provides a framework including general parameters of possible morphological change and a generative system specifying possible and/or likely changes. As the model is elaborated in the future, it should become possible for it to provide accounts of potential modifications of morphology in even more diverse circumstances. Logically speaking, the model provides a framework for interpretation and prediction that is already applicable to extremely varied settings of morphological change and is logically independent of any particular circumstance. Thus the approach specifies a natural logic, in accord with my usage.

A key idea of Piaget's "genetic epistemology" (Piaget, 1969) is that the mind acquires its cognitive structure in part by adapting to natural possibilities. The structures do not in all cases have to be preprogrammed because a moderately intelligent mind without very domain-specific gifts can acquire domain-specific structures as it adapts to the limited range of possible solutions to domain-specific cognitive problems.

Modern connectionist thinkers often cite Piaget as a progenitor to important aspects of their approach. Piaget was an interactionist who assumed that innate and environmental forces had to work in concert to produce knowledge, in the context of "epistemological" constraints (which I take to be similar in many cases to what I call naturally logical constraints). And in connectionist thinking there is also a strong hint of interest in naturally logical systems (Elman et al., 1996) that might supplement the influences of innate and environmental forces in the development of mental capabilities, especially in the area of language.

Within the connectionist approach, empirical evidence can in some cases be developed through simulations that utilize computer models of neural networks. Architecturally simple neural networks have been shown to self-organize in surprising ways, providing an existence proof of the idea that simple learning systems can, under some circumstances, yield outcomes mimicking languagelike units, outcomes that seem unpredictable based upon the simple nature of the inputs to the networks. With small changes in timing of simulated events or other simple architecture constraints on networks (such as changing the size of a memory buffer with time in a way that might be thought to mimic the growth of memory capabilities in children), even more elaborate structural learning of languagelike units can be obtained.

Such results offer encouragement for the hypothesis that highly structured systems, functioning in such a way as to suggest rich languagelike modularity (even if the system shows little or no modularity in its physical pattern of connections), might be attainable through self-organizational development in the context of a few minimal initial architectural constraints. It seems unlikely that such constraints would have to take the form of hardwired grammatical principles, but could instead perhaps be the results of more general predisposition principles such as, perhaps, active search, analysis, storage, reanalysis, self-correction, cross-element evaluation, and a few general selectional principles such as efficiency and completeness of solutions.

In *Rethinking Innateness: A Connectionist Perspective on Development* (Elman et al., 1996), an outline is provided of innate constraints that might be posited to occur in language development. The great bulk of the discussion is devoted to consideration of

(1) representational constraints that might be implemented in specific innately determined microcircuitry

(2) architectural constraints that might be implemented at

   (a) the level of individual neural units in terms of cytoarchitecture type, firing threshold, and so on

   (b) the local level in terms of number of cell layers in the brain, packing density, and so on

   (c) the global level, reflected in connections between brain regions and so on

(3) timing constraints that might be implemented in terms of

   (a) number of cell divisions in neurogenesis

   (b) waves of synaptic growth and pruning

   (c) temporally adjusted development of sensory or motor systems.

These sorts of constraints of course apply to the neural implementation of mental systems. But in an intriguing suggestion of just a few lines, the authors invoke constraints outside the neural system, constraints associated with the nature of possibilities of linguistic growth:

... we note that there is one more very important source of constraint on development, namely the constraints which arise from the problem space itself ... this source is entirely external to the developing organism ... we have not included it in our taxonomy of 'ways to be innate' ...

We refer here to the fact that certain problems may have intrinsically good (or sometimes, unique) solutions. For example, the logical function Exclusive OR ... readily decomposes into two simpler functions, OR and AND. Given a range of possible architectures, networks will typically 'choose' this solution without being explicitly instructed to find it ... the solutions ... need not be internally specified. These outcomes are immanent in the problems themselves. (Elman et al., 1996, 34)

I might be inclined to modify the statement slightly. The outcomes are immanent in the problems only insofar as there are limited solutions to them. The specification of the solutions pertains to the domain of natural logic as I propose it. The above quotation represents most of what is said about the potential importance of natural logic in explaining the chaotic mysteries of development and self-organization in simulated systems in Elman et al. My suggestion is that we need to develop this idea richly and soon.

Given my inclination to criticize the Chomskyan approach, the recent "minimalist" program in linguistics (Chomsky, 1993b) deserves comment because in some regards the program has a flavor of natural logic. Consider again Neil Smith's review:

The aim of Minimalism is to show that "what is empirically unavoidable is imposed by 'design specifications' of external systems," so that all constraints on grammatical processes are motivated by either perceptual or conceptual considerations. Binding theory for instance, described here as a module of the grammar, is now suspected to be just "an external interpretive system": i.e., a set of conditions imposed from outside the grammar proper. Similarly, the reason for the ubiquitous existence of movement, which is clearly not conceptually necessary, may be that it is a response to requirements of communication: putting old information before new, for instance. (Smith, 1999, 87)

Further on in the review, Smith notes that the minimalist approach has sought to characterize human language as:

[approximating] a "perfect system": an optimal solution to the basic problem of relating sound and meaning. It is putatively optimal in the sense that it manifests properties of simplicity, naturalness, symmetry, elegance and economy. (p. 90)

After decades of insistence from Chomsky and his followers that primary linguistic capabilities must reside as strong predispositions in the mind (and further, in the genes), it seems odd that now we are encouraged, by the same camp, to consider the possibility that such capabilities may be as they are because they yield optimal solutions to the problem of relating sound to meaning. Such reasoning dovetails with the natural logic approach because it emphasizes natural constraints on the functions of language. But the minimalist program differs from my approach in that it requires that the "optimal solutions" be innate. In contrast, the natural logic approach leaves open the question of how much of language capability is innate and how much is self-organized by individual learners.

We hear from Smith that Chomsky has recently referred to the minimalist reasoning as "maybe the most interesting thing I've thought of" (Smith, 1999, 86). In fact there are quite a number of people who have for a long time entertained an important role for "external constraints" on language. In order to pursue the reasoning related to external constraints thoroughly, I contend that we need to develop a general theory of natural logic, and that this theory should assume a role among the primary domains of study in language along with the predispositions of the organism (nature) and the environments of learning (nurture).

A Naturally Logical Approach to Vocal Communicative Development in the Human Infant

Consider a case that illustrates the idea of a natural logic-based solution to accounts of development, a solution that contrasts sharply with preformationist approaches. The example is drawn from empirical/theoretical work on infant vocal development (Oller, 1980, 1995, 2000). In the account, the steps of the progression toward well-formed syllables in infancy are described and the naturally logical requirements that appear to guide the steps are specified.

The Landscape of Infant Vocal Development By way of background it is necessary to start with some observations regarding the landscape of infant vocal development. Human infants come to control the production of canonical or well-formed syllables typically by the time they are 5–8 months of age. Canonical syllable development, usually accompanied by the production of reduplicated sequences such as *bababa*, or *nanana*, is extremely recognizable to parents and others, and reveals that the infant has come to command a variety of properties of the infrastructure of phonology (consequently we call the study of these properties "infraphonology"). Prior to the onset of canonical babbling, infants produce a wide variety of "protophones," such as squeals, growls, isolated vowellike sounds, "raspberries," and so on. The protophones are presumably precursors to speech that differ in utilization pattern from the less speechlike vegetative sounds (coughs, burps, etc.) and fixed signals (cries, laughs, etc.).

Now it might be assumed that canonical babbling would be the first sort of event to be noted in scientific endeavors regarding infant vocalizations. But not so. The recognition of the onset of canonical babbling by psycholinguists was very late in coming because for the first few centuries of interest in infant vocalizations, only preformationist description was utilized. Observers made phonetic transcriptions of infant sounds, even long before the sounds were canonical (Fisichelli, 1950; Grégoire, 1948; Lewis, 1936). These phonetic transcriptions shoehorned the sounds of infants into mature categories of phonological function long before the sounds met the structural requirements of such categorization. Such categorizations were nonsensical in the same way that categorizing the fertilized ovum in terms of mature anatomical categories (by using the homunculus idea) was nonsensical. And because the categorization of infant sounds was garbled and rendered useless by shoehorning, it was not possible to notice the onset of canonical babbling—every infant sound was treated as if it were already canonical, and so the description begged the question of how the system of canonical speech sound production had developed.

In order to escape the trap of preformationism, protophones have to be described on their own terms, without shoehorning. This approach to description is much like that of ethologists when they describe the vocal system of some newly discovered mammal. They watch creatures in natural environments, take note of vocalizations and other apparent displays, and describe both the displays and the consistent circumstances under which they occur (Elowson et al., 1998; Hauser, 1996; Sutton, 1979). Ethologists then attempt to characterize the functions of vocalizations and other displays.

Observers of protophones do essentially the same thing: they do not try to fit the infant vocalizations into adult categories; rather, they treat the sounds on their own terms, as squeals, growls, yells, and so on, and they note how the sounds are used rather than attempting to force them into mature "meaning" categories. Special terms have to be made up for some of the protophones that babies produce. For instance, infants produce a category we now call "quasivowels," sounds that consist of normal phonation (the kind that is used in speech) produced when the infant is in a comfort state and with the supraglottal vocal tract at rest (not postured). These sounds have been called "comfort sounds," "small vowels," and sometimes "grunts" in a prior literature (although the term "quasivowels" refers to "grunts" that do not occur in conjunction with straining movements).

Similarly, we have to devise a term to refer to the "vowellike" normally phonated sounds (we call them "full vowels" or "fully resonant nuclei") that are produced with the vocal tract in any of a wide variety of special postures (for example, with spread or rounded lips, with tongue tip raised and fronted, and so on). Also important are the "marginal babbles" that consist of full vowels plus slowly articulated consonant-like elements. As for the reduplicated, well-formed syllable utterances that occur once the canonical stage is in place, we simply call them "Canonical Babbling." Many of the other protophones can be referred to with common parlance terms in this ethological description.

The Application of the Natural Logic   The occurrence and nature of protophones can be seen to be guided by a natural logic specifying the principles by which well-formed syllables are composed. The natural logic is independent of the human vocal system and brain, and should (according to my reasoning) constrain any self-organizing system to develop its system of syllables in the logically specified manner, as long as the system possesses or is simulated to possess a vocal tract and brain vaguely similar to those of humans. According to the reasoning, if another species of primate, for example, were to develop a communicative system requiring well-formed syllables, then the steps to be taken in that evolution would necessarily be very similar (at least in the broad outlines of the logic) to the ones the human infant takes.

The simplified natural logic account of vocal development (Oller, 1986) begins by recognizing that any vocal communication system, temporally organized, that would incorporate an open-ended number of units of transmission (a clear requirement of a powerful communication system), must have some way of chaining transmission units together. To be practically usable, chainable units must be constrained in time, and must be identifiable as distinct from other events that might be produced by the same vocal tract. These practical factors can be addressed optimally in the human case with a vocalization type that utilizes normal phonation (the most efficient kind of vocalization for humans, from the standpoint of the ratio of energy expenditure to discernible acoustic output) and that produces such phonation in brief chunks with a mean duration ranging from 200 to 400 ms, the duration of minimal rhythmic units in stereotypic repetitive movements of humans (drumming, finger tapping, etc.). (For a perspective on such rhythmicities, see Thelen, 1981.)

The normally phonated brief quasivowels that are specified by these constraints constitute the starting point on the journey toward well-formed syllables. They constitute a beginning minimal rhythmic unit for a system that can grow from that point, although it should be clear that quasivowels are to well-formed syllables as a bud is to a flower. The bud is not a segmentable part of the flower, but rather a precursor to it with a structure of its own, a structure that must be reconstituted in a complex mapping of precursors to mature forms as the flower blooms.

Next we recognize that it is necessary to string units together rapidly if there is to be rapid transmission of information through the vocal system. How are the units to be differentiated in chains? The simplest way (and this simplicity constraint should apply to temporally organized vocal systems in general, not just to the human one) is for the speaker to vary amplitude abruptly: to drop the amplitude of the quasivowels and then raise that amplitude systematically in accord with the optimal rhythmic patterning of the organism. This pattern of ups and downs of amplitude results in an identifiable sequence of quasivowels, bounded by the low-amplitude "margins" adjacent to each syllabic nucleus, where the nuclei are the amplitudinal high points of the quasivowels.

There are two broad ways to drop amplitude. One is to halt phonation at the glottis, and the other is to halt or inhibit the flow of phonated energy by closing off the supraglottal tract. If one inhibits flow with the back of the tongue (a maneuver that is particularly common in sucking anyway), moving from quasivowels to a lower-intensity sound inhibited by the tongue gesture, one obtains a sequence of quasivowel and consonant-like margin. In the early infant, we call sounds produced by normal phonation along with such suckinglike articulated tongue gestures "gooing." The addition of articulated margins possessing this "gooing" characteristic makes it possible for there to be a chain of quasivowels, each of which is identifiable, and each of which is a minimal rhythmic unit (a precursor that has some, but lacks other, properties of the mature syllable).

The next step is to make it possible for the individual nuclei of the potential rhythmic sequence to be differentiated (thus allowing contrasts of different nuclear-vowel types). The natural logic specifies, then, that the character of nuclei must be modifiable, and this can be done relatively easily by posturing the vocal tract in differentiable ways during normal phonation without closing the tract (as in the case of creation of margins). So, for example, if the infant opens the mouth and postures the tongue forward during normal phonation, a vowellike sound is produced that can be contrasted with the quasivowel. With additional postures, additional vowellike contrasts can be constructed.

The infant who is capable of producing sequences of quasivowels and full vowels, bounded by low-amplitude margins, is capable of producing a series of differentiated syllables, and when these different types are paired with meanings, a wide variety of different word types can be produced. So it is a logical step of development (in order to elaborate the vocal communicative system) to sequence the newly formed full vowels (just as sequencing for quasivowels is a logical step), with vocal tract closures bounding the nuclei thus formed in similar rhythm to that utilized previously, thereby allowing the infant to produce sequences of either quasivowels or full vowels or both, with each nucleus being identifiable as distinct.

To make the system yet more efficient, there is another logical step that can be introduced, consistent with both vocal tract movement capabilities of humans and auditory propensities of both humans and other mammals. The movements of the supraglottal tract that create the margins between syllable nuclei can themselves be refined. If they are made to occur rapidly, so as to maintain the 200–400 ms duration of syllables as a whole, then the spectral transitions between margins and nuclei (corresponding to the movements of the vocal tract) can become sources of information themselves, embedded within the syllable complex. Since the movements are necessary to create efficient chaining (and ultimately hierarchical rhythm), the exploitation of the opportunity that is afforded by speeding movements up to a pace that naturally conforms to the timing of rapid rhythmic stereotypies can yield a new source of information that rapidly multiplies the bit rate of transmission.

The increase in bit rate, of course, depends on the possibility of differentiating the margins themselves (i.e., producing a variety of auditorily differentiable margin types), but by this point in development, the infant has already mastered at least the rudiments of two ways of producing margins (glottal stopping and supraglottal closure). A series of additional types of supraglottal closure (or consonant types) is possible, and all these types become efficiently available to raise the bit rate of the potential system as soon as the rapid spectral transition is efficiently in place.

So at this point of our story, the natural logic (with reference along the way to a variety of special conditions of the organism) has specified a sequence of steps that lead systematically toward a transmission system that is rich by being practically open-ended (since these developments produce a wide variety of syllable types and no strict limit on how long chains of syllables can be) and efficient. The system depends on the use of contrastive nuclei and margins that are embedded in each minimal rhythmic unit, strung together in rapid succession in a manner that is consistent with production, perception, and other mental capabilities of the organism.

Let me emphasize at this point that the organismic limitations or predispositions specify what the organism might be able to do or be especially inclined to do, but they do not, in this way of thinking, constitute necessary adaptations for language. Furthermore, the naturally logical steps specified above, according to this reasoning, need not be preprogrammed in any way into the organism because they are aspects of the possibilities that would present themselves to any evolving or developing organism that might move toward a vocal communication system of substantial power (i.e., one with the full power of syllabicity).

The natural logic specifies that there are only certain solutions to the problem of creating a communication tool that will work well. The class of solutions may be indefinitely large (for example, because there are many kinds of nuclei and margins that can be selected by individual infants), but it is not entirely unconstrained because the general steps of the logic must be observed, or else the resulting system will lose ground either in completeness (it will not be open-ended enough to make it possible to form new words in a large and always growing lexicon) or in efficiency (it will fail to be able to transmit information at a bit rate that would make it competitive in the jungle of natural selection).

The fact that the selectional criteria of completeness and efficiency are related to each other in a dynamic tension suggests that there will always be room for adaptation of a natural vocal communication system and that, consequently, the number of natural communication systems that can evolve is extremely large, perhaps infinite. My suspicion is that even for individual children acquiring individual human languages, the steps in the process involve some flexibility all along the way and that the concrete capabilities of each infant will be unique. Still, the gross plan of development must be guided by the goal of the adaptation, which is a communication system with completeness and efficiency. These criteria, along with the inherent physical and mental limitations of the organism, and the constraints of the natural logic of potential communication systems, appear to create a circumstance that constrains the class of optimal solutions to adaptation of the communicative tool.

The Explanation of Stages of Infant Vocal Development

The natural logic of canonical syllable formation outlined above provides a clear basis for explaining how and, more important, why (at least in broad strokes) infant vocalizations emerge as they do. The well-recognized stages are four:

1. The *phonation stage*, in which quasivowels are prominent

2. The *primitive articulation stage*, where gooing emerges

3. the *expansion stage*, where first we see the systematic use of full vowels, and later of marginal babbling, where full vowels are combined with consonant-like margins in slowly articulated sequences

4. The *canonical stage*, where canonical babbling first occurs under systematic control with transitions between margins and full vowels, transitions that are quickened to the point that maximum efficiency of syllable production is attained.

The sequentially organized principles (they represent modules of capability) of the natural logic for canonical syllables specify the following requirements for a powerful and efficient vocal transmission system:

1. A *production capability for nuclei* that are identifiable and distinguishable from other vocal types, produced at a pace that conforms to the general rhythmic inclinations of the organism

2. An *articulation capability allowing production of chains* or sequences of nuclei with margins that separate syllables of the chains, in accord with the physical abilities of the organism

3. An *ability to produce contrasts among nuclei*

4. An *ability to chain contrastive nuclei* through articulation of margins that separate syllables of fully articulated chains

5. An *ability to produce fully articulated chains rapidly*, again in accord with the optimized rhythmic capability of the organism, exploiting the production and perceptual opportunities afforded by rapid transitions.

Note that the elements of the logic all refer to *capabilities* that must be attained and that the capabilities are naturally ordered in a hierarchy of presupposition (figure 18.2). Articulation between nuclei (criterion 2) requires that there exist nuclei. Similarly, contrasts among nuclei (criterion 3) require that nuclei exist. So criterion 1 is logically presupposed by both 2 and 3. It is not clear that 2 needs to precede 3 for logical reasons. But 3 clearly precedes 4, and 4 precedes 5, yielding the sequence 1 > 2, 3 > 4 > 5.

In general the process of development in normal human infants appears to proceed in accord with the logic, and there is fairly good evidence to support the posited sequence. The empirical evidence suggests there are some children in whom the primitive articulation and expansion stages merge (i.e., gooing and full vowels seem to emerge simultaneously). This pattern is logically possible—2 and 3 do not appear to be related in logical presupposition. But the other aspects of the ordering predicted by the natural logic appear to be observed uniformly, as logic seems to demand, by infants all over the world.

My suggestion, of course, is that the reason the natural logic is observed all over the world is not so much a question of human nature as it is a matter of necessity. Any other creature that might have evolved to possess such a powerful communication system as ours (and whose communication system was temporally ordered and vocal in nature) would also have had to proceed in development of

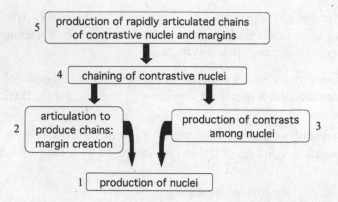

Note: Arrows indicate presuppositional relations

Figure 18.2
A sketch of naturally logical infraphonological capabilities (principles) required by the property of well-formed "syllabicity."

transmission units in accord with the natural logic. This is not a matter of innateness, but a matter of natural possibility.

How Infraphonology Has Introduced Practical as Well as Theoretical Advances
The natural logic approach has had fundamental impact in theoretical consideration regarding infant vocalizations, but it has also supported bountiful empirical products. An extensive review of the results is provided in Oller (2000). For a highlight, consider the fact that profoundly deaf infants are now known to differ dramatically from hearing infants in their pattern of vocal development. In particular, they are known to begin the canonical babbling stage late, and not just a little late. For practical purposes the distribution of ages of onset for normally developing hearing infants and for profoundly deaf infants is nearly disjunct (Eilers and Oller, 1994). Normally developing hearing infants virtually always begin the canonical stage by 10 months of age, and profoundly deaf infants rarely begin that early.

This empirical result is a direct product of the infraphonological, natural logic approach. Prior to the existence of the approach, the scientific tradition in infant vocal development was entirely preformationist in character, and infant protophones were routinely treated as if they could be fully characterized in terms of mature phonetic features—they were transcribed alphabetically as if they constituted well-formed canonical syllables, even in cases where the protophones were precanonical. The preformationist characterization begged the question as to whether children were or were not in the canonical stage because the very method of observation presumed all infant protophones were canonical.

Thus it was impossible in the preformationist era of infant vocalizations (prior to the 1970s) to recognize differences between hearing and deaf infants with regard to canonical babbling, because it was impossible to recognize canonical babbling as distinct from any other kind of vocalization. And it was claimed, in one of the most widely cited myths of child development, that deaf and hearing infants babbled similarly, except perhaps with regard to amount of babbling produced (Lenneberg et al., 1965; Mavilya, 1969). With the advent of the infraphonological approach, the difference between deaf and hearing infants with regard to one of the most important developments of infancy, canonical babbling, immediately became apparent.

Infrastructure and Modularity in Infant Vocal Capabilities

My approach to developing an understanding of infant vocal development and why it has the form it does, is guided by the assumption that the necessities of the task of developing a powerful system are at the root of every step in the progression of stages in infant sounds. The steps are discernible and isolable, and every aspect of the progression is constrained by the necessities. Furthermore, each of the steps can be recognized to represent a kind of capability that must be attained, with successive capabilities representing natural steps of growth beyond prior ones, steps that in most cases apparently cannot be taken without prior foundations.

Evolution and development do not typically allow leaps. If an infant were to go straight to the expansion stage at birth, for example, all the logically necessary components or capabilities inferior to those presupposed by expansion stage vocalization would have to be commanded at once. Skipping a stage entails the incorporation of the capability that the stripped stage requires. Component is built upon component, and some components must precede others.

A theory of macroscopic modularity must, in my view, begin at the level of these infrastructural components. The modules are literally the capabilities, and the presuppositional relationships among them specify developmental sequences as well as inherent relationships among capabilities. This idea suggests that there might be concrete structures (presumably distributed networks) in the brain to implement each of the steps, and that the networks might develop self-organizationally in sequence as specified by the logic.

The ability to produce and control normal phonation should (according to the reasoning) be an early brain development and an early evolved characteristic in humans—indeed, the fact that nonhuman primates are not able to phonate at will (at least not to the same extent as humans) is correlated with differences in brain structure between human and nonhuman primates (Jürgens, 1992). Similarly, the ability to control articulation should develop as a separate, modular system (and the control of movement of the vocal tract is managed by very different mechanisms, even to the point of different cranial nerve transmission lines, than those that manage phonation), and the timing of control should similarly be largely isolable as an additional neurophysiological system.

All this may seem obvious to those of a biological orientation, but the surprising fact is that in linguistics and psycholinguistics the search for modularity does not typically begin at such a point. Instead, the effort commonly begins by looking at mature systems, and this narrowly defined approach has led to the attempt to designate mature system properties as innate and available from birth (or at least from the point at which talking begins). In the approach of rationalistic linguists, the steps of vocal development specified in the natural logic above are ignored altogether. Instead, phonological capability is seen as composed of abilities to manage phonological features of syllables; both features and syllables are seen as well-formed from the beginning. The featural hierarchy that is posited for mature natural languages is typically assumed within markedness or optimality theory (see Archangeli and Langendoen, 1997) to be innately given. But such an approach offers no explanation as to how the mature capabilities got to be the way they are.

To assert innateness is to push the problem away rather than to offer a solution to it. The typical innateness claim about natural phonological constraints provides no specification as to the nature of the infrastructure that underlies such constraints. Consequently, the modules of function in rationalist modularity theory are developmentally unworkable, and they provide little help to those who would wish to evaluate the self-organization of communicative systems in individual children or the evolution of communicative systems in species.

The Distinction Between Signal and Value

The infraphonological example presented above provides a brief account of the potential role of natural logic in explaining the nature of vocal development and in specifying modules of function that will of necessity obtain in any rich, temporally organized communication system. The example offers only a sketch of one property from a natural logic in the domain of infraphonology. The property in this case is called syllabicity. There are a number of others that relate to each other in the presuppositional system (see Oller, 2000). All of the infraphonological properties pertain to ways that a *transmission* or "signal" *system* for communication might be structured.

There is, of course, another broad realm that is always involved in communication, and that realm concerns the messages or "values" of communication. In every communicative act, there is always a distinction to be drawn between signal and value. As a result, communicative systems that become elaborate and powerful develop modular signal systems that are clearly distinguishable from their modular value systems. This, in and of itself, represents a high-order modularity in communicative systems.

In specifying the nature of *signal* systems and the natural logic that underlies them, I have used the term "infraphonology" to designate the entire domain of study. In specifying the nature of *value* systems, I have conversely used the term "infrasemiotics" to encompass the study of referential meanings, sentential meanings, illocutionary forces, and so on.

An Example on the Infrasemiotic Side

As an example of infrasemiotic development, illuminated by a naturally logical scheme, consider the problem that faces the human child learning to speak at a very early stage, say around 18 months of age. The child already knows how to utilize a number of words or wordlike sequences of sounds. But each word tends to be utilized in limited circumstances. This is what we would expect from the vocalizations or other communicative displays of nonhuman primates, although in the latter case, learning new vocal displays appears to be extremely limited if not impossible. Our focus here is on the fact that for nonhuman primates, as for the example child, individual displays (and words for the child) do have one thing in common: each has one and only one communicative function, or, in the terminology of J. L. Austin, one "illocutionary force" (Austin, 1962). One word or display may be utilized as, for example, a social affiliative act and another as a complaint, but neither word changes roles at this stage of development (or at least such early words do not change roles very freely).

So the child might say "mommy" but the word at this stage is not analytically referential, since it does not necessarily designate a person for the child. Instead, the word is limited to being said by the child when engaged in certain events, for example cuddling or seeking of the same. For a child at this stage of development, the word would not be used to express other functions such as displeasure or complaint. "No-like-it," on the other hand, could be produced as a single word of complaint or refusal, but would not be used in other ways (for example, as a mere comment). This is a pattern of early speech that suggests that the child has no distinction between infrastructural components pertaining to referential meaning and illocutionary force. Both usages, "mommy" and "no-like-it," have illocutionary functions, but these cannot be altered until the child develops further.

At a later stage, the same child would produce the word "Mommy," now with referential meaning, to designate to the mother in a variety of circumstances and with a variety of illocutionary forces. The child could say "Mommy" when requesting the attention of the mother, when refusing to accompany the mother, when questioning whether the mother is or is not in the next room, and so on. The referential meaning of the word (the fact that it refers to a particular person) is constant across these circumstances, but the illocutionary force varies. In fact, at this later point in development the entire lexicon of nouns (which grows rapidly during this period) comes to be utilized with multiple illocutionary forces. For many reasons, not the least of which is storage efficiency, it obviously makes sense to modularize the system of values into a meaning (or semantic) component and an illocutionary force component that are related freely in individual acts of communication. From this point on, there is no need to store each semantic value along with each of its potential forces—the forces can be one system and the semantic values another. This restructuring does not just save space and time. It also conforms to the natural distinction between what one refers to in saying something about something, and what one intends to do in the act of so saying, to paraphrase Austin.

I posit the existence of a naturally logical property, which might be called "semanticity," denoting a capacity that children must master to differentiate meaning and illocutionary force.

Consider now the logical foundations for semanticity. At an early stage of communication development (or in an early stage of evolution for a creature moving toward a languagelike capacity) an organism may have the ability to produce a wide range of socially acquired, vocal illocutionary forces, some of which consist of mere referential acts and others of which consist of complaints, requests, and so on—all social acts of one sort or another. The ability to produce words in this illocutionarily limited way is based upon a variety of communicative properties (to be specified in natural logic). For example, such a child must command the property of conventionality (corresponding to the ability to acquire new pairings of vocalizations with forces, in social contexts), as well as the presupposed property of imitation (corresponding to the ability to replicate particular phonological sequences modeled [whether intentionally or not] by other speakers of the community). The three properties of infrasemiotic natural logic here noted are portrayed, with presuppositional relations indicated, in figure 18.3.[1]

As the number of the illocutionarily limited acts increases, there develops a naturally logical opportunity to reorganize the system in terms of references and forces,

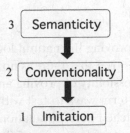

Note: Arrows indicate presuppositional relations

Figure 18.3
Three naturally logical properties of infrasemiotics.

a reorganization that can instantiate the distinction between meaning and illocutionary force in such a way that any referential expression can be free to express any illocutionary force. Both the need for completeness and the need for efficiency (in terms of mental storage, if nothing else) of communicative capability encourage this reorganization, which yields a new property of semanticity. The new property presupposes that the organism already commands the above-mentioned properties of conventionality and imitation (as well as a variety of others not discussed here), as indicated in figure 18.3.

So how is this proposal different from that of rationalistic linguistics? In the radical rationalist's approach, the ability to function semantically must be presumed to be present innately as a set of lexical frames to be filled merely by triggering when children are learning the pairings between meanings (values) and sounds (signals). It is not built upon a prior infrastructure of more primitive communicative acts, but is fully preformed in the infant's mind. The preformationist descriptive bias of the radical rationalist approach (which always focuses on the mature system's fully formed structures) imposes this rigidity, and renders its theoretical treatment impotent with regard to rich developmental description and devoid of content with regard to evolution.

It should be noted that there are numerous nonrationalistic proposals about early development of vocabulary, proposals that work directly with empirical information about child language and that are compatible with the infrastructural approach (see, e.g., Bates et al., 1979) even if they do not, in many cases, expressly address the naturally logical constraints I am invoking.

The Need for a General Theory of Natural Logic

One might object that there is no proving or disproving the natural logic proposal. And if one has in mind a simple experiment, it is true that there is no proving this proposal at present any more than there is the possibility of proving an innateness-of-language proposal. But in fact there are long-term ways to test with converging empirical studies the sort of proposal I have sketched here. Simulations of the sort that are currently being conducted by connectionists offer some of the most fruitful immediate possibilities because they allow rapid seeking of existence proofs for the idea that relatively modest architectural and timing constraints in a neural network are sufficient to allow dramatic self-organization that is not predictable on the basis of the stimulus set presented to the net. My contention is that natural logic can explain much of what happens in these mysterious simulation outcomes, for the simple reason that if the system is programmed to seek solutions, and if there is any sort of selection mechanism that prunes away inefficient nets and allows efficient and powerful ones to survive, then they may well evolve toward rich capabilities without much in the way of complex initial architecture. The directions in which they will evolve, according to my reasoning, will not, however, be entirely unpredictable, because they will abide by the constraints of natural logic.

This is not an argument that innateness should be left out of our characterization of language or cognition. It is an argument instead that natural logic needs to be added to that characterization.

Notes

1. As discussed in the present chapter, the "minimalist program" currently being pursued by a variety of Chomskyan scholars may have opened a door to evolutionary considerations. One indication of a shift is seen in Hauser, Chomsky, and Fitch (2002), an article that appeared after the present chapter was originally composed. The basic innateness arguments do not, however, appear to have been substantively altered in the minimalist program.

2. Each of the properties, of course, represents a major domain of development, and not all the *principles* corresponding to each *property* must be implemented before all the principles of any more advanced property. (The distinction between properties and principles is maintained throughout this formulation of a natural logic of communication systems—as an example of the distinction, note that the sequence depicted in figure 18.2 constitutes *principles* of implementation for the infraphonological *property* of Syllabicity). Principles corresponding to each property must themselves be implemented in logical sequence; regarding such presuppositional relationships as those indicated in figure 18.3, it can be said that the primitive principles for any more primitive property must be developed before any principles for any more advanced property. As more advanced principles are implemented for primitive properties, the groundwork is laid for the more advanced principles of more advanced properties.

References

Archangeli D, Langendoen DT (1997) Optimality Theory: An Overview. Malden, MA: Blackwell.

Austin JL (1962) How to Do Things with Words. London: Oxford University Press.

Bates E, Benigni L, Bretherton I, Camaioni L, Volterra V (1979) The Emergence of Symbols: Cognition and Communication in Infancy. New York: Academic Press.

Bates E, MacWhinney B (1982) Functionalist approaches to grammar. In: Language Acquisition: The State of the Art (Wanner E, Gleitman LR, eds), 173–218. Cambridge: Cambridge University Press.

Bonnet C (1762) Considérations sur les Corps Organisés. Amsterdam: Marc-Michel Rey.

Chomsky N (1957) Syntactic Structures. The Hague: Mouton.

Chomsky N (1965) Aspects of the Theory of Syntax. Cambridge, MA: MIT Press.

Chomsky N (1966) Cartesian Linguistics. New York: Harper & Row.

Chomsky N (1968) Language and Mind. New York: Harcourt.

Chomsky N (1986) Knowledge of Language: Its Nature, Origin and Use. New York: Praeger.

Chomsky N (1993a) Language and Thought. Wakefield, RI: Moyer Bell.

Chomsky N (1993b) A minimalist program for linguistic theory. In: The View from Building 20: Essays in Linguistics in Honor of Sylvain Bromberger (Hale K, Keyser SJ, eds), 1–52. Cambridge, MA: MIT Press.

Eilers RE, Oller DK (1994) Infant vocalizations and the early diagnosis of severe hearing impairment. J Pediatrics 124: 199–203.

Elman JL, Bates, EA, Johnson MH, Karmiloff-Smith A, Parisi, D, Plunkett K (1996) Rethinking Innateness: A Connectionist Perspective on Development. Cambridge, MA: MIT Press.

Elowson AM, Snowdon CT, Lazaro-Perea C (1998) "Babbling" and social context in infant monkeys: Parallels to human infants. Trends Cog Sci 2: 31–37.

Fisichelli RM (1950) An Experimental Study of the Prelinguistic Speech Development of Institutionalized Infants. Unpublished doctoral dissertation, Fordham University.

Grégoire A (1948) L'Apprentissage du Langage. Originally published in Lingua 1 (1948): 162–164, 168–169, 170–172. Reprinted in: Child Language: A Book of Readings (Bar-Adon A, Leopold WF, eds), 91–95. Englewood Cliffs, NJ: Prentice-Hall, 1971.

Griffin D (1992) Animal Minds. Chicago: University of Chicago Press.

Haley MC, Lunsford RF (1994) Noam Chomsky. New York: Twain.

Hauser M (1996) The Evolution of Communication. Cambridge, MA: MIT Press.

Hauser M, Chomsky N, Fitch WT (2002) The faculty of language: What is it, who has it, and how did it evolve? Science 298: 1569–1579.

Jürgens U (1992) On the neurobiology of vocal communication. In: Nonverbal Vocal Communication (Papousek H, Jürgens U, Papousek M, eds), 31–42. New York: Cambridge University Press.

Koffka K (1928) The Growth of the Mind. London: Kegan Paul, Trench and Trubner.

Lenneberg E, Rebelsky FG, Nichols IA (1965) The vocalizations of infants born to deaf and hearing parents. Hum Devel 8: 23–37.

Lewis MM (1936) Infant Speech. New York: Harcourt Brace.

Mavilya M (1969) Spontaneous Vocalizations and Babbling in Hearing Impaired Infants. Unpublished doctoral dissertation, Columbia University. University Microfilms no. 70-12879.

Montague R (1970) Universal grammar. Theoria 36: 373–398.

Newmeyer FJ (1996) Generative Linguistics: A Historical Perspective. London: Routledge.

Oller DK (1980) The emergence of the sounds of speech in infancy. In: Child Phonology, vol. 1: Production (Yeni-Komshian G, Kavanagh J, Ferguson C, eds), 93–112. New York: Academic Press.

Oller DK (1986) Metaphonology and infant vocalizations. In: Precursors of Early Speech (Lindblom B, Zetterstrom R, eds), 21–35. New York: Stockton Press.

Oller DK (1995) Development of vocalizations in infancy. In: Human Communication and Its Disorders: A Review, vol. IV (Winitz W, ed), 1–30. Timonium, MD: York Press.

Oller DK (2000) The Emergence of the Speech Capacity. Mahwah, NJ: Erlbaum.

Partee BH (1979) Constraining transformational Montague grammar: A framework and a fragment. In: Linguistics, Philosophy, and Montague Grammar (Davis S, Mithun M, eds), 51–102. Austin: University of Texas Press.

Piaget J (1969) Genetic Epistemology. New York: Columbia Forum.

Smith N (1999) Chomsky: Ideas and Ideals. Cambridge: Cambridge University Press.

Sutton D (1979) Mechanisms underlying learned vocal control in primates. In: Neurobiology of Social Communication in Primates: An Evolutionary Perspective (Steklis HD, Raleigh MJ, eds), 45–67. New York: Academic Press.

Thelen E (1981) Rhythmical behavior in infancy: An ethological perspective. Dev Psych 17: 237–257.

Contributors

Lee Altenberg
Department of Information and
Computer Sciences
University of Hawaii at Manoa
Honolulu, Hawaii

Lauren Ancel Meyers
Department of Integrative Biology
University of Texas at Austin
Austin, Texas

Carl Anderson
IcoSystem Corporation
Cambridge, Massachusetts

Robert N. Brandon
Departments of Philosophy and Biology
Duke University
Durham, North Carolina

Angela D. Buscalioni
Unit of Paleontology
Department of Biology
Autonomous University of Madrid
Madrid, Spain

Raffaele Calabretta
Department of Neural Systems and
Artificial Life
Institute of Cognitive Sciences and
Technologies
Italian National Research Council
Rome, Italy

Werner Callebaut
Konrad Lorenz Institute for Evolution
and Cognition Research
Altenberg, Austria
and
Faculty of Sciences
Limburgs Universitair Centrum
Diepenbeek, Belgium

Anne Dejoan
Department of Aeronautics
Imperial College of Science,
Technology and Medicine
London, United Kingdom

Rafael Delgado-Buscalioni
Centre for Computational Science

Department of Chemistry
University College, London,
United Kingdom

Gunther J. Eble
Centre National de la Recherche
Scientifique
CNRS UMR 5561
BIOGEOSCIENCES
Dijon, France

Walter Fontana
Department of Systems Biology
Harvard Medical School
Boston, Massachusetts

Fernand Gobet
Department of Human Sciences
Brunel University
Uxbridge, United Kingdom

Alicia de la Iglesia
IES Mirasierra
Madrid, Spain

Slavik V. Jablan
Mathematical Institute
University of Belgrade
Belgrade, Federal Republic of
Yugoslavia

Luigi Marengo
DSGSS
Università di Teramo
Teramo, Italy

Daniel W. McShea
Department of Biology
Duke University
Durham, North Carolina

Jason Mezey
Department of Biological Sciences
University of Florida
Tallahassee, Florida

D. Kimbrough Oller
School of Audiology and
Speech–Language Pathology
University of Memphis
Memphis, Tennessee

Domenico Parisi
Department of Neural Systems and
Artificial Life
Institute of Cognitive Sciences and
Technologies
Italian National Research Council
Rome, Italy

Corrado Pasquali
DSGSS
Università di Teramo
Teramo, Italy

Diego Rasskin-Gutman
Gene Expression Laboratory
The Salk Institute
La Jolla, California

Gerhard Schlosser
Brain Research Institute
University of Bremen
Bremen, Germany

Herbert A. Simon †
Department of Psychology
Carnegie Mellon University
Pittsburgh, Pennsylvania

Roger D. K. Thomas
Department of Geosciences
Franklin & Marshall College
Lancaster, Pennsylvania

Marco Valente
DSIE
Università dell'Aquila
L'Aquila, Italy

Boris M. Velichkovsky
Department of Psychology
Dresden University of Technology
Dresden, Germany

Günter P. Wagner
Department of Ecology and
Evolutionary Biology
Yale University
New Haven, Connecticut

Rasmus G. Winther
Department of History and Philosophy
of Science
Indiana University
Bloomington, Indiana

Index

Printed in the United States
by Baker & Taylor Publisher Services